Peroxisomes: Biogenesis, Function, and Role in Human Disease

Tsuneo Imanaka • Nobuyuki Shimozawa
Editors

Peroxisomes: Biogenesis, Function, and Role in Human Disease

Springer

Editors
Tsuneo Imanaka
Faculty of Pharmaceutical Sciences
Hiroshima International University
Kure, Hiroshima, Japan

University of Toyama
Toyama, Japan

Nobuyuki Shimozawa
Division of Genomics Research,
Life Science Research Center
Gifu University
Gifu, Japan

Department of Pediatrics
Gifu University School of Medicine
Gifu, Japan

ISBN 978-981-15-1168-4 ISBN 978-981-15-1169-1 (eBook)
https://doi.org/10.1007/978-981-15-1169-1

This Springer imprint is published by the registered company Springer Nature Singapore Pte Ltd.
The registered company address is: 152 Beach Road, #21-01/04 Gateway East, Singapore 189721, Singapore

Foreword

Peroxisomes are subcellular organelles found in almost all eukaryotic cells. They are involved in a variety of metabolic processes, including the β-oxidation of very long chain fatty acids as well as the synthesis of ether-phospholipids and bile acids in mammals. Peroxisomal disorders represent disease conditions caused by defects in the biogenesis and function of the peroxisome.

This time, Prof. Shimozawa, Prof. Imanaka, and other distinguished clinicians and researchers who lead peroxisomal research in Japan have published the textbook *Peroxisomes*. This book includes current progress in basic research of peroxisomes and important roles of peroxisomes for genetic diseases as well as cancer and age-related diseases. In the textbook, Prof. Shimozawa organized the clinical aspect of peroxisomal disorders. Prof. Kaga described the neurophysiology and neuropsychology in adrenoleukodystrophy (ALD), and Prof. Yokoyama described the lipidomics of peroxisomal disorders as new aspects of peroxisomal disorders. On the other hand, Prof. Imanaka organized the basic science of peroxisomes.

Peroxisomal disorders including ALD are sometimes difficult to diagnose clinically, as these disorders present in a variety of symptoms. To make a definitive diagnosis of peroxisomal disorders requires various clinical tests based on biochemical and molecular genetics. Prof. Shimozawa, Gifu University, has screened many patients with peroxisomal disorders for more than 30 years. There are few laboratories in the world that can diagnose peroxisomal disorders. Prof. Shimozawa and his coworkers have a lot of experiences in this area and made many achievements.

The Ministry of Health, Labor and Welfare of Japan has conducted national research projects for intractable diseases. I have been worked extensively a project related to lysosomal and peroxisomal diseases as Principle Investigator for more than 10 years. Prof. Shimozawa's group including above-mentioned professors has achieved many successful works in the field of peroxisomal research. I sincerely express my thanks for their great contribution to this research project.

I hope that many clinicians and researchers including postgraduate students in the world would read the wonderful textbook published by Springer-Nature to understand the basics of peroxisomes and its clinical importance for the diagnosis and treatment of peroxisomal disorders.

Lysosomal and Peroxisomal Research Team
supported by a Ministry of Health
Labor and Welfare Refractory Disease Policy Project in Japan
Tokyo, Japan

The Jikei University School of Medicine
Tokyo, Japan
October 2, 2019

Yoshikatsu Eto

Preface

Peroxisomes are organelles that are present in almost all eukaryotic cells. The organelles were first described as "microbodies" by Rhodin in 1954. Subsequently, de Duve et al. found that they contained various H_2O_2-producing oxidases along with H_2O_2-degrading catalase and named them peroxisomes in 1965. At present, it is well known that peroxisomes are essential organelle to maintain cellular function in lipid metabolism, redox homeostasis, and intracellular signaling. Recently, it has been shown that physical contact between peroxisomes and other organelles plays an important role in the transporting or shuttling of metabolites for regulation of cellular functions.

With regard to the biogenesis of the peroxisome, several models have been proposed. It is now considered most likely that peroxisomes are endoplasmic reticulum-derived organelles. The biosynthesis of peroxisomes in mammals involves three different processes: the formation of the pre-peroxisome from the ER, the import of the peroxisomal membrane and matrix proteins to the pre-peroxisome, and the growth and division of the peroxisome. It has been revealed that many proteins called "peroxins" encoded by *PEX* genes are involved in peroxisome biogenesis.

Mutation of the genes encoding peroxisomal proteins and/or proteins involved in peroxisome biogenesis causes congenital metabolic disorders. Brown first described Zellweger's cerebro-hepato-renal syndrome (Zellweger syndrome) in 1964. Then, Goldfisher discovered that peroxisomes were absent from tissues in patients with Zellweger syndrome. This syndrome contributed to our understanding of peroxisomal assembly and the metabolic importance of peroxisomes. In 1992, Shimozawa discovered the first gene responsible for Zellweger syndrome, which led to the elucidation of many pathogenic PEX genes.

Our understanding of peroxisome functions, biogenesis, and diseases is deepened through continuous progress of peroxisome research. Recently, many excellent reviews have been published about peroxisome and related diseases. These reviews are targeted at experts who are interested in research on peroxisomes. On the other hand, we provide here the whole picture of peroxisome, based on present knowledge, for postgraduate students, young scientists, and clinicians in the field of internal medicine, pediatrics, and neurology who want to understand peroxisomes.

Fig. 1 Prof. Takashi Hashimoto in the 54th Annual Meeting of the Japanese Society for Inherited Metabolic Diseases at Gifu, Japan, in November 2012. Prof. Hashimoto makes a speech at the welcome party

This book includes the history of peroxisome research, biogenesis, and function of peroxisomes, and peroxisomal disorders. In addition, the book covers specific subjects such as isolation of peroxisomes, structural biology of peroxisomal proteins involved in peroxisome biogenesis, lipidomics of peroxisomal disorders, and neurophysiology and psychology for adrenoleukodystrophy. The book also shows the future direction of peroxisome research and therapeutic development for peroxisomal disorders. We hope the book is helpful for young people to understand peroxisomes.

We would like to express our admiration for the contribution of Professor Emeritus Takashi Hashimoto at Shinshu University in peroxisomal functions and disorders (Fig. 1). He was one of the great pioneers in the field. He has successfully characterized peroxisomal β-oxidation system. He purified the enzymes involved in the β-oxidation and cloned the genes encoding the enzymes. He also prepared antibodies against the enzymes and distributed them to many researchers around the world. The antibodies were of help to demonstrate several peroxisomal disorders caused by deficiency of peroxisomal β-oxidation enzymes. He passed away on April 18, 2019, at the age of 86, and we dedicate this book to him and are deeply grateful to him for his great contribution to the field of peroxisome and peroxisomal disorders.

In addition, we appreciate Ms. Kripa Guruprasad and Drs. Sue Lee and Emmy Lee of Springer Nature for the continuous encouragement and support in preparing this book.

Hiroshima, Japan; Toyama, Japan — Tsuneo Imanaka
Gifu, Japan — Nobuyuki Shimozawa
September 2019

Contents

Part III Topics in Peroxisome Research

Contributors

Yuko Fujiwara, Ph.D. Faculty of Pharma-Sciences, Teikyo University, Tokyo, Japan

Kotaro Hama, Ph.D. Faculty of Pharma-Sciences, Teikyo University, Tokyo, Japan

Tsuneo Imanaka, Ph.D. Faculty of Pharmaceutical Sciences, Hiroshima International University, Kure, Hiroshima, Japan
University of Toyama, Toyama, Japan

Makiko Kaga, M.D., Ph.D. Tokyo Metropolitan Tobu Medical Center for Children with Developmental Disabilities, Tokyo, Japan
National Institute of Mental Health, National Center of Neurology and Psychiatry, Tokyo, Japan

Hiroaki Kato, Ph.D. Graduate School of Pharmaceutical Sciences, Kyoto University, Kyoto, Japan

Kosuke Kawaguchi, Ph.D. Graduate School of Medicine and Pharmaceutical Sciences, University of Toyama, Toyama, Japan

Masashi Morita, Ph.D. Graduate School of Medicine and Pharmaceutical Sciences, University of Toyama, Toyama, Japan

Masahide Oku, Ph.D. Division of Applied Life Sciences, Graduate School of Agriculture, Kyoto University, Kyoto, Japan

Yasuyoshi Sakai, Ph.D. Division of Applied Life Sciences, Graduate School of Agriculture, Kyoto University, Kyoto, Japan

Nobuyuki Shimozawa, M.D., Ph.D. Division of Genomics Research, Life Science Research Center, Gifu University, Gifu, Japan
Department of Pediatrics, Gifu University School of Medicine, Gifu, Japan

Shigeo Takashima, Ph.D. Division of Genomics Research, Life Science Research Center, Gifu University, Gifu, Japan

Kazuaki Yokoyama, Ph.D. Faculty of Pharma-Sciences, Teikyo University, Tokyo, Japan

Part I
Biogenesis and Function of Peroxisome

Chapter 1
The History of Peroxisomal Research

Tsuneo Imanaka

Abstract Peroxisomes are subcellular organelles bounded by a single membrane. They are involved in a variety of metabolic processes, including the β-oxidation of very long chain fatty acids, as well as the synthesis of ether-phospholipids and bile acid in mammals. These organelles were first described in 1954 in the cytoplasm of the proximal tubule cells of the mouse kidney by Rhodin and were first known as "microbodies". Subsequently, in 1965 de Duve et al. isolated microbodies from the rat liver and defined them as membrane-bound organelles containing various H_2O_2-producing oxidases along with H_2O_2-degrading catalase, and named them peroxisomes. The fatty acid β-oxidation system was identified in rat liver peroxisomes in 1976. Goldfischer discovered that peroxisomes were absent from the tissues of patients with Zellweger syndrome in 1973, and the metabolic defects that characterize this disease contributed to the elucidation of the metabolic roles of peroxisomes in humans. With regard to the biogenesis of the peroxisome in mammals, several models have been proposed and the following process is generally accepted. Pre-peroxisomes bud from the endoplasmic reticulum, and peroxisomal membrane and matrix proteins are then imported into these pre-peroxisomes. The mature peroxisomes grow by division. Here, I look back the history of peroxisomal research based on the investigation of the biogenesis and function of peroxisomes, along with peroxisomal diseases.

Keywords Peroxisome · Biogenesis · Function · Lipid metabolism · Peroxisomal diseases

Abbreviations

ER	Endoplasmic reticulum
PBD	Peroxisome biogenesis disorder

T. Imanaka (✉)
Faculty of Pharmaceutical Sciences, Hiroshima International University, Kure, Hiroshima, Japan

University of Toyama, Toyama, Japan
e-mail: imanaka@hirokoku-u.ac.jp

T. Imanaka, N. Shimozawa (eds.), *Peroxisomes: Biogenesis, Function, and Role in Human Disease*, https://doi.org/10.1007/978-981-15-1169-1_1

PMP	Peroxisomal membrane protein
RCDP	Rhizomelic chondrodysplasia punctata
VLCFA	Very long chain fatty acid
X-ALD	X-linked adrenoleukodystrophy

1.1 Introduction

Peroxisomes are single membrane-bound organelles existing in almost all eukaryotes and are ubiquitous in mammalian cells. Most cells contain spherical or spheroidal peroxisomes with a diameter between 0.1 and 1 μm (Fig. 1.1), although the peroxisomes vary considerably in shape and size in different tissues. Peroxisomes are small organelles comprising only a few percent of the volume of the human liver and kidney cells, where peroxisomes are most abundant. Nonetheless, despite such low abundance, the function of peroxisomes is indispensable in mammals, including humans.

These organelles were first described in 1954 by Rhodin while he was a postgraduate student at the Karolinska Institute working in the laboratory of Sjöstrand, one of the pioneers of electron microscopy. Rhodin closely examined the proximal tubule cells of the mouse kidney and observed a small organelle that had not been previously described. It was surrounded by a single membrane and filled with a fine granular matrix. He must have been excited about the appearance of an unknown cell compartment. He called these compartments “microbodies” (Rhodin 1954; de Duve 1983). Subsequently Rouiller and Bermhard reported similar structures in rat hepatic cells (Bernhard and Rouiller 1956). However, cell biologists did not express much interest in these microbodies since their function was unknown.

The microbodies were first characterized biochemically by de Duve and his colleagues (Baudhuin et al. 1965; de Duve and Baudhuin 1966). de Duve had always maintained that all members of a given organelle family have the same enzymatic composition (Fig. 1.2). He and his colleagues undertook efforts to separate the microbodies from other organelles, especially lysosomes. Initially, the distribution of the microbodies and lysosomes were found to be in tight proximity when normal rat liver cells were fractionated. In contrast, when rats were injected with the detergent Triton WR-1339, the lysosome density was selectively lowered by the accumulation of lipids derived from plasma lipoproteins, and the microbodies were then clearly separated from the lysosomes (Leighton et al. 1968). They also found that microbodies contained various H_2O_2-producing oxidases and the H_2O_2-degrading enzyme catalase (de Duve and Baudhuin 1966; Baudhuin 1969). They named these organelles “peroxisomes”, but the physiological function was still poorly understood.

In 1967 Beevers and his colleagues discovered particles that were characterized by a high equilibrium density in a sucrose gradient, which is a typical property of peroxisomes. These particles from castor been endosperm had glyoxylate cycle enzymes, so he named them “glyoxysomes” (Breidenbach and

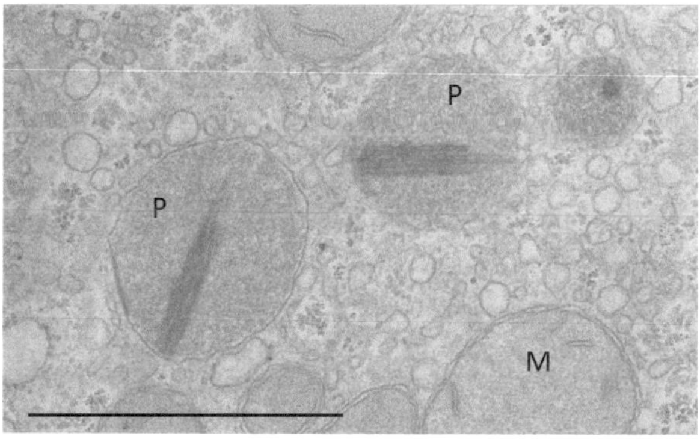

Fig. 1.1 Electron micrograph of rat liver peroxisomes. P peroxisome, M mitochondrion. The bar, 1 μm. [Courtesy of Sadaki Yokota (Professor emeritus at Yamanashi University)]

Fig. 1.2 Prof. de Duve in his laboratory of Biochemical Cytology at Rockefeller University in 1987. Prof. de Duve is standing close to the centrifuge that he developed. (Figure from Imanaka 2018 with the permission of the Pharmaceutical Society of Japan)

Beevers 1967). Soon after they reported that glyoxysomes contain catalase, glycol oxidase and urate oxidase (Breidenbach et al. 1968). This finding supports the hypothesis that there is a phylogenetic relationship between peroxisomes and glyoxysomes in the plant and animal kingdoms.

1.2 Peroxisome Function

In 1969 Beevers's group found glyoxysomes have all the enzymes necessary for the β-oxidation of fatty acids (Cooper and Beevers 1969). The β-oxidation enzymes allow efficient conversion of fatty acid to carbohydrate via the glyoxylate cycle. Interestingly, the first step of fatty acid β-oxidation was found to be a hydrogen-producing oxidation that is different from that of the mitochondrial fatty acid β-oxidation system. Surprisingly in hindsight, the significance of this observation failed to be recognized for almost a decade.

On the other hand, it was shown in mammals at that time that ethyl-*p*-chlorophenoxybutyrate (CPIB or clofibrate) decreased the concentration of serum neutral lipids in patients with hyperlipidemia (Thorp and Waring 1962). Then rodent studies showed that clofibrate increased liver size and induced a remarkable proliferation of peroxisomes (Hess et al. 1965; Svoboda and Azarnoff 1966). However, the metabolic implications of these findings did not immediately receive attention. In the 1970s, based on the evidence that clofibrate and other hypolipidemic drugs induced the proliferation of hepatic peroxisomes, peroxisomes came to be understood to play an important role in lipid metabolism (Reddy and Krishnakantha 1975). Nonetheless, there was still no direct corroboration.

In 1976, Lazarow and de Duve found that rat liver peroxisomes were capable of palmitoyl-CoA oxidation, with reduction of O_2 to H_2O_2 in a cyan-insensitive manner, and that this oxidation was enhanced in rats given clofibrate (Lazarow and de Duve 1976). This was the first report on the physiological function of peroxisomes in mammals. They also suggested that the hepatic fatty acid β-oxidation system plays a role in reducing lipid serum levels with hypolipidemic drugs. It was subsequently shown that treating rodents with these drugs also induced certain peroxisomal enzymes involved in fatty acid β-oxidation (Osumi and Hashimoto 1978).

With regard to peroxisome function, several different metabolic pathways after the β-oxidation system had been reported in mammalian peroxisomes. These include very long chain fatty acid (VLCFA) β-oxidation, as well as the synthesis of ether-phospholipids and bile acid. Elucidation of these metabolic roles of peroxisomes in humans was coupled to the research on Zellweger syndrome, in which no functional peroxisomes exist (Goldfischer et al. 1973). In the middle years of the 1980s, Moser and his colleagues demonstrated that VLCFAs were increased in the plasma and cultured skin fibroblasts from these patients, and VLCFA oxidation was impaired in the homogenates of cultured skin fibroblasts (Moser et al. 1984). The activity of dihydroxyacetone phosphate acyltransferase, a peroxisomal enzyme with a major role in ether-phospholipid synthesis, was found to be deficient in the skin fibroblasts of patients with Zellweger syndrome (Datta et al. 1984). Defective peroxisomal cleavage of the C27-steroid side chain in bile acid synthesis was also found in Zellweger syndrome (Kase et al. 1985). Later, antibodies against peroxisomal enzymes prepared by Hashimoto's group at Shinshu University helped to confirm a relationship to a deficiency in peroxisomal β-oxidation enzymes in several different peroxisomal diseases. In addition to lipid metabolism, peroxisomes play a role in several non-lipid metabolic pathways, including purine, polyamine, glyoxylate and D-amino acid metabolism as wells as reactive oxygen metabolism. Recently, it is suggested that peroxisomes are multifunctional organelles that interact with other organelles in various metabolic and signaling pathways.

Concerning proliferation of peroxisomes and their function, it must be pointed out that the peroxisome proliferator-activated receptor (PPAR) was discovered in 1990 (Issemann and Green 1990). PPAR is a general transcriptional regulator of lipid homeostasis, and several hypolipidemic and antidiabetic agents have been shown to be effective ligands. PPARα is known to be able to control not only peroxisomal metabolism, but also lipid homeostasis (Vamecq et al. 2014).

1.3 Biogenesis of the Peroxisome

In the early days of this research when microbodies were first found in the rat liver, they were thought to be precursors of mitochondria based on morphological observations in the course of liver regeneration and under various pathological conditions (Bernhard and Rouiller 1956). In the 1960–1970s, peroxisomes were thought to be formed from the endoplasmic reticulum (ER), since peroxisomes often appeared in clusters surrounded by a smooth ER and seemed to be continuous in a number of places with the ER (Novikoff and Novikoff 1972). A pulse-chase study of catalase in the rat liver showed that labeled-catalase appeared first in the ER fraction and then was transported to the mitochondrial fraction (Higashi and Peters Jr. 1963), although peroxisomes had not yet been identified in tissues.

In the 1980s, it was demonstrated that many peroxisomal proteins, including membrane proteins, are synthesized on free polysomes. In vitro import studies were developed and showed that newly synthesized peroxisomal matrix proteins were directly transported into peroxisomes. We demonstrated that acyl-CoA oxidase, a peroxisomal matrix protein, was imported into purified rat liver peroxisomes and that the import required ATP hydrolysis, but not a change in the membrane potential (Imanaka et al. 1987). In addition, in 1987 Gould, a postgraduate student in the Subramani laboratory at the University of California (San Diego), showed a peroxisome targeting signal in the COOH-terminal 12 amino acids of firefly luciferase by the transfection of the cDNA in CV-1 monkey cells (the luciferase is present in peroxisome-like organelles in the cells of the firefly lantern organ) (Gould et al. 1987). He subsequently showed that the COOH-terminal amino acids Ser-Lys-Leu (SKL) were sufficient for the targeting of the protein to peroxisomes (Gould et al. 1988). At the time, it had become generally accepted that peroxisomes could form by growth and division from pre-existing organelles, like mitochondria (Lazarow and Fujiki 1985).

By the end of the 1980s, a search for the genes involved in peroxisome biogenesis had been initiated. Erdman in Prof. Kunau's laboratory at Ruhr-University in Germany isolated mutants of the yeast *Saccharomyces cerevisiae* that affected peroxisomal assembly (pas mutations) (Erdmann et al. 1989). The mutants were isolated based on the characteristics of the cells that were not able to grow on oleic acid as the sole carbon source. Subsequently, he isolated the *PAS1* gene required for peroxisome biogenesis. Interestingly, Pas1p is a member of the ATPases associated with diverse cellular activities (AAA) -ATPase (Erdmann et al. 1991). On the other hand, Fujiki and his colleagues in Meiji Institute of Life Science at that time isolated mutant CHO cells defective in peroxisome assembly by screening cells with dihydroxyacetone phosphate acyltransferase deficiency (Tsukamoto et al. 1990). They isolated a *peroxisome assembly factor (PAF)* gene encoding a peroxisomal membrane protein (PMP) that restored peroxisome biogenesis (Tsukamoto et al. 1991). Subsequently, many genes involved in the biogenesis of peroxisomes have been identified in different yeast strains and given names such as *PER*, *PAY*, and *PEM*. As the names became confusing in the course of multiple descriptions of the same gene in different species, the names were unified in 1996 as the gene "*PEX*" and the protein peroxin, "Pexp" for short (Distel et al. 1996). The genes are numbered according to the order in which they were identified.

During this time, it was shown that peroxisomes were not detected in cells with a mutation in either the *PEX3* or *PEX19* gene, and peroxisomes reappeared upon transfection with a normal cDNA of *PEX3* or *PEX19*, respectively. Furthermore, the newly synthesized Pex3p was shown to be inserted into ER membranes. Based on these observations, it is considered most likely at present that peroxisomes are ER-derived organelles in mammals (Agrawal and Subramani 2016), although there is a suggestion that pre-peroxisomes are formed by the fusion of vesicles derived from the ER and mitochondrial outer membranes (Sugiura et al. 2017). The biosynthesis of peroxisomes in mammals is proposed to involve three different processes, i.e. budding of pre-peroxisomes from the ER and the import of PMPs and matrix proteins into the pre-peroxisomes and peroxisomes, growth into mature peroxisomes and ultimately, division/proliferation (Fig. 1.3). However, it should be note that when peroxisomes already exist, newly synthesized PMPs are directly transported to peroxisomes.

In retrospect, a peroxisome biogenesis model was proposed based on the experimental results obtained by the latest technology of that time, and then the model was reevaluated and revised based on new evidence obtained by subsequently emerging technology. Tracing the evolving history of the developing conceptualization of the biogenesis of the peroxisome is considerable interest. The idea of where the peroxisome originates goes back to the ER hypothesis from over half a century ago. It is known that many of the so-called peroxins encoded by the *PEX* genes are involved in peroxisome biogenesis. To date, 36 peroxins have been identified. Currently, it is important to analyze the sites that newly synthesized endogenous Pex3p or Pex16p, which initiate *de novo* peroxisome biogenesis, target in cells completely lacking peroxisome remnants.

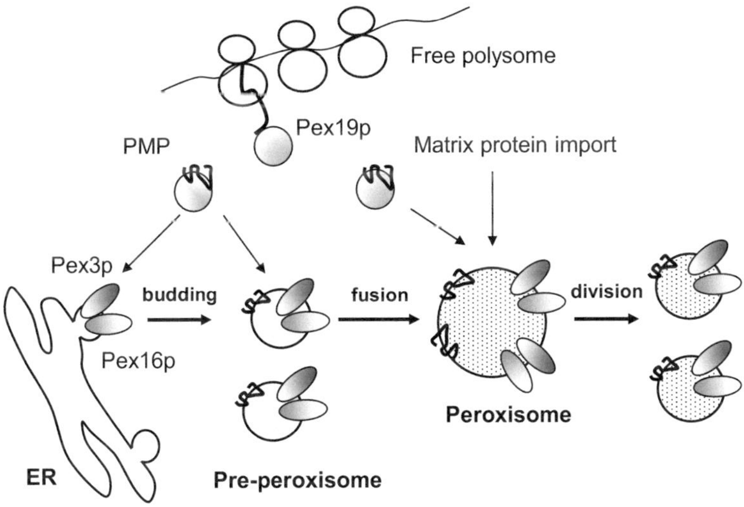

Fig. 1.3 Biogenesis of the peroxisome. A pre-peroxisome containing Pex3p and Pex16p buds from the ER. Newly synthesized PMP and matrix proteins are post-translationally imported into the pre-peroxisome. After fusion of the pre-peroxisomes, a mature peroxisome is formed. The number of peroxisomes increases by division. (Figure from Imanaka 2018 with the permission of the Pharmaceutical Society of Japan)

1.4 Peroxisomal Diseases

The peroxisomal diseases comprise a congenital metabolic disorder and mutation of the genes encoding peroxisomal proteins and/or proteins involved in peroxisome biogenesis results in a variety of human diseases. These diseases are generally classified into two groups: peroxisome biogenesis disorders (PBDs) and single peroxisomal enzyme deficiency disorders (Wanders et al. 2017).

Concerning the PBDs, Bowen first described Zellweger syndrome as Zellweger's cerebro-hepato-renal syndrome in 1964 (Bowen et al. 1964). As mentioned above, Goldfischer discovered that peroxisomes were absent from tissues in Zellweger's patients (Goldfischer et al. 1973). Initially, the importance of this finding was not well recognized. However, the syndrome came into the spotlight as a specific inheritable metabolic disease with a defect in organelle assembly through the emerging understanding of the metabolic importance of peroxisomes. In addition, as clinically similar but less severe phenotypes, neonatal adrenoleukodystrophy (Scotto et al. 1982) and infantile Refsum disease were reported (Ulrich et al. 1978). As another phenotype, rhizomelic chondrodysplasia punctata (RCDP) was reported by Spranger et al. in 1971 (Spranger et al. 1971). PCDP is clinically characterized by proximal shortening of the limbs and multiple punctuate epiphyseal calcifications.

These diseases were first described in the past, before the biochemical and molecular bases of this disease spectrum were fully determined. During discovery of the *PEX* genes, intensive investigation to identify the genes responsible for PBDs was undertaken. Shimozawa in 1992 first identified a point mutation in the *PAF1* (*PEX2*) gene in a patient that resulted in the premature termination of Pex2p (Shimozawa et al. 1992). Then many mutated genes were identified in Zellweger syndrome and surprisingly, in the case of RCDP, mutations of *PEX7* gene was independently reported by three different laboratories in the same issue of Nature Genetics (Braverman et al. 1997; Motley et al. 1997; Purdue et al. 1997).

The diseases due to single peroxisomal enzyme deficiency include defects in peroxisomal matrix enzymes as well as PMPs. Based on the discovery of metabolic pathways in peroxisomes, such as β-oxidation of VLCFA and branched chain of fatty acids, as well as the biosynthesis of ether-phospholipids and bile acids, many peroxisome diseases with a single enzyme deficiency have been discovered. Some of these diseases were described in the past and later the responsible proteins or genes were identified. The clinical phenotypes of single peroxisomal enzyme deficiency depend upon the specific function of a given protein in peroxisomal metabolism. These include mutation of the ABC transporters, ABCD1 and ABCD3, the enzymes involved in β or α-oxidation, ether-phospholipid synthesis, glyoxylate detoxification, and catalase (Wanders et al. 2017).

Among the cases of peroxisomal enzyme deficiency, X-linked adrenoleukodystrophy (X-ALD) is the most common peroxisomal disease caused by mutations in the *ABCD1* gene. The affected patients display progressive demyelination of the central nervous system (CNS), adrenal insufficiency, and testicular dysfunction as pathological characteristics. According to the literature (Engelen et al. 2012; Trompier and Savary 2013), in 1897 Heubner first described a young boy with rapidly progressive neurologic deterioration consistent with X-ALD who had "diffuse sclerosis" on autopsy. In 1970, the name adrenoleukodystrophy was proposed and the condition ultimately came to be known as X-ALD (Blaw 1970). In 1976, Igarashi established the biochemical basis for X-ALD by showing VLCFA accumulation in the brain and adrenal glands of patients (Igarashi et al. 1976). In 1993, Aubourg and his colleagues identified the *ALD* (*ABCD1*) gene by positional cloning as the gene in which mutations are responsible for X-ALD (Mosser et al. 1993). As the VLCFA-CoA synthetase gene had been suspected to be associated with X-ALD at that time, it was surprising that the responsible protein was this transporter.

The indispensable role of peroxisomes in human health and development is evidenced by the existence of a large number of inborn errors of peroxisome assembly and peroxisomal metabolism. At present, 23 genetic diseases caused by the mutation of 32 genes are classified as peroxisomal diseases. With regard to these peroxisomal diseases, many genes have been identified and an understanding of their pathogenesis is as yet still a work in progress. However, many of these diseases are severe, so progress is urgently required. At present, therapeutic approaches to X-ALD are under investigation. I hope in the near future therapeutic protocols for X-ALD will become established and serve as a model approach to other peroxisomal diseases.

Additionally, glyoxysomes, glycosomes and Woronin bodies are known to be organelles related to the peroxisomes. They are formed by the same mechanism as peroxisomes and carry out specific functions in different species. Glyoxysomes, as mentioned above, include fatty acid β-oxidation and glyoxylate cycle enzymes, and are involved in plant germination (Graham 2008). The glycosomes, a protozoan microbody-like organelle, were discovered in *Trypanosome brucei* by Opperdoes and Borst in 1977 (Opperdoes and Borst 1977). Glycosomes possess glycolytic enzymes but no catalase, and play an important role in survival of protozoans in the host animal blood stream, including humans (Haanstra et al. 2016). The Woronin bodies in ascomycete fungi contain the structural protein, hexagonal peroxisome protein (Hex1). The hyphal cells are separated by perforated septa and by Woronin bodies. They seal the pore to protect the remaining cells when extensive wound-induced damage occurs (Steinberg et al. 2017). This chapter was written based on the review article (Imanaka 2019).

Acknowledgments The publication is supported in part by a Grant-in-Aid for Intractable Diseases from the Ministry of Health, Labour and Welfare of Japan. Pacific Edit reviewed the manuscript prior to submission.

References

Agrawal G, Subramani S (2016) *De novo* peroxisome biogenesis: evolving concepts and conundrums. Biochim Biophys Acta 1863(5):892–901

Baudhuin P (1969) Liver peroxisomes, cytology and function. Ann N Y Acad Sci 168(2):214–228

Baudhuin P et al (1965) Combined biochemical and morphological study of particulate fractions from rat liver. Analysis of preparations enriched in lysosomes or in particles containing urate oxidase, D-amino acid oxidase, and catalase. J Cell Biol 26(1):219–243

Bernhard W, Rouiller C (1956) Microbodies and the problem of mitochondrial regeneration in liver cells. J Biophys Biochem Cytol 2(4 Suppl):355–360

Blaw M (1970) Melanodermic type leukodystrophy (adrenoleukodystrophy). In: Vinken PJ, Bruyn GW (eds) Handbook of clinical neurology, vol 10. American Elsevier, New York, pp 128–133

Bowen P et al (1964) A familial syndrome of multiple congenital defects. Bull Johns Hopkins Hosp 114:402–414

Braverman N et al (1997) Human PEX7 encodes the peroxisomal PTS2 receptor and is responsible for rhizomelic chondrodysplasia punctata. Nat Genet 15(4):369–376

Breidenbach RW, Beevers H (1967) Association of the glyoxylate cycle enzymes in a novel subcellular particle from castor bean endosperm. Biochem Biophys Res Commun 27(4):462–469

Breidenbach RW et al (1968) Characterization of glyoxysomes from castor bean endosperm. Plant Physiol 43(5):705–713

Cooper TG, Beevers H (1969) β oxidation in glyoxysomes from castor bean endosperm. J Biol Chem 244(13):3514–3520

Datta NS et al (1984) Deficiency of enzymes catalyzing the biosynthesis of glycerol-ether lipids in Zellweger syndrome. A new category of metabolic disease involving the absence of peroxisomes. N Engl J Med 311(17):1080–1083

de Duve C (1983) Microbodies in the living cell. Sci Am 248(5):74–84

de Duve C, Baudhuin P (1966) Peroxisomes (microbodies and related particles). Physiol Rev 46(2):323–357

Distel B et al (1996) A unified nomenclature for peroxisome biogenesis factors. J Cell Biol 135(1):1–3

Engelen M et al (2012) X-linked adrenoleukodystrophy (X-ALD): clinical presentation and guidelines for diagnosis, follow-up and management. Orphanet J Rare Dis 7:51

Erdmann R et al (1989) Isolation of peroxisome-deficient mutants of *Saccharomyces cerevisiae*. Proc Natl Acad Sci U S A 86(14):5419–5423

Erdmann R et al (1991) *PAS1*, a yeast gene required for peroxisome biogenesis, encodes a member of a novel family of putative ATPases. Cell 64(3):499–510

Goldfischer S et al (1973) Peroxisomal and mitochondrial defects in the cerebro-hepato-renal syndrome. Science 182(4107):62–64

Gould SG et al (1987) Identification of a peroxisomal targeting signal at the carboxy terminus of firefly luciferase. J Cell Biol 105(6 Pt 2):2923–2931

Gould SJ et al (1988) Identification of peroxisomal targeting signals located at the carboxy terminus of four peroxisomal proteins. J Cell Biol 107(3):897–905

Graham IA (2008) Seed storage oil mobilization. Annu Rev Plant Biol 59:115–142

Haanstra JR et al (2016) Biogenesis, maintenance and dynamics of glycosomes in trypanosomatid parasites. Biochim Biophys Acta 1863(5):1038–1048

Hess R et al (1965) Nature of the hepatomegalic effect produced by ethyl-chlorophenoxy-isobutyrate in the rat. Nature 208(5013):856–858

Higashi T, Peters T Jr (1963) Studies on rat liver catalase. I. Combined immunochemical and enzymatic determination of catalase in liver cell fractions. J Biol Chem 238:3945–3951

Igarashi M et al (1976) Fatty acid abnormality in adrenoleukodystrophy. J Neurochem 26(4):851–860

Imanaka T et al (1987) Translocation of acyl-CoA oxidase into peroxisomes requires ATP hydrolysis but not a membrane potential. J Cell Biol 105(6 Pt 2):2915–2922

Imanaka T (2018) Biogenesis, the function of peroxisomes, and their role in genetic disease: with a focus on the ABC transporter. Yakugaku Zasshi 138:1067–1083

Imanaka T (2019) Biogenesis and function of peroxisomes with a focus on the ABC transporter. Biol Pharm Bull 42(5):649–665. https://doi.org/10.1248/bpb.b18-00723

Issemann I, Green S (1990) Activation of a member of the steroid hormone receptor superfamily by peroxisome proliferators. Nature 347(6294):645–650

Kase BF et al (1985) Defective peroxisomal cleavage of the C27-steroid side chain in the cerebro-hepato-renal syndrome of Zellweger. J Clin Invest 75(2):427–435

Lazarow PB, de Duve C (1976) A fatty acyl-CoA oxidizing system in rat liver peroxisomes; enhancement by clofibrate, a hypolipidemic drug. Proc Natl Acad Sci U S A 73(6):2043–2046

Lazarow PB, Fujiki Y (1985) Biogenesis of peroxisomes. Annu Rev Cell Biol 1:489–530

Leighton F et al (1968) The large-scale separation of peroxisomes, mitochondria, and lysosomes from the livers of rats injected with triton WR-1339. Improved isolation procedures, automated analysis, biochemical and morphological properties of fractions. J Cell Biol 37(2):482–513

Moser AE et al (1984) The cerebrohepatorenal (Zellweger) syndrome. Increased levels and impaired degradation of very-long-chain fatty acids and their use in prenatal diagnosis. N Engl J Med 310(18):1141–1146

Mosser J et al (1993) Putative X-linked adrenoleukodystrophy gene shares unexpected homology with ABC transporters. Nature 361(6414):726–730

Motley AM et al (1997) Rhizomelic chondrodysplasia punctata is a peroxisomal protein targeting disease caused by a non-functional PTS2 receptor. Nat Genet 15(4):377–380

Novikoff PM, Novikoff AB (1972) Peroxisomes in absorptive cells of mammalian small intestine. J Cell Biol 53(2):532–560

Opperdoes FR, Borst P (1977) Localization of nine glycolytic enzymes in a microbody-like organelle in *Trypanosoma brucei*: the glycosome. FEBS Lett 80(2):360–364

Osumi T, Hashimoto T (1978) Enhancement of fatty acyl-CoA oxidizing activity in rat liver peroxisomes by di-(2-ethylhexyl)phthalate. J Biochem 83(5):1361–1365

Purdue PE et al (1997) Rhizomelic chondrodysplasia punctata is caused by deficiency of human PEX7, a homologue of the yeast PTS2 receptor. Nat Genet 15(4):381–384

Reddy JK, Krishnakantha TP (1975) Hepatic peroxisome proliferation: induction by two novel compounds structurally unrelated to clofibrate. Science 190(4216):787–789
Rhodin J (1954) Correlation of ultrastructural organization and function in normal and experimentally changed proximal convoluted tubule cells of the mouse kidney. PhD thesis.
Scotto JM et al (1982) Infantile phytanic acid storage disease, a possible variant of Refsum's disease: three cases, including ultrastructural studies of the liver. J Inherit Metab Dis 5(2):83–90
Shimozawa N et al (1992) A human gene responsible for Zellweger syndrome that affects peroxisome assembly. Science 255(5048):1132–1134
Spranger JW et al (1971) Heterogeneity of Chondrodysplasia punctata. Humangenetik 11(3):190–212
Steinberg G et al (2017) Woronin body-based sealing of septal pores. Fungal Genet Biol 109:53–55
Sugiura A et al (2017) Newly born peroxisomes are a hybrid of mitochondrial and ER-derived pre-peroxisomes. Nature 542(7640):251–254
Svoboda DJ, Azarnoff DL (1966) Response of hepatic microbodies to a hypolipidemic agent, ethyl chlorophenoxyisobutyrate (CPIB). J Cell Biol 30(2):442–450
Thorp JM, Waring WS (1962) Modification of metabolism and distribution of lipids by ethyl chlorophenoxyisobutyrate. Nature 194:948–949
Trompier D, Savary S (2013) X-linked adrenoleukodystrophy. Colloquium series on the genetic basis of human disease #4, Morgan and Claypool Life Sciences, Williston, USA
Tsukamoto T et al (1990) Isolation and characterization of Chinese hamster ovary cell mutants defective in assembly of peroxisomes. J Cell Biol 110(3):651–660
Tsukamoto T et al (1991) Restoration by a 35K membrane protein of peroxisome assembly in a peroxisome-deficient mammalian cell mutant. Nature 350(6313):77–81
Ulrich J et al (1978) Adrenoleukodystrophy. Preliminary report of a connatal case. Light- and electron microscopical, immunohistochemical and biochemical findings. Acta Neuropathol 43(1–2):77–83
Vamecq J et al (2014) The human peroxisome in health and disease: the story of an oddity becoming a vital organelle. Biochimie 98:4–15
Wanders RJ et al (2017) Clinical and laboratory diagnosis of peroxisomal disorders. Methods Mol Biol 1595:329–342

Chapter 2
Peroxisome Biogenesis

Kosuke Kawaguchi and Tsuneo Imanaka

Abstract It is considered most likely at present that peroxisomes are endoplasmic reticulum (ER)-derived organelles. The biosynthesis of peroxisomes in mammals involves three different processes, the formation of the pre-peroxisome from the ER, the import of the peroxisomal membrane and matrix proteins to the pre-peroxisome, and the growth and division of the peroxisome. Very recently a new process was reported, i.e. that pre-peroxisomes are formed by the fusion of vesicles derived from the ER and outer mitochondrial membranes. Based on recent findings, we discuss where the pre-peroxisomes are formed and become organized into mature peroxisomes. It is known that many proteins, called "peroxins", are encoded by *PEX* genes and involved in peroxisome biogenesis, including the targeting of peroxisomal matrix and membrane proteins. To date, 36 peroxins have been identified. Here, recent progress in the mechanisms by which peroxisomal matrix and membrane proteins are targeted to the peroxisome are discussed. In addition, the selective targeting of ATP-binding cassette (ABC) transporter subfamily D to peroxisomes as well as lysosomes is also covered.

Keywords Peroxisome · Biogenesis · Endoplasmic reticulum · Mitochondria · Targeting · ABC transporter

Abbreviations

ABC	ATP-binding cassette
ER	Endoplasmic reticulum
PMP	Peroxisomal membrane protein

K. Kawaguchi
Graduate School of Medicine and Pharmaceutical Sciences, University of Toyama, Toyama, Japan
e-mail: kkawa@pha.u-toyama.ac.jp

T. Imanaka (✉)
Faculty of Pharmaceutical Sciences, Hiroshima International University, Kure, Hiroshima, Japan

University of Toyama, Toyama, Japan
e-mail: imanaka@hirokoku-u.ac.jp

T. Imanaka, N. Shimozawa (eds.), *Peroxisomes: Biogenesis, Function, and Role in Human Disease*, https://doi.org/10.1007/978-981-15-1169-1_2

PPV	Pre-peroxisomal vesicle
PTS	Peroxisomal targeting signal
TA	Tail-anchored
VLCFA	Very long chain fatty acid

2.1 Introduction

Peroxisomes are single-membrane bound organelles that are ubiquitous in almost all eukaryotic cells. These organelles are involved in a variety of metabolic processes, including the β-oxidation of fatty acids, especially very long chain fatty acids (VLCFA), and, in mammals, the synthesis of ether-phospholipids and bile acids (Waterham et al. 2016; Honsho and Fujiki 2017; Walker et al. 2018). Early studies suggested that peroxisomes, like mitochondria and chloroplasts, are autonomous organelles exclusively derived from the growth and division of pre-existing peroxisomes, because peroxisomal matrix and membrane proteins had been shown to be synthesized on cytosolic free polysomes and imported to pre-existing peroxisomes (de Duve 1982; Lazarow and Fujiki 1985). However, it was subsequently found that there are no peroxisomes in certain mutant cells lacking the proteins required for peroxisome biogenesis, and peroxisomes appear in these cells upon the introduction of the genes responsible for peroxisome biogenesis (Hohfeld et al. 1991; Wiemer et al. 1996). These findings indicate that peroxisome biogenesis occurs *de novo*. Furthermore, it has been shown that endoplasmic reticulum (ER) is involved in *de novo* peroxisome biogenesis in *Saccharomyces cerevisiae* as well as mammals (Tabak et al. 2003; Hoepfner et al. 2005).

It is currently thought that peroxisomes arise *de novo* from the ER as pre-peroxisome vesicles. During the fusion of these vesicles as well as the post-translational transport of peroxisomal membrane and matrix proteins, pre-peroxisomes grow into mature peroxisomes and divide in two. However, it should be note that when peroxisomes already exist, newly synthesized peroxisomal membrane proteins (PMPs) are directly transported to peroxisomes, as in a "growth and division" model.

The proteins involved in peroxisome biogenesis are known as peroxins, referred to as PEX or Pex proteins in mammals and yeast, respectively (Fujiki 2016; Farre et al. 2019), and 36 peroxins are presently identified across a variety of species (Table 2.1). Most of these peroxins are essential for the import of peroxisomal matrix proteins, and the rest are involved in the sorting of PMPs and peroxisome proliferation.

Among the PMPs, ATP-binding cassette (ABC) transporter subfamily D exists on peroxisomal membranes. ABCD1–4 belong to subfamily D in mammals and ABCD3 is one of the major components of the mammalian peroxisome (Imanaka et al. 1999). ABCD1–3 are involved in the transport of the various fatty acids including VLCFA, branched chain fatty acids and bile acid intermediates, and di- and tri-hydroxycholestanoyl-CoA (DHCA and THCA) for β-oxidation (see Chap. 4). The

Table 2.1 List of peroxins

		Identified in		
Peroxin	Property	Mammal	Yeast	Fungi
Pex1p	AAA-ATPase complex	○	○	○
Pex2p	Ubiquitin ligase (E3), RING-subcomplex	○	○	○
Pex3p	Membrane anchor of Pex19p	○	○	○
Pex4p	Ubiquitin conjugating enzyme (E2)		○	○
Pex5p	PTS1-receptor, PTS2-co-receptor	○	○	○
Pex6p	AAA-ATPase complex	○	○	○
Pex7p	PTS2-receptor	○	○	○
Pex8p	might be involved in cargo release		○	○
Pex9p	PTS1-receptor under a strictly defined growth condition		○[a]	
Pex10p	Ubiquitin ligase (E3), RING-subcomplex	○	○	○
Pex11p	Membrane elongation, recruits the fission machinery	○	○	○
Pex12p	Ubiquitin ligase (E3), RING-subcomplex	○	○	○
Pex13p	PTS-receptor docking complex at the peroxisomal membrane	○	○	○
Pex14p	PTS-receptor docking complex at the peroxisomal membrane	○	○	○
Pex15p	Membrane anchor of AAA-ATPase complex		○[a]	
Pex16p	RecruitsPMPs into ER	○	○[b]	○
Pex17p	PTS-receptor docking complex at the peroxisomal membrane		○	
Pex18p	PTS2-co-receptor		○[a]	
Pex19p	Chaperon and receptor of PMPs	○	○	○
Pex20p	PTS2-co-receptor		○[b,c]	○
Pex21p	PTS2-co-receptor		○[a]	
Pex22p	Subcomplex of ubiquitin conjugating enzyme (E2)		○	○
Pex23p	Comprise a family of dysferlin domain-containing peroxin	○	○	○
Pex24p	Required for peroxisome assembly		○	○
Pex25p	Membrane elongation, recruits the fission machinery		○	
Pex26p	Membrane anchor of AAA-ATPase complex	○	○	○
Pex27p	Negatively affects fission		○[a]	
Pex28p	Controlsperoxisome size and proliferation		○[a]	
Pex29p	Forms ER subdomain for PPV exit site		○	
Pex30p	Forms ER subdomain for PPV exit site		○[a]	
Pex31p	Forms ER subdomain for PPV exit site		○[a]	
Pex32p	Controls peroxisome size and number		○[a,c]	
Pex33p	PTS-receptor docking complex at the peroxisomal membrane			○
Pex34p	Functional ortholog of mammalian Pex16p		○[a]	
Pex35p	Regulates peroxisome size and abundance		○[a]	
Pex36p	Functional ortholog of mammalian Pex16p		○[c]	

Identified in [a]*Saccharomyces cerevisiae*, [b]*Yarrowia lipolytica* or [c]*Komagataella phaffii*

ABCD1–3 are targeted to peroxisomes in a Pex19p-dependent manner. In contrast, ABCD4 is targeted to lysosomes via the ER since ABCD4 lacks a domain for the interaction of Pex19p, and is co-translationally inserted into the ER membrane. We discuss the organelle selective translocation of ABCD1–4.

This chapter will focus on the formation of the pre-peroxisome from the ER, the targeting of peroxisomal matrix and membrane proteins, and the detailed mechanisms underlying the targeting of ABCD proteins to peroxisomes.

2.2 The Location Where the Pre-peroxisomes Are Formed and How They Mature into Peroxisomes

Since the discovery of peroxisomes, their biogenesis has been much debated. An ER origin of the peroxisome was initially proposed based on electron microscopy studies in mammalian cells (Novikoff and Novikoff 1972), as well as subcellular fractionation studies in plants (Gonzalez and Beevers 1976). Subsequently, it was revealed that PMPs are synthesized on free polysomes in the cytosol and post-translationally inserted into the peroxisomal membrane (Fujiki et al. 1984; Suzuki et al. 1987; Imanaka et al. 1996). Therefore, peroxisomes came to be considered as autonomous organelles that arise exclusively by the growth and division of pre-existing peroxisomes (Lazarow and Fujiki 1985).

In the process of a characterization of mammalian cells having a *PEX* gene mutation, it was revealed that peroxisomes were not detected in cells with a mutation in either the *PEX3* or *PEX19* gene, later the *PEX16* gene, and that peroxisomes appeared upon transfection with a normal cDNA of these genes. It is difficult to explain this observation from the growth and division model. Furthermore, the newly synthesized Pex3p was shown to be inserted into the ER membranes in mammals and *S. cerevisiae*. It is considered most likely at present that peroxisomes are ER-derived organelles. Three peroxins, Pex3p, Pex19p, and Pex16p, play a critical role in peroxisome membrane biogenesis in mammals. The functional orthologs of Pex16p are Pex34p and Pex36p in *S. cerevisiae* and *Komagataella phaffii* (formerly called *Pichia pastoris*), respectively. The steps for *de novo* peroxisome formation are described below (Fig. 2.1).

2.2.1 Insertion of PMPs into the ER Membrane and Their Sorting to the Specific Domain

Not only Pex3p and Pex16p, but also several PMPs, transit to peroxisomes via the ER, and the pathway is conserved across species from yeast to mammals (Elgersma et al. 1997; Titorenko and Rachubinski 1998; Lisenbee et al. 2003; Karnik and Trelease 2005; Kragt et al. 2005; Kim et al. 2006; van der Zand et al. 2010; Schrul and Kopito 2016). For protein targeting to and insertion into the ER, two pathways are highly conserved. One is the classical co-translational pathway, utilized by most membrane proteins, involving targeting by the signal recognition particle (SRP) followed by insertion via the Sec61 translocon (Rapoport 2007). The other is post-

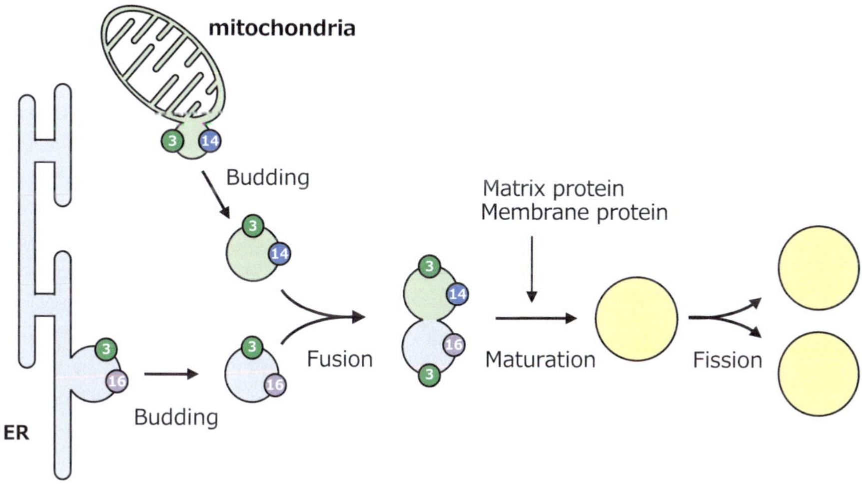

Fig. 2.1 Schematic representation of *de novo* peroxisome biogenesis in mammalian cells. The targeting of mammalian Pex16p to the ER was co-translational but not in a Pex3p- and Pex19p-dependent manner. In contrast, Pex3p requires Pex16p for ER sorting. After the insertion of PMPs into the ER, the PMPs must to be sorted to specific sites in the ER where pre-peroxisomal vesicles (PPVs) exit. Then the ER-derived PPVs might be fused with the mitochondria-derived PPVs. Pre-peroxisomes are post-translationally received PMPs and matrix proteins that become mature peroxisomes. The peroxisomes are then divided into two peroxisomes through elongation, constriction, and scission

translational insertion of tail-anchored (TA) proteins mediated by cytosolic receptor proteins, Get3 in yeast and ASNA-1/TRC40 in mammals (Borgese and Fasana 2011; Hegde and Keenan 2011).

The first direct evidence for involvement of the ER in peroxisome biogenesis was shown in *Yarrowia lipolytica* without any overexpression of PMPs (Titorenko and Rachubinski 1998). Two peroxins, Pex2p and Pex16p, were shown to be delivered to peroxisomes via the ER depending on Srp54p, a subunit of SRP. In mammalian cells, Pex16p expressed at endogenous levels is located mostly on peroxisomes. However, when expressed in Pex16p-deficient cells lacking peroxisomes, Pex16p localizes to the ER (Kim et al. 2006). The targeting of mammalian Pex16p to the ER was co-translational, similar to Pex3p, and in a SRP-dependent but not Pex3p- or Pex19p-dependent manner. Meanwhile Pex3p itself requires Pex16p for ER sorting (Aranovich et al. 2014). In *S. cerevisiae*, Pex3p, Pex8p, Pex13p and Pex14p were inserted into the ER in a Sec61 complex dependent manner (van der Zand et al. 2010; Thoms et al. 2012).

The trafficking of TA-PMPs via the ER is not consistent in yeast and mammalian cells. In *S. cerevisiae*, a TA peroxin, Pex15p, interacts with Get3 and is then inserted into the ER before it is finally translocated to the peroxisomes. Pex15p requires the Get3 complex, since Pex15p mis-localizes to the mitochondria in its absence (Schuldiner et al. 2008). In contrast to yeast, the TA peroxin Pex26p, the mammalian functional ortholog of yeast Pex15p, does not require ASNA-1 for peroxisomal targeting, but rather is directly transported to peroxisomes in a manner dependent on Pex19p, a peroxin responsible for the targeting of PMPs to peroxisomes (see below) (Yagita et al. 2013).

After the insertion of PMPs into the ER, they must to be sorted to specific ER sites where pre-peroxisomal vesicles (PPVs) exit. The signals required for the ER insertion and for intra-ER sorting have been characterized in detail in yeast Pex3p (Fakieh et al. 2013). The NH_2-terminal 17-amino acid region, conserved in its human and *Drosophila* homologs, has two signals that are each sufficient for sorting to the exit site. Two other types of intra-ER sorting of PMPs to the exit site have been revealed in *K. phaffii* (Agrawal et al. 2016, 2017). One is Pex3p- and Pex19p-independent sorting with the docking subcomplex proteins (Pex13p, Pex14p and Pex17p), and Pex3p. The other is Pex3p and Pex19p dependent sorting of RING (really interesting new gene) domain PMPs (Pex2p, Pex10p and Pex12p), and Pex11pγ. In mammalian cells, Pex16p is essential for the Pex19p-independent recruitment of both Pex3p and Pmp34p to the ER or for their intra-ER sorting (Hua et al. 2015). Mutational analysis of Pex16p identified the molecular targeting signals involved in its ER-to-peroxisome trafficking and the domain essential for PMP recruitment at the ER. The specific domain (AA.66–103) and adjacent transmembrane region are necessary for PMP recruitment at the ER, and a separate region (AA.71–81) serves as the ER-to-peroxisome signal.

2.2.2 *Budding and Fusion of PPV*

Little is known about the PPV exit site, but it seems to be a stable ER subdomain like the ER exit site. In *S. cerevisiae*, Pex30p and its paralog, Pex31p, were shown to be enriched in ER subdomains that serve as the PPV exit site (Joshi et al. 2016). Moreover, PPVs and even lipid droplets (LDs) bud from similar ER domains through a process requiring cooperation between Pex30p and seipin, a widely conserved protein frequently mutated in familial forms of lipodystrophy. Simultaneous deletion of these components significantly inhibits the budding of PPVs and LDs, indicating their partially redundant roles in organelle biogenesis (Joshi et al. 2018). Many other proteins, such as secretory proteins, ESCRT-III proteins, Pex19p and Pex29p in yeast (Titorenko and Rachubinski 1998; Perry et al. 2009; Agrawal et al. 2011; Mast et al. 2016, 2018), and Sec16B in mammals (Tani et al. 2011; Yonekawa et al. 2011), are implicated in PPV budding.

It is particularly important that peroxisomal matrix protein import should not occur into PPV buds while PMPs reside in the ER. This requirement appears to be met be segregating PMPs into distinct ER-derived PPVs (Agrawal et al. 2016; van der Zand et al. 2012). In *S. cerevisiae*, one type of PPV contains the docking subcomplex (Pex13p, Pex14p and Pex17p), while the other has components of the RING-subcomplex (Pex2p, Pex10p and Pex12p) (van der Zand et al. 2012). It was shown in *K. phaffii* that one PPV contains, in addition to the docking complex, the

PMPs, Pex3p and two RING proteins, Pex10p and Pex12p, whereas the other contains Pex2p, Pex3p and Pex11pγ (Agrawal et al. 2016; Farre et al. 2017).

Moreover, a recent study in mammals suggests that some PPVs might also originate from mitochondria (Sugiura et al. 2017). The translocation of exogenously expressed Pex3p-YFP or Pex16p-YFP in Zellweger syndrome fibroblasts lacking peroxisomes by a mutation in either the *PEX3* or *PEX16* gene was investigated. Interestingly, Pex3p-YFP was targeted to the outer mitochondrial membranes with endogenous Pex14p in the cells lacking Pex3p, and Pex16p-YFP was targeted to the ER in a Pex3p- and Pex19p-dependent manner. These observations show that Pex3p and Pex14p are released into vesicles and fuse with ER-derived vesicles containing Pex16p. In early studies, it was shown that many overexpressed PMPs are targeted to mitochondria in mutant cell lines lacking peroxisomes. In addition, it has been suggested that human Pex3p is co-translationally integrated into ER and exits the ER in budding vesicles in an *in vitro* assay (Mayerhofer et al. 2016). The expression levels of Pex3p and Pex16p might affect their targeting. Detailed analysis of the targeting of Pex3p and Pex16p using endogenous expression levels is required.

2.2.3 The Elongation and Fission of Peroxisomes

Pre-peroxisomes are post-translationally received PMPs and matrix proteins and become mature peroxisomes. The peroxisomes are then divided into two peroxisomes through elongation, constriction, and scission. Pex11p is involved in the elongation step. In mammals, peroxisomes are associated with microtubules through Ras GTPase and MIRO1, an adaptor linking mammalian peroxisomes to microtubules (Castro et al. 2018). As MIRO1 has been shown to mediate pulling forces in several cell types, it is believed to contribute to peroxisome elongation through microtubules together with Pex11p, especially Pex11pβ. However, peroxisomes can elongate in the presence of microtubule-depolymerizing drugs, suggesting that other pulling forces are involved in peroxisome elongation with Pex11p (Passmore et al. 2017). How peroxisomes constrict is not well understood at present. However, as peroxisomes seem to share the components of a common division machinery with mitochondria, the multiple constriction steps involved in mitochondrial division might be helpful in guiding peroxisome research.

Peroxisome fission starts in *K. phaffii* with the phosphorylation of Pex11p(Ser173) that stimulates its interaction with the adaptor, Fis1p (Joshi et al. 2012). Fis1p then recruits the peripheral receptors, Mdv1p and/or Caf4p (Motley et al. 2008). They assemble a Dnm1p ring around the peroxisome constriction site. Mammals do not have homologs of these proteins, and Dlp1p is recruited to peroxisomes by Mffp and Fis1p (Loson et al. 2013). Yeast Dnm1p interacts with Fis1p and two Pex11p helices named B1 and B3. The hydrolysis of GTP by Dnm1p, enhanced by the interaction with the B3 helix of Pex11p, leads to a constriction that divides the peroxisome. Dnm1p is required for scission (Williams et al. 2015). Dnm1p forms a ring-like structure around membranes, and the hydrolysis of GTP leads to a constriction that divides the organelle.

2.3 Peroxisomal Matrix Protein Targeting

The peroxisomal matrix proteins are synthesized in the cytosol and imported to peroxisomes in a post-translational manner. Protein import into peroxisomes differs from protein import into other organelles such as mitochondria and chloroplasts (Leon et al. 2006a). Proteins may be imported in a fully folded, oligomeric and cofactor bounded state. The import process can be divided into four steps: (1) cargo recognition in the cytosol, (2) docking of the receptor-cargo complex at the peroxisomal membrane, (3) cargo translocation across the membrane, and (4) receptor recycling (Fig. 2.2).

2.3.1 Cargo Recognition in the Cytosol

The majority of the matrix proteins possess a COOH-terminal type 1 peroxisomal targeting signal (PTS1), which is defined as the tripeptide SKL and its variants in the consensus (S/A/C)-(K/R/H)-(L/A) (Gould et al. 1987, 1990; Lametschwandtner et al. 1998). PTS1 is recognized by the receptor Pex5p, which contains two separate domains: a COOH-terminal tetratricopeptide repeat (TPR) domain providing a high affinity PTS1-binding site (Stanley and Wilmanns 2006) and NH_2-terminal domain binding with other peroxins (Stanley et al. 2006).

PTS type 2 (PTS2) conforms to the motif R-(L/V/I/Q)-xx-(L/V/I/H)-(L/S/G/A)-x-(H/Q)-(L/A) and is localized close to the NH_2-terminal of the protein (Swinkels et al. 1991; Lazarow 2006). Few proteins (only four in human and three in *S. cerevisiae*) are known to use PTS2 (Grunau et al. 2009; Jung et al. 2010), whereas in plants, approximately one third of peroxisomal matrix proteins possess PTS2 (Lingner et al. 2011), The PTS2-dependent pathway is absent in *Caenorhabditis elegans*, *Drosophila melanogaster* and diatoms (Motley et al. 2000; Faust et al. 2012; Fodor et al. 2012). PTS2-cargo is recognized by Pex7p, which contains several tryptophan-aspartic acid (WD) repeats that mediate the binding (Marzioch et al. 1994). Whereas Pex5p is sufficient for targeting PTS1-containing proteins to the peroxisomal membrane, Pex7p requires the assistance of species-specific auxiliary proteins. These PTS2-co-receptors are the orthologous redundant Pex18p and Pex21p in *S. cerevisiae*, Pex20p in most other yeast and fungi, and Pex5pL, a longer splice variant of the PTS1-receptor Pex5p, in mammals and plants (Schliebs and Kunau 2006).

In addition to the PTS1- or PTS2-dependent pathways, several matrix proteins may be co-imported in association with PTS-cargo proteins. This mechanism is known as "piggy-back import" and has been reported for the enoyl-CoA isomerases

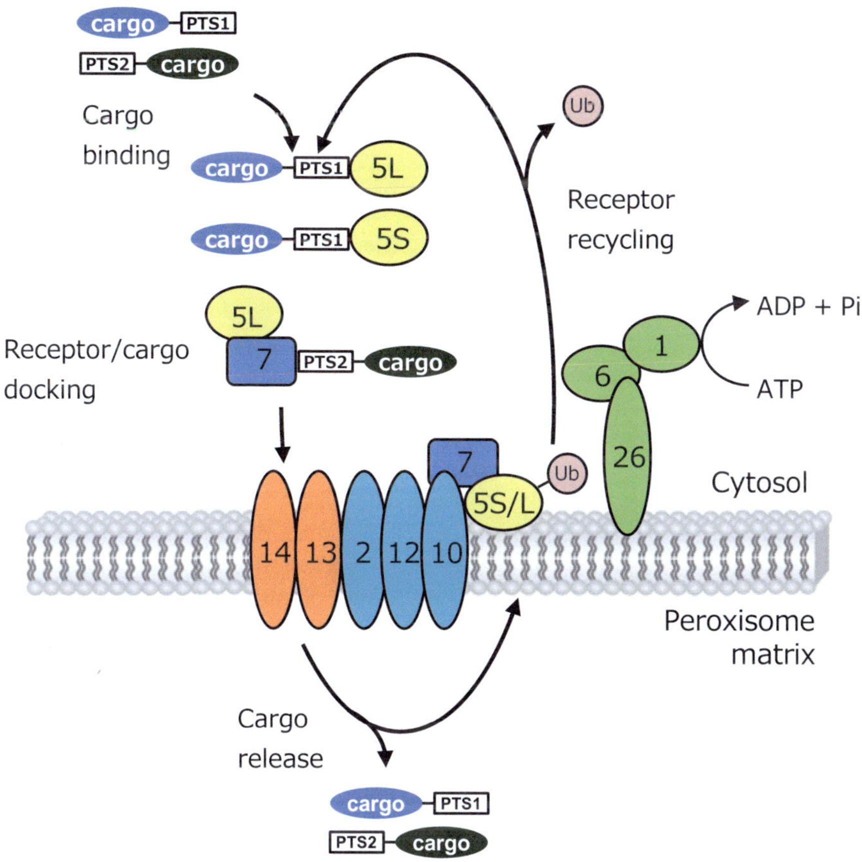

Fig. 2.2 Peroxisomal matrix protein import cycle in mammalian cells. Peroxisome matrix proteins harboring PTS1 or PTS2 are recognized by the receptors Pex5p and Pex7p, respectively, in the cytosol. The cargo-bound receptors associate with the peroxisomal membrane via a docking complex that consists of the core components Pex13p and Pex14p. After docking of the cargo-loaded receptors, the cargo must be translocated across the membrane and released inside the peroxisome. The mechanism by which cargo is released into the peroxisomal lumen is still unknown. After cargo release into the peroxisomal lumen, the receptors are mono-ubiquitinated at the peroxisomal membrane by the RING protein subcomplex, Pex2p, Pex10p and Pex12p. After mono-ubiquitination, PTS-receptors are transported to the cytosol, a process mediated by ATPase associated with diverse cellular activities (AAA)-ATPase complex composed of Pex1p and Pex6p. After being exported back to the cytosol, PTS-receptors are deubiquitinated and then reused in the next round of import

Eci1p and Dci1p in *S. cerevisiae* (Yang et al. 2001), the five acyl-CoA oxidase isoforms in *Y. lipolytica* (Titorenko et al. 2002) and Cu/Zn superoxide dismutase in mammals (Islinger et al. 2009). Furthermore, it has been reported that other proteins, like acyl-CoA oxidase from *S. cerevisiae* (Klein et al. 2002) and alcohol oxidase from *Ogataea polymorpha* (formerly called *Hansenula polymorpha*) (Gunkel et al. 2004), interact directly with Pex5p in a PTS1-independent manner. These proteins bind to the NH_2-terminal region of Pex5p without any need of the TPR domain required for PTS1 recognition. This process is called "non-PTS import" (Gunkel et al. 2004).

2.3.2 Docking of Receptor-Cargo Complex at the Peroxisomal Membrane

After the recognition of cargo in the cytosol, the cargo-bound receptors associate with the peroxisomal membrane via a docking complex comprised of the core components Pex13p and Pex14p. Pex13p is an integral membrane protein, which interacts with Pex14p via both the COOH-terminal Src-homology-3 (SH3) domain and an intraperoxisomal binding site, and also with PTS1-receptor Pex5p and PTS2-receptor Pex7p via the SH3-domain and NH_2-terminus, respectively (Pires et al. 2003; Schell-Steven et al. 2005; Williams and Distel 2006). Pex14p harbors a proline-rich region that associates with the SH3-domain of Pex13p, and also interacts with both PTS-receptors at distinct sites (Niederhoff et al. 2005). Pex14p is generally considered as the initial binding partner of cargo-bound PTS1-receptor Pex5p via two hydrophobic cavities, which interact with the characteristic WxxxF/Y motifs of Pex5p (Schliebs et al. 1999; Saidowsky et al. 2001).

In yeast, Pex17p is involved in the docking complex along with Pex13p and Pex14p. Yeast Pex17p is a peripheral membrane protein of unknown function, which associates with peroxisomes via Pex14p but not with Pex5p. A deficiency of Pex17p results in a failure to import matrix proteins into peroxisomes via either the PTS1- or PTS2-dependent pathways for unknown reasons (Huhse et al. 1998). A homolog of Pex17p has not yet been identified in higher eukaryotes. These docking complexes associate with another subcomplex comprised of three conserved RING domain proteins, Pex2p, Pex10p, and Pex12p, which possess E3 ligase activities. Together, the docking and RING subcomplexes form the importomer complex (Rayapuram and Subramani 2006; Meinecke et al. 2010).

2.3.3 Cargo Translocation Across the Membrane

After docking with the cargo-loaded receptors, the cargo must be translocated across the membrane and released inside the peroxisome. However, these processes are still elusive. Over the years, several observations provided insights into how folded and oligomeric proteins are transported across the membrane. The most convincing model is that the translocation occurs through a transiently opened pore. This pore is proposed to be composed of Pex14p and Pex13p (Grou et al. 2009) or Pex14p and the PTS1-receptor Pex5p (Erdmann and Schliebs 2005). Indeed, Pex5p possesses certain unique features. Pex5p changes its membrane topology during protein import cascade, adjusts to a carbonate-resistant and a partial protease-protected state, and thereby behaves like an integral membrane protein (Gouveia et al. 2000, 2003a; Platta et al. 2005; Kerssen et al. 2006). Furthermore, Pex5p and Pex14p together constitute the minimal translocon required for translocation of intraperoxisomal Pex8p across the peroxisomal membrane in *K. phaffii* (Ma et al. 2009). *In vitro*, the Pex5p-Pex14p complex of *S. cerevisiae* is able to form an ion-conducting channel with a variable pore size of up to 9 nm according to the size of the cargo (Meinecke ct al. 2010). However, the actual architecture of the pore, as well as the driving force for translocation, remains unexplained.

The mechanism of how cargo is released into the peroxisomal lumen is also unknown. An *in vitro* study in *O. polymorpha* suggested that Pex8p might be responsible for this process, because this peroxin is able to dissociate the receptor-cargo complex, and that a change in pH might resolve Pex5p-oligomers into the monomeric form and thereby also lead to the dissociation of the cargo from this receptor (Wang et al. 2003). Another study in *K. phaffii* suggests that Pex8p is concerned with disassembly of Pex5p oligomers and cargo depending on the redox-state (Ma et al. 2013). However, Pex8p is a less widely conserved yeast protein and is not conserved in higher eukaryotes (Kiel et al. 2006). Therefore, it is not easy to establish a consensus model. In mammalian cells, it was demonstrated that the NH_2-terminus of Pex14p takes part in the release of Pex5p-bound PTS1 protein from the translocon into the peroxisomal lumen (Freitas et al. 2011).

2.3.4 Receptor Recycling

After cargo release into the peroxisomal lumen, the receptor and co-receptor are exported to the cytosol and recycled. During these processes, the receptors are mono-ubiquitinated and transported in an ATP-dependent manner (Gouveia et al. 2003b; Oliveira et al. 2003). In general, ubiquitination is a highly conserved post-translational protein modification, resulting in covalent attachment of a 76 amino acid polypeptide, ubiquitin (Goldstein et al. 1975). This attachment is archived by an enzyme-cascade, i.e., a ubiquitin-activating E1 enzyme, a ubiquitin-conjugating E2 enzyme and a ubiquitin-ligating E3 enzyme. During this modification, ubiquitin is activated in an ATP-dependent process by the E1 enzyme, which forms a thioester with the COOH-terminal amino acid (Gly) of ubiquitin. Subsequently, the activated ubiquitin is transferred to the active-site cysteine of the E2 enzyme to form an E2-ubiquitin thioester intermediate. Finally, the E3 ligase transfers ubiquitin from the E2, generally to a lysine residue or NH_2-terminus of the substrate, but also, although rarely, to a cysteine (Kerscher et al. 2006). This mono-ubiquitination is required prior to the export of Pex5p back to the cytosol and is conserved across species from yeast to mammals, and occurs at a conserved NH_2-terminal cysteine residue of Pex5p. Pex4p, anchored to peroxisomal membrane via Pex22p, functions as the E2-conjugating enzyme in yeast (Platta et al. 2007; Williams et al. 2007, 2012). Its association with Pex22p stimulates the full E2 activity of Pex4p

(El Magraoui et al. 2014). As mammalian cells lack Pex4p- and Pex22p-orthologs, the functional related isoforms E2D1/2/3 (UbcH5a/b/c) function as E2 for cysteine mono-ubiquitination (Grou et al. 2008). The RING protein subcomplex, comprised of Pex2p, Pex10p and Pex12p, functions at the peroxisomal membrane as E3 ligase and is responsible for proper mono-ubiquitination of Pex5p (Platta et al. 2009; El Magraoui et al. 2012). The PTS2-coreceptor Pex18p of *S. cerevisiae* and its ortholog Pex20p of *K. phaffii* were found to be ubiquitinated in a manner similar to Pex5p (Purdue and Lazarow 2001; Leon et al. 2006b). Both Pex18p and Pex20p are poly-ubiquitinated on lysine residues and mono-ubiquitinated on cysteines. The mono-ubiquitination is essential for matrix protein import, while the poly-ubiquitination regulates the stability of the PTS2-co-receptors (Leon et al. 2006b; Hensel et al. 2011; Liu and Subramani 2013) and renders the Pex5p a substrate ready for proteasome-mediated degradation (Francisco et al. 2014; Schwartzkopff et al. 2015).

After mono-ubiquitination, PTS-receptors or co-receptors are transported to the cytosol mediated by ATPase associated with diverse cellular activities (AAA)-ATPase complex composed of Pex1p and Pex6p (Platta et al. 2005; Miyata and Fujiki 2005). Pex1p and Pex6p display a dual localization in the cytosol and at the peroxisomal membrane. They are anchored to the peroxisomal membrane by associating with the TA proteins Pex15p, Pex26p and APEM9 in yeast, mammals and plants, respectively (Birschmann et al. 2003; Matsumoto et al. 2003; Goto et al. 2011; Nashiro et al. 2011). It is supposed that the binding and hydrolysis of ATP by AAA-ATPases induce conformational changes which generate the force for pulling the receptor out of the membrane (Fujiki et al. 2012; Grimm et al. 2012). However, the exact molecular mechanisms of substrate recognition and extraction from the membrane are still unclear.

After being exported back to cytosol, PTS-receptors and co-receptors are deubiquitinated by Ubp15p for the mono-ubiquitinated Pex5p in yeast or by USP9X in mammals (Debelyy et al. 2011; Grou et al. 2012). Subsequently, the PTS receptors are reused in the next round of import.

2.4 Targeting of the Peroxisomal Membrane Proteins

As mentioned above, certain PMPs are inserted into the ER membrane prior to their transit to the peroxisome. Here, the direct targeting pathway of PMPs to peroxisomes is described (Fig. 2.3). It has been demonstrated that only three peroxins in mammals, Pex3p, Pex19p and Pex16p, are essential for peroxisomal membrane biogenesis (Lazarow and Fujiki 1985; Fujiki et al. 2006). In the absence of these, there is no peroxisome remnant or "ghost" and PMPs are mistargeted or degraded (Kim et al. 2006).

PMPs are synthesized on cytosolic free polysomes and inserted post-translationally into peroxisomal membranes. The PMP import pathway is distinct from that of matrix protein. Therefore, PMPs harbor neither PTS1 nor PTS2, but

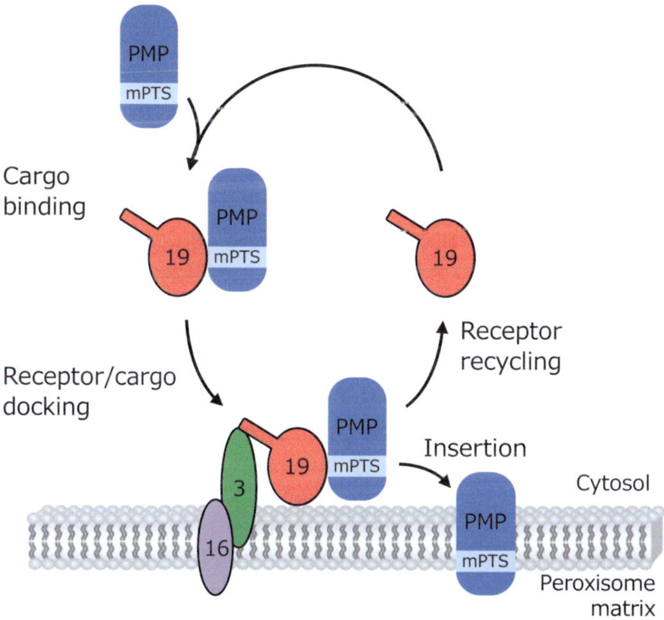

Fig. 2.3 The peroxisomal membrane protein import cycle in mammalian cells. After being translated on cytosolic free polysomes, PMPs are recognized by Pex19p via mPTS. Pex19p functions as both a shuttling PMP receptor and chaperone, with a capacity for self-recovery after PMP release. After the formation of the Pex19p-cargo complex in the cytosol, Pex19p docks on Pex3p at the peroxisomal membrane. Subsequently, PMP is inserted into the peroxisomal membrane and Pex19p recycles back to the cytosol for another round of PMP import

rather, contain a membrane peroxisome targeting signal (mPTS) consisting of a Pex19p binding motif and membrane anchor sequence (Jones et al. 2001). The PMPs that are sorted to the peroxisomal membrane in a Pex19p-depending manner are annotated as class I. The identified mPTS sequences are characterized by a cluster of basic residues or a mixture of basic and hydrophobic amino acids, as well as the presence of one or more short and usually α-helix segments, in addition to at least one transmembrane segment (Jones et al. 2001; Fransen et al. 2001; Rottensteiner et al. 2004). In spite of these similarities in the mPTS motif, a clear consensus cannot be deduced. A subclass of peroxisomal PMPs belonging to TA proteins possesses a common COOH-terminal hydrophobic sequence as the mPTS that is responsible for both targeting and membrane insertion. TA proteins are inserted into the ER membranes by the ER-associated and ATP-dependent Get pathway (Borgese and Fasana 2011; Hegde and Keenan 2011). However, at least two TA proteins, Pex26p and Fis1p, which are involved in peroxisome biogenesis, are sorted to peroxisomes in a Pex19p-dependent manner (Chen et al. 2014). In addition to the Pex19p-dependent pathway, class II PMPs do not require Pex19p-binding for peroxisomal targeting. The Pex19p-independent pathway was initially identified for Pex3p in mammalian cells (Kim et al. 2006; Jones et al. 2004), which traffics via the

ER (Fakieh et al. 2013; Mayerhofer et al. 2016). Since the early studies, many PMPs have been revealed to traffic to peroxisomes via the ER. However, it should also be note that the same PMP may traffic to peroxisomes directly or indirectly. For instance, mammalian Pex3p can also be transported directly to peroxisomes in a Pex16p- and Pex19p-dependent manner (Matsuzaki and Fujiki 2008).

Pex19p is ubiquitous in peroxisome-containing organisms and is a hydrophobic and acidic protein exhibiting a broad binding specificity for PMPs. Among vertebrates, Pex19p are highly conserved and most versions are 299 amino acids in length. In contrast, Pex19p from plants, fungi and amoebozoa are highly diverse in sequence and length, suggesting functional diversity. Pex19p is predominantly cytosolic, but a small yet significant portion of Pex19p interacts with the peroxisomal membrane (Jones et al. 2004; Matsuzono et al. 1999). A farnesylation consensus motif exists at the COOH-terminus of all of the known forms of Pex19p except for the trypanosomatid (Banerjee et al. 2005). Pex19p possesses the ability to interact with most PMPs and stabilizes newly synthesized ones before sorting to the peroxisomal membrane in a Pex3p-dependent manner (Jones et al. 2004; Halbach et al. 2006). Based on these findings, Pex19p functions as both a shuttling PMP receptor and chaperone, with a capacity for self-recovery after PMP release. In addition, it is suggested that Pex19p may function in the regulation of the docking process between cargo-loaded PTS1-receptor Pex5p and Pex14p (Knoops et al. 2014; Veenhuis and van der Klei 2014), and in peroxisomal inheritance by binding directly to the myosin motor protein Myo2p in a Pex19p-farnesylation dependent manner (Otzen et al. 2012).

Pex19p has multiple protein-protein interaction regions: the NH_2-terminus binds to Pex3p and thus mediates peroxisomal docking, a central domain is suggested to play a role in the formation of the PTS-receptor docking complex, and the COOH-terminal domain binds to the mPTS of several PMPs (Shibata et al. 2004; Matsuzono et al. 2006; Sato et al. 2008). The binding affinity of the Pex3p-Pex19p complex has been determined to be less than 10 nM. The high affinity of this complex seems to be caused by an intricate set of interactions beyond a simple binary binding site. The affinity of the NH_2-terminal binding site was determined structurally to be approximately five to ten times less than that full-length Pex3p, but this is still sufficient to establish a stable and constitutive protein/protein complex (Sato et al. 2010). The finding of a second low affinity Pex3p binding site within the central part of Pex19p could help explain the difference in binding affinity, but does not appear to be sufficient to establish binding on its own (Schmidt et al. 2010, 2012). Little is known on the mechanism of the mPTS-mediated binding between PMPs and Pex19p. A crystal structure of the COOH-terminal segment revealed that it forms a largely α-helical bundle, thereby molding the mPTS-binding domain (Schueller et al. 2010). PMP binding data from truncated Pex19p indicated a crucial contribution of the first α-helical bundle in the COOH-terminal domain to PMP binding. The binding affinity corresponds to a dissociation constant K_D of approximately 10 μM and is thus moderate. Furthermore, replacement of a number of exposed hydrophobic residues from this helix leads to loss of mPTS binding, indicating the non-specific hydrophobic interactions to be crucial (Schueller et al. 2010). After the formation of

Pex19p-cargo complexes in the cytosol, Pex19p docks on Pex3p at the peroxisomal membrane (Muntau et al. 2003; Fang et al. 2004), thereby forming a trimeric Pex3p-Pex19p-cargo complex (Pinto et al. 2006). The crucial interaction between Pex3p and Pex19p was reported initially in *S. cerevisiae* (Gotte et al. 1998). The importance of the Pex3p-Pex19p interplay is further highlighted by the observation that an impaired interaction between these proteins contributes to a defective peroxisomal membrane biogenesis in the fibroblasts of a Zellweger Syndrome patient (Muntau et al. 2000).

Pex3p is anchored in the peroxisomal membrane by its NH_2-terminal transmembrane segment, so that it exposes a large COOH-terminal domain towards the cytosol. Similar to Pex19p, vertebrate Pex3p is highly conserved and is 372 or 373 amino acids in length. In contrast, Pex3p from more distantly related organisms such as fungi and plants generally share less than 35% sequence identity with human Pex3p and are divergent in length. Despite these differences, the targeting information is still contained within the NH_2-terminus 46 amino acids (Wiemer et al. 1996; Kammerer et al. 1998; Haan et al. 2002; Tam et al. 2005). The cytosolic region of Pex3p that is comprised of a "twisted six-helix bundle" binds to an NH_2-terminal site of Pex19p that is different from the PMP binding site of Pex19p. Based on this, Pex3p has been proposed to function as a Pex19p-dependent receptor and membrane docking factor (Fang et al. 2004). The Pex19p binding site on Pex3p is distant from the anticipated NH_2-terminal transmembrane anchor. Three Pex3p loop regions contribute to this Pex19p binding site. Several residues, responsible for the Pex3p-Pex19p interaction in both Pex3p and Pex19p, especially hydrophobic residues, are universally conserved among the respective sequences from various species, indicating that the interaction is conserved as well. A cluster of conserved hydrophobic residues near the NH_2-terminal membrane insertion site seems to play a role in facilitating PMP membrane insertion, possibly by deforming the peroxisomal membrane (Huhse et al. 1998; Chen et al. 2014; Schmidt et al. 2012).

Although the structure of the fragments of the Pex3p-Pex19p complex have been determined, little is known about the architecture and conformational transitions this complex may undergo during the mPTS-PMP delivery process into the peroxisomal membrane.

2.5 Organelle Selective Targeting of ABC Transporters

The ATP-binding cassette (ABC) transporters comprise a superfamily of membrane-bound proteins found in almost all organisms from eubacteria to mammals. The architecture of the ABC transporters is composed of two nucleotide-binding domains (NBDs) that catalyze the hydrolysis of ATP and two transmembrane domains (TMDs) that form a translocation pathway. There are 48 ABC transporters in humans that are classified into seven subfamilies, A–G, based on structural organization and amino acid homology (Dassa and Bouige 2001; Vasiliou et al. 2009). The ABC transporters exist on plasma membranes and intracellular compartments such as the mitochondria, peroxisomes, ER, Golgi apparatus and lysosomes, and play important roles in transporting a wide variety of substrates across both extra- and intracellular membranes in order to sustain cellular homeostasis. Defects in their functions are related to various diseases (Ueda 2011).

To date, four ABCD transporters have been identified: adrenoleukodystrophy protein (ALDP/ABCD1), ALDP-related protein (ALDRP/ABCD2), the 70-kDa peroxisomal membrane protein (PMP70/ABCD3) and the PMP70-related protein (P70R/ABCD4) (Kamijo et al. 1990; Mosser et al. 1993; Lombard-Platet et al. 1996; Shani et al. 1997). The ABCD transporters have a predicted ABC half-transporter structure, with one TMD comprised of six transmembrane helixes (TM1–6) and one NBD. ABCD1–3 are known to be peroxisomal proteins and predominantly function as a homodimer (Morita and Imanaka 2012), but a heterodimeric structure has also been suggested (Geillon et al. 2014). It is reported that ABCD1 and ABCD2 are involved in the transport of long and very long chain fatty acids (VLCFAs) or their CoA-derivatives into peroxisomes (van Roermund et al. 2008, 2011). ABCD3 is thought to play an important role in the transport of branched chain acyl-CoA and the bile acid intermediates DHCA and THCA (van Roermund et al. 2014). It is known that ABCD1 and ABCD3 defects are the cause of X-linked adrenoleukodystrophy (X-ALD), a neurodegenerative disease, and hepatosplenomegaly, a liver disease, respectively (Mosser et al. 1993; Ferdinandusse et al. 2015).

The ABCD transporters are translated on free polysomes. ABCD1–3 are sorted to peroxisomes in a Pex19p-dependent manner. An NH_2-terminal hydrophilic region containing an H0 motif, a hydrophobic segment adjacent to the NH_2-terminal portion of TM1, plays an important role in this translocation (Lee et al. 2014). In contrast, ABCD4 hardly interacts with Pex19p because of its lack of the NH_2-terminal H0 motif, and as a result, ABCD4 is recognized by a certain SRP and integrated into the ER membrane (Kashiwayama et al. 2009). Subsequently, ABCD4 is translocated to lysosomes in a manner dependent on the lysosomal targeting ability of lysosomal membrane protein, LMBD1 (Kawaguchi et al. 2016).

In terms of the trafficking of newly synthesized peroxisomal ABCD transporters, ABCD3 has been studied in the greatest detail (Sacksteder et al. 2000; Biermanns and Gartner 2001; Kashiwayama et al. 2005, 2007). As ABCD3 is a hydrophobic integral membrane protein, binding with Pex19p is indispensable for ABCD3 to retain its soluble form and proper conformation for targeting to peroxisomes. In the absence of Pex19p, ABCD3 mostly remains in the cytosol and is rapidly degraded (Kinoshita et al. 1998). Various truncated or mutated ABCD3 were prepared to investigate the region of ABCD3(AA.1–659) that is critical for binding with Pex19p (Kashiwayama et al. 2005). The truncation of the NH_2-terminal 61 amino acids of ABCD3 severely reduced the interaction of ABCD3 with Pex19p. COOH-terminal truncation of TM6 also decreased the interaction, but further deletion of TM3–5 did not result in any additional decrease. In these conditions, the stoichiometry for the binding of ABCD3 to Pex19p is one-to-one. Furthermore, chimeric proteins of the NH_2-terminal 61 amino acids as well as TM5 and TM6 fused with *Escherichia coli* dihydrofolate reductase (DHFR) associate with Pex19p. Therefore, both the NH_2-terminal 61 amino acids and the regions around TM5 and TM6 are required for the efficient interaction of ABCD3 with Pex19p, and the lack of either region leads to a decrease in the interaction. It is deduced that the amino acid sequence xxx-(CFILTVW)-xx-(ACFILQVWY)-(CILV)-xx-(ACFILVWY)-(ILQRV)-xxx is the common Pex19p binding motif in PMPs (Rottensteiner et al. 2004). Eight corresponding regions are found in the ABCD3 sequence. Among them, three sites are found in the TM5–6 region and one overlaps with the isolated Pex19p binding site of ABCD3. However, none of these sites are found in NH_2-terminal 61 amino acids region.

In terms of mPTS, there are no consensus primary amino acid sequences or common structural properties. To characterize the regions responsible for the peroxisomal targeting of ABCD3, various COOH-terminally or NH_2-terminally deleted constructs of ABCD3 fused with GFP were prepared and their intracellular localization determined (Kashiwayama et al. 2007). In the COOH-terminally truncated ABCD3, ABCD3(AA.1–144)-GFP, including TM1 and TM2 of ABCD3, was still sorted to peroxisomes. However, further deletion of TM2 (AA.125–144) led to mislocalization of the ABCD3(AA.1–124)-GFP. The substitution of TM2 in ABCD3(AA.1–144)-GFP for TM4 (AA.224–259) or TM6 (AA.314–347) did not affect the peroxisomal localization. Meanwhile, ABCD3(TM2)-GFP, ABCD3(TM4)-GFP and ABCD3(TM6)-GFP did not localize to peroxisomes by themselves. These facts indicate that the important targeting information of ABCD3 is included within the NH_2-terminal 124-amino acid region and that at least two TM are necessary for proper peroxisomal targeting. In the NH_2-terminal 124-amino acid region, there are three clusters of basic amino acids which are suggested to be the peroxisomal targeting motif of other PMPs (Dyer et al. 1996; Baerends et al. 2000; Honsho and Fujiki 2001; Wang et al. 2001). However, the substitution of these basic amino acids to alanine does not affect the peroxisomal targeting of ABCD3(AA.1–144)-GFP. This finding revealed that these basic amino acid clusters are not essential forw the targeting of ABCD3. In the case of Pmp22p, it is reported that Y-xxx-L-xxx-P-xxx-(K/Q/N) comprise the core of the mPTS of Pmp22p orthologs, the basic cluster in the first peroxisomal matrix loop is not essential for targeting, and the hydrophobic sequence is required for the targeting and/or insertion of Pmp22p (Pause et al. 2000). Based on the hydropathy profile, all peroxisomal ABC proteins possess two hydrophobic segments adjacent to the NH_2-terminal side of TM1. When hydrophobic segments of ABCD3, i.e. Leu^{21}-Leu^{22}-Leu^{23} and Ile^{70}-Leu^{71}, were substituted to hydrophilic amino acids, both ABCD3(AA.1–144 L21Q/L22Q/L23Q)-GFP and ABCD3(AA.1–144 I70N/L71Q)-GFP were degraded by proteasomes and did not exhibit any peroxisomal localization even in the presence of a proteasome inhibitor. As mentioned above, it is important for ABCD3 to interact with Pex19p to maintain its solubility and proper conformation. ABCD3(AA.1–659 L21Q/L22Q/L23Q) was reduced in solubility and the efficiency of the interaction with Pex19p. On the other hand, ABCD3(AA.1–659 I70N/L71Q) existed in a soluble form in the presence of Pex19p and interacted with Pex19p at almost the same level as wild type ABCD3, but this mutant ABCD3 did not localize to peroxisomes. These data indicate that the first set of the NH_2-terminal hydrophobic residues, Leu^{21}-Leu^{22}-Leu^{23}, is essential for the interaction with Pex19p in order to keep ABCD3 soluble and in proper conformation in the sorting process, and the second set, Ile^{70}-Leu^{71}, is necessary for the targeting step after the interaction with Pex19p. Further analysis of the subcellular localization of the NH_2-terminal truncated ABCD3 fused with the COOH-terminus of GFP revealed that ABCD3 possesses a second mPTS in the region of TM5–6 (AA.263–375). In this region, a hydrophobic motif, Ile^{307}-Leu^{308}, also exists. The substitution of Ile^{307} and Leu^{308} to Asn^{307} and Gln^{308}, respectively, led to the failure in peroxisomal localization of ABCD3(AA.1–659 I307N/L308Q)-GFP, although ABCD3(AA.1–659 I307N/L308Q) interacted with Pex19p and was solubilized by Pex19p. The hydrophobic amino acid pairs Ile^{70}-Leu^{71} and Ile^{307}-Leu^{308} adjacent to TMD1 and TMD5, respectively, are essential for the targeting to peroxisomes, but not for binding with Pex19p. This indicates that the targeting element and Pex19p-binding site are functionally separated.

As discussed above, the NH_2-terminal region of ABCD3 is important for binding with Pex19p and targeting to peroxisomes. Furthermore, it has been shown that the NH_2-terminal region of ABCD3 possesses multiple organelle-targeting signals (Iwashita et al. 2010). The TM1–2 region of ABCD3 (AA.81–150) are targeted and inserted into the ER membrane, since a hydrophobic transmembrane segment is recognized by a signal recognition particle and cotranslationally targeted to the ER (Egea et al. 2005). The NH_2-terminal 80 amino acid residues suppress the ER targeting of the TM1–2 regions. Furthermore, the NH_2-terminal nine amino acid residues are essential for the suppression of ER-targeting. Among these residues, Ser^5 is indispensable for the peroxisomal targeting of ABCD3 (Sakaue et al. 2016). It has also been demonstrated that specific binding factors, 50-kDa and 20-kDa proteins, are involved in the ER-targeting suppression system.

A hypothetical scenario for the peroxisomal targeting of ABCD3 is as follows (Fig. 2.4). After being synthesized on free cytosolic ribosomes, the NH_2-terminal short motif is recognized by specific binding factors. This interaction may suppress the function of signal recognition particles and their receptor. Subsequently, ABCD3 interacts with Pex19p at the NH_2-terminal hydrophobic motif constituted by Leu^{21}-Leu^{22}-Leu^{23} and the region of TM5–6. Through this interaction, ABCD3 is retained in a soluble state and proper conformation in the cytosol. Then the ABCD3-Pex19p complex is transported to peroxisomes by the mPTSs located in the NH_2-terminal 124-amino acid region as well as the region of TM5–6. In this transporting process, the hydrophobic amino acid pairs Ile^{70}-Leu^{71} and Ile^{307}-Leu^{308} adjacent to TM1 and TM5, respectively, are essential. Finally, ABCD3 is inserted into the peroxisomal membranes through a putative proteinaceous receptor on the peroxisomal membrane. In this process, at least two TMs are required for proper insertion.

The targeting of ABCD1 to peroxisomes has also been characterized (Halbach et al. 2005). It was shown that the 14-amino acid motif (F-(F/L)-x-(R/Q/K)-(L/F)-(L/I/K)-x-LLKIL-(F/I/V)-P) adjacent to TM1 functions as an mPTS for ABCD1, and it was demonstrated that the substitution or deletion of these hydrophobic residues significantly reduced the targeting efficiency. In particular, the three amino acids Leu^{78}-Leu^{79}-Arg^{80} were shown to be critical for peroxisomal targeting of ABCD1. This region corresponds to Ile^{70}-Leu^{71}-Lys^{72} in ABCD3, which was identified as an mPTS in ABCD3. In ABCD2, a potential Pex19p binding

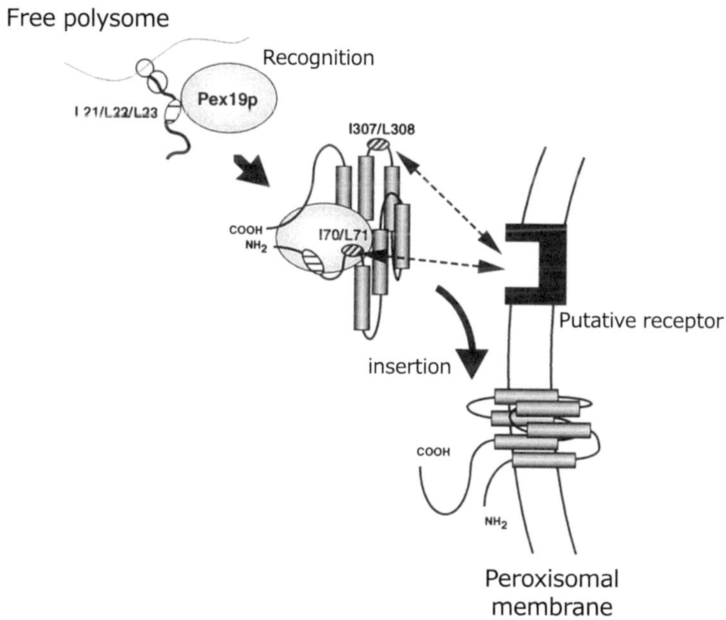

Fig. 2.4 A hypothetical model for the targeting of ABCD3 to the peroxisome. After being synthesized in the cytosol, ABCD3 is recognized by and bound to Pex19p at the NH_2-terminal hydrophobic motif constituted by Leu^{21}-Leu^{22}-Leu^{23} and the region of TM5–TM6. The ABCD3-Pex19p complex is transported to peroxisomes by the mPTSs located in the NH_2-terminal 124-amino acid region and the region of ABCD3(AA.263–375) (the hydrophobicity of Ile^{70}-Leu^{71} and Ile^{307}-Leu^{308} might be essential). Finally, ABCD3 is inserted into peroxisomal membranes through certain unidentified proteinaceous components on these peroxisomal membranes. In this process, at least two TMs are required for correct insertion

site was also identified as ABCD1 (Halbach et al. 2005). This corresponds to AA.84–97, which are localized in proximity to the putative TM1. However, no experimental data are available as yet to support the functionality of this putative Pex19p binding site.

2.6 Concluding Remarks

Since the *PEX* genes were first identified in the early 1990s, our understanding of biogenesis of peroxisome has deepened. It has been revealed that there is a unique mechanism by which matrix proteins are imported into peroxisomes in a complex with their receptors, such as Pex5p, and these receptors are recycled through the ubiquitination system. With regard to PMPs targeting, an indirect pathway via the ER has been elucidated and it is well accepted that the ER-derived vesicles significantly contribute to *de novo* peroxisome biogenesis. In addition, it has been recognized that PMPs targets peroxisomal membranes in a complex with Pex19p through interaction with the Pex3p on the peroxisomes.

Yet, despite such progress, there are unresolved questions which remain. Where pre-peroxisomes are formed is one question. It is important to analyze the sites that newly synthesized endogenous Pex3p or Pex16p, which initiate *de novo* peroxisome biogenesis, target in cells completely lacking peroxisome remnants. We think PPV formation is initiated anywhere that newly synthesized Pex3p or Pex16p has reached. Considering the fact that overexpressed PMPs are mistargeted to mitochondria in mammalian cells in the absence of peroxisomes, where PPV forms might depend on the expression levels of Pex3p and Pex16p. It might also vary depending on where PPV forms among different species and cell types. Recently, it was revealed that peroxisomal membranes (probably PPV) dynamically come into contact with other organelles such as the ER and mitochondria. This contact might affect the place where pre-peroxisome is formed as well. In addition, with regard to targeting of matrix proteins and PMPs to peroxisomes, the precise mechanisms by which receptor proteins such as Pex5p and Pex19p release the cargo into the peroxisomes or their membranes and then return to recycling processes need to be elucidated.

Peroxisomes are highly dynamic organelles and can rapidly change in size, number and protein content in response to alterations in the prevailing environmental conditions. It is of critical importance to determine the precise mechanism of peroxisome maturation through elongation, constriction and scission. Furthermore, it is also important to understand how control is exerted over the number of peroxisomes in cells. It is to be hoped that a better understanding of peroxisome biogenesis will flow from new evidence obtained by ever more sophisticated new experiments in the near future.

Acknowledgments The publication is supported in part by a Grant-in-Aid for Intractable Diseases from the Ministry of health, Labour and Welfare of Japan, and by Early-Career Scientists from the Ministry of Education, Culture, Sports, Science and Technology of Japan. Pacific Edit reviewed the manuscript prior to submission.

References

Agrawal G et al (2011) Cell-free sorting of peroxisomal membrane proteins from the endoplasmic reticulum. Proc Natl Acad Sci U S A 108(22):9113–9118

Agrawal G et al (2016) Distinct requirements for intra-ER sorting and budding of peroxisomal membrane proteins from the ER. J Cell Biol 212(3):335–348

Agrawal G et al (2017) Functional regions of the peroxin Pex19 necessary for peroxisome biogenesis. J Biol Chem 292(27):11547–11560

Aranovich A et al (2014) PEX16 contributes to peroxisome maintenance by constantly trafficking PEX3 via the ER. J Cell Sci 127(Pt 17):3675–3686

Baerends RJ et al (2000) A stretch of positively charged amino acids at the N terminus of *Hansenula polymorpha* Pex3p is involved in incorporation of the protein into the peroxisomal membrane. J Biol Chem 275(14):9986–9995

Banerjee SK et al (2005) Identification of trypanosomatid PEX19: functional characterization reveals impact on cell growth and glycosome size and number. Mol Biochem Parasitol 142(1):47–55

Biermanns M, Gartner J (2001) Targeting elements in the amino-terminal part direct the human 70-kDa peroxisomal integral membrane protein (PMP70) to peroxisomes. Biochem Biophys Res Commun 285(3):649–655

Birschmann I et al (2003) Pex15p of *Saccharomyces cerevisiae* provides a molecular basis for recruitment of the AAA peroxin Pex6p to peroxisomal membranes. Mol Biol Cell 14(6):2226–2236

Borgese N, Fasana E (2011) Targeting pathways of C-tail-anchored proteins. Biochim Biophys Acta 1808(3):937–946

Castro IG et al (2018) A role for mitochondrial Rho GTPase 1 (MIRO1) in motility and membrane dynamics of peroxisomes. Traffic 19(3):229–242

Chen Y et al (2014) Hydrophobic handoff for direct delivery of peroxisome tail-anchored proteins. Nat Commun 5:5790

Dassa E, Bouige P (2001) The ABC of ABCS: a phylogenetic and functional classification of ABC systems in living organisms. Res Microbiol 152(3–4):211–229

de Duve C (1982) Peroxisomes and related particles in historical perspective. Ann N Y Acad Sci 386:1–4

Debelyy MO et al (2011) Ubp15p, a ubiquitin hydrolase associated with the peroxisomal export machinery. J Biol Chem 286(32):28223–28234

Dyer JM et al (1996) The sorting sequence of the peroxisomal integral membrane protein PMP47 is contained within a short hydrophilic loop. J Cell Biol 133(2):269–280

Egea PF et al (2005) Targeting proteins to membranes: structure of the signal recognition particle. Curr Opin Struct Biol 15(2):213–220

El Magraoui F et al (2012) The RING-type ubiquitin ligases Pex2p, Pex10p and Pex12p form a heteromeric complex that displays enhanced activity in an ubiquitin conjugating enzyme-selective manner. FEBS J 279(11):2060–2070

El Magraoui F et al (2014) The cytosolic domain of Pex22p stimulates the Pex4p-dependent ubiquitination of the PTS1-receptor. PLoS One 9(8):e105894

Elgersma Y et al (1997) Overexpression of Pex15p, a phosphorylated peroxisomal integral membrane protein required for peroxisome assembly in *S. cerevisiae*, causes proliferation of the endoplasmic reticulum membrane. EMBO J 16(24):7326–7341

Erdmann R, Schliebs W (2005) Peroxisomal matrix protein import: the transient pore model. Nat Rev Mol Cell Biol 6(9):738–742

Fakieh MH et al (2013) Intra-ER sorting of the peroxisomal membrane protein Pex3 relies on its luminal domain. Biol Open 2(8):829–837

Fang Y et al (2004) PEX3 functions as a PEX19 docking factor in the import of class I peroxisomal membrane proteins. J Cell Biol 164(6):863–875

Farre JC et al (2017) A new yeast peroxin, Pex36, a functional homolog of mammalian PEX16, functions in the ER-to-peroxisome traffic of peroxisomal membrane proteins. J Mol Biol 429(23):3743–3762

Farre JC et al (2019) Peroxisome biogenesis, membrane contact sites, and quality control. EMBO Rep 20(1)

Faust JE et al (2012) An inventory of peroxisomal proteins and pathways in *Drosophila melanogaster*. Traffic 13(10):1378–1392

Ferdinandusse S et al (2015) A novel bile acid biosynthesis defect due to a deficiency of peroxisomal ABCD3. Hum Mol Genet 24(2):361–370

Fodor K et al (2012) Molecular requirements for peroxisomal targeting of alanine-glyoxylate aminotransferase as an essential determinant in primary hyperoxaluria type 1. PLoS Biol 10(4):e1001309

Francisco T et al (2014) Ubiquitin in the peroxisomal protein import pathway. Biochimie 98:29–35

Fransen M et al (2001) Human pex19p binds peroxisomal integral membrane proteins at regions distinct from their sorting sequences. Mol Cell Biol 21(13):4413–4424

Freitas MO et al (2011) PEX5 protein binds monomeric catalase blocking its tetramerization and releases it upon binding the N-terminal domain of PEX14. J Biol Chem 286(47):40509–40519

Fujiki Y (2016) Peroxisome biogenesis and human peroxisome-deficiency disorders. Proc Jpn Acad Ser B Phys Biol Sci 92(10):463–477

Fujiki Y et al (1984) Synthesis of a major integral membrane polypeptide of rat liver peroxisomes on free polysomes. Proc Natl Acad Sci U S A 81(22):7127–7131

Fujiki Y et al (2006) Import of peroxisomal membrane proteins: the interplay of Pex3p- and Pex19p-mediated interactions. Biochim Biophys Acta 1763(12):1639–1646

Fujiki Y et al (2012) New insights into dynamic and functional assembly of the AAA peroxins, Pex1p and Pex6p, and their membrane receptor Pex26p in shuttling of PTS1-receptor Pex5p during peroxisome biogenesis. Biochim Biophys Acta 1823(1):145–149

Geillon F et al (2014) Structure-function analysis of peroxisomal ATP-binding cassette transporters using chimeric dimers. J Biol Chem 289(35):24511–24520

Goldstein G et al (1975) Isolation of a polypeptide that has lymphocyte-differentiating properties and is probably represented universally in living cells. Proc Natl Acad Sci U S A 72(1):11–15

Gonzalez E, Beevers H (1976) Role of the endoplasmic reticulum in glyoxysome formation in castor bean endosperm. Plant Physiol 57(3):406–409

Goto S et al (2011) Arabidopsis ABERRANT PEROXISOME MORPHOLOGY9 is a peroxin that recruits the PEX1-PEX6 complex to peroxisomes. Plant Cell 23(4):1573–1587

Gotte K et al (1998) Pex19p, a farnesylated protein essential for peroxisome biogenesis. Mol Cell Biol 18(1):616–628

Gould SG et al (1987) Identification of a peroxisomal targeting signal at the carboxy terminus of firefly luciferase. J Cell Biol 105(6 Pt 2):2923–2931

Gould SJ et al (1990) Peroxisomal protein import is conserved between yeast, plants, insects and mammals. EMBO J 9(1):85–90

Gouveia AM et al (2000) Characterization of peroxisomal Pex5p from rat liver. Pex5p in the Pex5p-Pex14p membrane complex is a transmembrane protein. J Biol Chem 275(42):32444–32451

Gouveia AM et al (2003a) Insertion of Pex5p into the peroxisomal membrane is cargo protein-dependent. J Biol Chem 278(7):4389–4392

Gouveia AM et al (2003b) Characterization of the peroxisomal cycling receptor, Pex5p, using a cell-free *in vitro* import system. J Biol Chem 278(1):226–232

Grimm I et al (2012) The AAA-type ATPases Pex1p and Pex6p and their role in peroxisomal matrix protein import in *Saccharomyces cerevisiae*. Biochim Biophys Acta 1823(1):150–158

Grou CP et al (2008) Members of the E2D (UbcH5) family mediate the ubiquitination of the conserved cysteine of Pex5p, the peroxisomal import receptor. J Biol Chem 283(21):14190–14197

Grou CP et al (2009) The peroxisomal protein import machinery—a case report of transient ubiquitination with a new flavor. Cell Mol Life Sci 66(2):254–262

Grou CP et al (2012) Identification of ubiquitin-specific protease 9X (USP9X) as a deubiquitinase acting on ubiquitin-peroxin 5 (PEX5) thioester conjugate. J Biol Chem 287(16):12815–12827

Grunau S et al (2009) Peroxisomal targeting of PTS2 pre-import complexes in the yeast *Saccharomyces cerevisiae*. Traffic 10(4):451–460

Gunkel K et al (2004) Routing of *Hansenula polymorpha* alcohol oxidase: an alternative peroxisomal protein-sorting machinery. Mol Biol Cell 15(3):1347–1355

Haan GJ et al (2002) *Hansenula polymorpha* Pex3p is a peripheral component of the peroxisomal membrane. J Biol Chem 277(29):26609–26617

Halbach A et al (2005) Function of the PEX19-binding site of human adrenoleukodystrophy protein as targeting motif in man and yeast. PMP targeting is evolutionarily conserved. J Biol Chem 280(22):21176–21182

Halbach A et al (2006) Targeting of the tail-anchored peroxisomal membrane proteins PEX26 and PEX15 occurs through C-terminal PEX19-binding sites. J Cell Sci 119(Pt 12):2508–2517

Hegde RS, Keenan RJ (2011) Tail-anchored membrane protein insertion into the endoplasmic reticulum. Nat Rev Mol Cell Biol 12(12):787–798
Hensel A et al (2011) Cysteine-dependent ubiquitination of Pex18p is linked to cargo translocation across the peroxisomal membrane. J Biol Chem 286(50):43495–43505
Hoepfner D et al (2005) Contribution of the endoplasmic reticulum to peroxisome formation. Cell 122(1):85–95
Hohfeld J et al (1991) *PAS3*, a *Saccharomyces cerevisiae* gene encoding a peroxisomal integral membrane protein essential for peroxisome biogenesis. J Cell Biol 114(6):1167–1178
Honsho M, Fujiki Y (2001) Topogenesis of peroxisomal membrane protein requires a short, positively charged intervening-loop sequence and flanking hydrophobic segments. study using human membrane protein PMP34. J Biol Chem 276(12):9375–9382
Honsho M, Fujiki Y (2017) Plasmalogen homeostasis—regulation of plasmalogen biosynthesis and its physiological consequence in mammals. FEBS Lett 591(18):2720–2729
Hua R et al (2015) Multiple domains in PEX16 mediate its trafficking and recruitment of peroxisomal proteins to the ER. Traffic 16(8):832–852
Huhse B et al (1998) Pex17p of *Saccharomyces cerevisiae* is a novel peroxin and component of the peroxisomal protein translocation machinery. J Cell Biol 140(1):49–60
Imanaka T et al (1996) Insertion of the 70-kDa peroxisomal membrane protein into peroxisomal membranes *in vivo* and *in vitro*. J Biol Chem 271(7):3706–3713
Imanaka T et al (1999) Characterization of the 70-kDa peroxisomal membrane protein, an ATP binding cassette transporter. J Biol Chem 274(17):11968–11976
Islinger M et al (2009) Hitchhiking of Cu/Zn superoxide dismutase to peroxisomes—evidence for a natural piggyback import mechanism in mammals. Traffic 10(11):1711–1721
Iwashita S et al (2010) Multiple organelle-targeting signals in the N-terminal portion of peroxisomal membrane protein PMP70. J Biochem 147(4):581–590
Jones JM et al (2001) Multiple distinct targeting signals in integral peroxisomal membrane proteins. J Cell Biol 153(6):1141–1150
Jones JM et al (2004) PEX19 is a predominantly cytosolic chaperone and import receptor for class 1 peroxisomal membrane proteins. J Cell Biol 164(1):57–67
Joshi S et al (2012) Phosphorylation-dependent Pex11p and Fis1p interaction regulates peroxisome division. Mol Biol Cell 23(7):1307–1315
Joshi AS et al (2016) A family of membrane-shaping proteins at ER subdomains regulates pre-peroxisomal vesicle biogenesis. J Cell Biol 215(4):515–529
Joshi AS et al (2018) Lipid droplet and peroxisome biogenesis occur at the same ER subdomains. Nat Commun 9(1):2940
Jung S et al (2010) Dynamic changes in the subcellular distribution of Gpd1p in response to cell stress. J Biol Chem 285(9):6739–6749
Kamijo K et al (1990) The 70-kDa peroxisomal membrane protein is a member of the Mdr (P-glycoprotein)-related ATP-binding protein superfamily. J Biol Chem 265(8):4534–4540
Kammerer S et al (1998) Cloning and characterization of the gene encoding the human peroxisomal assembly protein Pex3p. FEBS Lett 429(1):53–60
Karnik SK, Trelease RN (2005) *Arabidopsis* peroxin 16 coexists at steady state in peroxisomes and endoplasmic reticulum. Plant Physiol 138(4):1967–1981
Kashiwayama Y et al (2005) Role of Pex19p in the targeting of PMP70 to peroxisome. Biochim Biophys Acta 1746(2):116–128
Kashiwayama Y et al (2007) Hydrophobic regions adjacent to transmembrane domains 1 and 5 are important for the targeting of the 70-kDa peroxisomal membrane protein. J Biol Chem 282(46):33831–33844
Kashiwayama Y et al (2009) 70-kDa peroxisomal membrane protein related protein (P70R/ABCD4) localizes to endoplasmic reticulum not peroxisomes, and NH_2-terminal hydrophobic property determines the subcellular localization of ABC subfamily D proteins. Exp Cell Res 315(2):190–205

Kawaguchi K et al (2016) Translocation of the ABC transporter ABCD4 from the endoplasmic reticulum to lysosomes requires the escort protein LMBD1. Sci Rep 6:30183
Kerscher O et al (2006) Modification of proteins by ubiquitin and ubiquitin-like proteins. Annu Rev Cell Dev Biol 22:159–180
Kerssen D et al (2006) Membrane association of the cycling peroxisome import receptor Pex5p. J Biol Chem 281(37):27003–27015
Kiel JA et al (2006) *PEX* genes in fungal genomes: common, rare or redundant. Traffic 7(10):1291–1303
Kim PK et al (2006) The origin and maintenance of mammalian peroxisomes involves a *de novo* PEX16-dependent pathway from the ER. J Cell Biol 173(4):521–532
Kinoshita N et al (1998) Newly identified Chinese hamster ovary cell mutants are defective in biogenesis of peroxisomal membrane vesicles (Peroxisomal ghosts), representing a novel complementation group in mammals. J Biol Chem 273(37):24122–24130
Klein AT et al (2002) *Saccharomyces cerevisiae* acyl-CoA oxidase follows a novel, non-PTS1, import pathway into peroxisomes that is dependent on Pex5p. J Biol Chem 277(28):25011–25019
Knoops K et al (2014) Preperoxisomal vesicles can form in the absence of Pex3. J Cell Biol 204(5):659–668
Kragt A et al (2005) Endoplasmic reticulum-directed Pex3p routes to peroxisomes and restores peroxisome formation in a *Saccharomyces cerevisiae pex3*Δ strain. J Biol Chem 280(40):34350–34357
Lametschwandtner G et al (1998) The difference in recognition of terminal tripeptides as peroxisomal targeting signal 1 between yeast and human is due to different affinities of their receptor Pex5p to the cognate signal and to residues adjacent to it. J Biol Chem 273(50):33635–33643
Lazarow PB (2006) The import receptor Pex7p and the PTS2 targeting sequence. Biochim Biophys Acta 1763(12):1599–1604
Lazarow PB, Fujiki Y (1985) Biogenesis of peroxisomes. Annu Rev Cell Biol 1:489–530
Lee A et al (2014) Role of NH_2-terminal hydrophobic motif in the subcellular localization of ATP-binding cassette protein subfamily D: common features in eukaryotic organisms. Biochem Biophys Res Commun 453(3):612–618
Leon S et al (2006a) Uniqueness of the mechanism of protein import into the peroxisome matrix: transport of folded, co-factor-bound and oligomeric proteins by shuttling receptors. Biochim Biophys Acta 1763(12):1552–1564
Leon S et al (2006b) Dynamics of the peroxisomal import cycle of PpPex20p: ubiquitin-dependent localization and regulation. J Cell Biol 172(1):67–78
Lingner T et al (2011) Identification of novel plant peroxisomal targeting signals by a combination of machine learning methods and in vivo subcellular targeting analyses. Plant Cell 23(4):1556–1572
Lisenbee CS et al (2003) Peroxisomal ascorbate peroxidase resides within a subdomain of rough endoplasmic reticulum in wild-type Arabidopsis cells. Plant Physiol 132(2):870–882
Liu X, Subramani S (2013) Unique requirements for mono- and polyubiquitination of the peroxisomal targeting signal co-receptor, Pex20. J Biol Chem 288(10):7230–7240
Lombard-Platet G et al (1996) A close relative of the adrenoleukodystrophy (*ALD*) gene codes for a peroxisomal protein with a specific expression pattern. Proc Natl Acad Sci U S A 93(3):1265–1269
Loson OC et al (2013) Fis1, Mff, MiD49, and MiD51 mediate Drp1 recruitment in mitochondrial fission. Mol Biol Cell 24(5):659–667
Ma C et al (2009) The peroxisomal matrix import of Pex8p requires only PTS receptors and Pex14p. Mol Biol Cell 20(16):3680–3689
Ma C et al (2013) Redox-regulated cargo binding and release by the peroxisomal targeting signal receptor, Pex5. J Biol Chem 288(38):27220–27231
Marzioch M et al (1994) *PAS7* encodes a novel yeast member of the WD-40 protein family essential for import of 3-oxoacyl-CoA thiolase, a PTS2-containing protein, into peroxisomes. EMBO J 13(20):4908–4918

Mast FD et al (2016) Peroxins Pex30 and Pex29 dynamically associate with reticulons to regulate peroxisome biogenesis from the endoplasmic reticulum. J Biol Chem 291(30):15408–15427
Mast FD et al (2018) ESCRT-III is required for scissioning new peroxisomes from the endoplasmic reticulum. J Cell Biol 217(6):2087–2102
Matsumoto N et al (2003) The pathogenic peroxin Pex26p recruits the Pex1p-Pex6p AAA ATPase complexes to peroxisomes. Nat Cell Biol 5(5):454–460
Matsuzaki T, Fujiki Y (2008) The peroxisomal membrane protein import receptor Pex3p is directly transported to peroxisomes by a novel Pex19p- and Pex16p-dependent pathway. J Cell Biol 183(7):1275–1286
Matsuzono Y et al (1999) Human PEX19: cDNA cloning by functional complementation, mutation analysis in a patient with Zellweger syndrome, and potential role in peroxisomal membrane assembly. Proc Natl Acad Sci U S A 96(5):2116–2121
Matsuzono Y et al (2006) Functional domain mapping of peroxin Pex19p: interaction with Pex3p is essential for function and translocation. J Cell Sci 119(Pt 17):3539–3550
Mayerhofer PU et al (2016) Human peroxin PEX3 is co-translationally integrated into the ER and exits the ER in budding vesicles. Traffic 17(2):117–130
Meinecke M et al (2010) The peroxisomal importomer constitutes a large and highly dynamic pore. Nat Cell Biol 12(3):273–277
Miyata N, Fujiki Y (2005) Shuttling mechanism of peroxisome targeting signal type 1 receptor Pex5: ATP-independent import and ATP-dependent export. Mol Cell Biol 25(24):10822–10832
Morita M, Imanaka T (2012) Peroxisomal ABC transporters: structure, function and role in disease. Biochim Biophys Acta 1822(9):1387–1396
Mosser J et al (1993) Putative X-linked adrenoleukodystrophy gene shares unexpected homology with ABC transporters. Nature 361(6414):726–730
Motley AM et al (2000) Caenorhabditis elegans has a single pathway to target matrix proteins to peroxisomes. EMBO Rep 1(1):40–46
Motley AM et al (2008) Dnm1p-dependent peroxisome fission requires Caf4p, Mdv1p and Fis1p. J Cell Sci 121(Pt 10):1633–1640
Muntau AC et al (2000) Defective peroxisome membrane synthesis due to mutations in human *PEX3* causes Zellweger syndrome, complementation group G. Am J Hum Genet 67(4):967–975
Muntau AC et al (2003) The interaction between human PEX3 and PEX19 characterized by fluorescence resonance energy transfer (FRET) analysis. Eur J Cell Biol 82(7):333–342
Nashiro C et al (2011) Recruiting mechanism of the AAA peroxins, Pex1p and Pex6p, to Pex26p on the peroxisomal membrane. Traffic 12(6):774–788
Niederhoff K et al (2005) Yeast Pex14p possesses two functionally distinct Pex5p and one Pex7p binding sites. J Biol Chem 280(42):35571–35578
Novikoff PM, Novikoff AB (1972) Peroxisomes in absorptive cells of mammalian small intestine. J Cell Biol 53(2):532–560
Oliveira ME et al (2003) The energetics of Pex5p-mediated peroxisomal protein import. J Biol Chem 278(41):39483–39488
Otzen M et al (2012) Pex19p contributes to peroxisome inheritance in the association of peroxisomes to Myo2p. Traffic 13(7):947–959
Passmore JB et al (2017) The respiratory chain inhibitor rotenone affects peroxisomal dynamics via its microtubule-destabilising activity. Histochem Cell Biol 148(3):331–341
Pause B et al (2000) Targeting of the 22 kDa integral peroxisomal membrane protein. FEBS Lett 471(1):23–28
Perry RJ et al (2009) Endoplasmic reticulum-associated secretory proteins Sec20p, Sec39p, and Dsl1p are involved in peroxisome biogenesis. Eukaryot Cell 8(6):830–843
Pinto MP et al (2006) The import competence of a peroxisomal membrane protein is determined by Pex19p before the docking step. J Biol Chem 281(45):34492–34502
Pires JR et al (2003) The ScPex13p SH3 domain exposes two distinct binding sites for Pex5p and Pex14p. J Mol Biol 326(5):1427–1435

Platta HW et al (2005) Functional role of the AAA peroxins in dislocation of the cycling PTS1 receptor back to the cytosol. Nat Cell Biol 7(8):817–822
Platta HW et al (2007) Ubiquitination of the peroxisomal import receptor Pex5p is required for its recycling. J Cell Biol 177(2):197–204
Platta HW et al (2009) Pex2 and Pex12 function as protein-ubiquitin ligases in peroxisomal protein import. Mol Cell Biol 29(20):5505–5516
Purdue PE, Lazarow PB (2001) Pex18p is constitutively degraded during peroxisome biogenesis. J Biol Chem 276(50):47684–47689
Rapoport TA (2007) Protein translocation across the eukaryotic endoplasmic reticulum and bacterial plasma membranes. Nature 450(7170):663–669
Rayapuram N, Subramani S (2006) The importomer—a peroxisomal membrane complex involved in protein translocation into the peroxisome matrix. Biochim Biophys Acta 1763(12):1613–1619
Rottensteiner H et al (2004) Peroxisomal membrane proteins contain common Pex19p-binding sites that are an integral part of their targeting signals. Mol Biol Cell 15(7):3406–3417
Sacksteder KA et al (2000) PEX19 binds multiple peroxisomal membrane proteins, is predominantly cytoplasmic, and is required for peroxisome membrane synthesis. J Cell Biol 148(5):931–944
Saidowsky J et al (2001) The di-aromatic pentapeptide repeats of the human peroxisome import receptor PEX5 are separate high affinity binding sites for the peroxisomal membrane protein PEX14. J Biol Chem 276(37):34524–34529
Sakaue H et al (2016) The N-terminal motif of PMP70 suppresses cotranslational targeting to the endoplasmic reticulum. J Biochem 159(5):539–551
Sato Y et al (2008) Characterization of the interaction between recombinant human peroxin Pex3p and Pex19p: identification of TRP-104 in Pex3p as a critical residue for the interaction. J Biol Chem 283(10):6136–6144
Sato Y et al (2010) Structural basis for docking of peroxisomal membrane protein carrier Pex19p onto its receptor Pex3p. EMBO J 29(24):4083–4093
Schell-Steven A et al (2005) Identification of a novel, intraperoxisomal Pex14-binding site in Pex13: association of Pex13 with the docking complex is essential for peroxisomal matrix protein import. Mol Cell Biol 25(8):3007–3018
Schliebs W, Kunau WH (2006) PTS2 co-receptors: diverse proteins with common features. Biochim Biophys Acta 1763(12):1605–1612
Schliebs W et al (1999) Recombinant human peroxisomal targeting signal receptor PEX5. Structural basis for interaction of PEX5 with PEX14. J Biol Chem 274(9):5666–5673
Schmidt F et al (2010) Insights into peroxisome function from the structure of PEX3 in complex with a soluble fragment of PEX19. J Biol Chem 285(33):25410–25417
Schmidt F et al (2012) The role of conserved PEX3 regions in PEX19-binding and peroxisome biogenesis. Traffic 13(9):1244–1260
Schrul B, Kopito RR (2016) Peroxin-dependent targeting of a lipid-droplet-destined membrane protein to ER subdomains. Nat Cell Biol 18(7):740–751
Schueller N et al (2010) The peroxisomal receptor Pex19p forms a helical mPTS recognition domain. EMBO J 29(15):2491–2500
Schuldiner M et al (2008) The GET complex mediates insertion of tail-anchored proteins into the ER membrane. Cell 134(4):634–645
Schwartzkopff B et al (2015) Cysteine-specific ubiquitination protects the peroxisomal import receptor Pex5p against proteasomal degradation. Biosci Rep 35(3):e00215
Shani N et al (1997) Identification of a fourth half ABC transporter in the human peroxisomal membrane. Hum Mol Genet 6(11):1925–1931
Shibata H et al (2004) Domain architecture and activity of human Pex19p, a chaperone-like protein for intracellular trafficking of peroxisomal membrane proteins. J Biol Chem 279(37):38486–38494
Stanley WA, Wilmanns M (2006) Dynamic architecture of the peroxisomal import receptor Pex5p. Biochim Biophys Acta 1763(12):1592–1598

Stanley WA et al (2006) Recognition of a functional peroxisome type 1 target by the dynamic import receptor pex5p. Mol Cell 24(5):653–663

Sugiura A et al (2017) Newly born peroxisomes are a hybrid of mitochondrial and ER-derived pre-peroxisomes. Nature 542(7640):251–254

Suzuki Y et al (1987) Biosynthesis of membrane polypeptides of rat liver peroxisomes. J Biochem 101(2):491–496

Swinkels BW et al (1991) A novel, cleavable peroxisomal targeting signal at the amino-terminus of the rat 3-ketoacyl-CoA thiolase. EMBO J 10(11):3255–3262

Tabak HF et al (2003) Peroxisomes start their life in the endoplasmic reticulum. Traffic 4(8):512–518

Tam YY et al (2005) Pex3p initiates the formation of a preperoxisomal compartment from a subdomain of the endoplasmic reticulum in *Saccharomyces cerevisiae*. J Biol Chem 280(41):34933–34939

Tani K et al (2011) Dual function of Sec16B: endoplasmic reticulum-derived protein secretion and peroxisome biogenesis in mammalian cells. Cell Logist 1(4):164–167

Thoms S et al (2012) Peroxisome formation requires the endoplasmic reticulum channel protein Sec61. Traffic 13(4):599–609

Titorenko VI, Rachubinski RA (1998) Mutants of the yeast *Yarrowia lipolytica* defective in protein exit from the endoplasmic reticulum are also defective in peroxisome biogenesis. Mol Cell Biol 18(5):2789–2803

Titorenko VI et al (2002) Acyl-CoA oxidase is imported as a heteropentameric, cofactor-containing complex into peroxisomes of *Yarrowia lipolytica*. J Cell Biol 156(3):481–494

Ueda K (2011) ABC proteins protect the human body and maintain optimal health. Biosci Biotechnol Biochem 75(3):401–409

van der Zand A et al (2010) Peroxisomal membrane proteins insert into the endoplasmic reticulum. Mol Biol Cell 21(12):2057–2065

van der Zand A et al (2012) Biochemically distinct vesicles from the endoplasmic reticulum fuse to form peroxisomes. Cell 149(2):397–409

van Roermund CW et al (2008) The human peroxisomal ABC half transporter ALDP functions as a homodimer and accepts acyl-CoA esters. FASEB J 22(12):4201–4208

van Roermund CW et al (2011) Differential substrate specificities of human ABCD1 and ABCD2 in peroxisomal fatty acid β-oxidation. Biochim Biophys Acta 1811(3):148–152

van Roermund CW et al (2014) A role for the human peroxisomal half-transporter ABCD3 in the oxidation of dicarboxylic acids. Biochim Biophys Acta 1841(4):563–568

Vasiliou V et al (2009) Human ATP-binding cassette (ABC) transporter family. Hum Genomics 3(3):281–290

Veenhuis M, van der Klei IJ (2014) A critical reflection on the principles of peroxisome formation in yeast. Front Physiol 5:110

Walker CL et al (2018) Redox regulation of homeostasis and proteostasis in peroxisomes. Physiol Rev 98(1):89–115

Wang X et al (2001) Discrete targeting signals direct Pmp47 to oleate-induced peroxisomes in *Saccharomyces cerevisiae*. J Biol Chem 276(14):10897–10905

Wang D et al (2003) Physical interactions of the peroxisomal targeting signal 1 receptor pex5p, studied by fluorescence correlation spectroscopy. J Biol Chem 278(44):43340–43345

Waterham HR et al (2016) Human disorders of peroxisome metabolism and biogenesis. Biochim Biophys Acta 1863(5):922–933

Wiemer EA et al (1996) Isolation and characterization of Pas2p, a peroxisomal membrane protein essential for peroxisome biogenesis in the methylotrophic yeast *Pichia pastoris*. J Biol Chem 271(31):18973–18980

Williams C, Distel B (2006) Pex13p: docking or cargo handling protein? Biochim Biophys Acta 1763(12):1585–1591

Williams C et al (2007) A conserved cysteine is essential for Pex4p-dependent ubiquitination of the peroxisomal import receptor Pex5p. J Biol Chem 282(31):22534–22543

Williams C et al (2012) Insights into ubiquitin-conjugating enzyme/co-activator interactions from the structure of the Pex4p:Pex22p complex. EMBO J 31(2):391–402

Williams C et al (2015) The membrane remodeling protein Pex11p activates the GTPase Dnm1p during peroxisomal fission. Proc Natl Acad Sci U S A 112(20):6377–6382

Yagita Y et al (2013) Tail-anchored PEX26 targets peroxisomes via a PEX19-dependent and TRC40-independent class I pathway. J Cell Biol 200(5):651–666

Yang X et al (2001) Eci1p uses a PTS1 to enter peroxisomes: either its own or that of a partner, Dci1p. Eur J Cell Biol 80(2):126–138

Yonekawa S et al (2011) Sec16B is involved in the endoplasmic reticulum export of the peroxisomal membrane biogenesis factor peroxin 16 (Pex16) in mammalian cells. Proc Natl Acad Sci U S A 108(31):12746–12751

Chapter 3
Peroxisome Degradation and Its Molecular Machinery

Masahide Oku and Yasuyoshi Sakai

Abstract Peroxisomal metabolism and its regulation play important roles in various cellular functions. The regulation of peroxisomal metabolism is controlled by modulation of peroxisome biogenesis as well as the degradation of intra-organellar components and the organelle itself. An accumulation of experimental findings demonstrate that the majority of organelle degradation is accomplished through autophagy, an important cellular process involving transport of cytoplasmic constituents into lysosomes for degradation. The first part of this chapter discusses several processes responsible for the degradation of peroxisomes in mammalian cells, including autophagy. Next, following a general description of the molecular machinery of autophagy, molecular details of selective autophagy of peroxisomes, termed pexophagy, are described based on studies conducted in yeast and mammalian cells. In the final section, expected medical applications associated with pexophagy are described along with potential future developments in this field.

Keywords Autophagy · Methylotrophic yeast · Peroxisome · PEX1 · Reactive oxygen species · Ubiquitin · Yeast genetics

3.1 Pathways for the Degradation of Mammalian Peroxisomes and Their Components

Feeding rodents a diet containing specific hypolipidemic agents or plasticizers, both of which are potent agonists of a group of nuclear hormone receptors (peroxisome proliferator-activated receptors, or PPARs), leads to an enlargement of the liver and an increase in the number of peroxisomes in the hepatocytes (Reddy et al. 1982).

M. Oku · Y. Sakai (✉)
Division of Applied Life Sciences, Graduate School of Agriculture, Kyoto University, Kyoto, Japan
e-mail: oku.masahide.6a@kyoto-u.ac.jp; ysakai@kais.kyoto-u.ac.jp

T. Imanaka, N. Shimozawa (eds.), *Peroxisomes: Biogenesis, Function, and Role in Human Disease*, https://doi.org/10.1007/978-981-15-1169-1_3

Subsequent removal of these chemicals from the diet results in a gradual decline in liver size and peroxisome number, affording an ideal experimental system for the morphological and molecular analyses of peroxisome degradation. Based on pioneering studies with the rat experimental model, three pathways were proposed to account for the degradation of peroxisomes: (1) Lon protease, (2) 15-lipoxygenase (15-LOX), and (3) autophagy pathways (Yokota and Dariush Fahimi 2009). The following section is a general introduction to these pathways.

3.1.1 Lon Protease

Peroxisomes from a wide range of organisms are endowed with an ATP-driven protease, known as Lon protease, located in the lumen of the organelle (Pomatto et al. 2017). The mammalian orthologue of this protein (peroxisomal Lon protease, or PSLP) has a molecular mass of 75 kDa and includes a C-terminal peroxisome targeting signal 1 (PTS1) (Kikuchi et al. 2004). The expression level of PSLP in rat hepatocytes was found to be up-regulated during treatment with di-(2-ethylhexyl) phthalate (DEHP), one of the PPAR agonists. Intriguingly, the expression level of PSLP continued to be induced 3 days after cessation of treatment with DEHP, in contrast to the other peroxisomal β-oxidation enzymes (Yokota et al. 2008). This observation implies PSLP plays a role in the clearance of peroxisome luminal contents (Yokota et al. 2008).

A study of peroxisome dynamics in a yeast species *Ogataea polymorpha* (previously called *Hansenula polymorpha*) gave supporting evidence to the role of PSLP in peroxisome degradation (Aksam et al. 2007). In this organism, accumulation of a misfolded protein in the peroxisome was enhanced by knockout of the PSLP homologue. Notably, further loss of an autophagy factor Atg1 (described below) along with blockage of the PSLP homologue resulted in severe toxicity to the yeast cells, concomitant with accumulation of intracellular reactive oxygen species (ROS). Toxicity resulting from the double knockout was more significant than that from a single knockout of either the PSLP homologue or a factor necessary for autophagy Atg1, suggesting a complementary role in the clearance of the peroxisomal proteins mediated by the Lon protease and autophagy.

3.1.2 15-LOX

15-LOX (also termed arachidonate 15-lipoxygenase, ALOX15) enzymes convert polyunsaturated fatty acids, such as linoleic acid and arachidonic acid, to their hydroperoxy derivatives via introduction of molecular dioxygen at specific positions (C13 for linoleic acid and C15 for arachidonic acid) (Colakoglu et al. 2018). To date, two types of human 15-LOX enzymes have been identified: (1) 15-LOX-1, which preferentially acts on linoleic acid, and (2) 15-LOX-2, which displays a preference for arachidonic acid. The hydroperoxy products of 15-LOX are bioactive, mediating multiple events in inflammation or tumorigenesis.

While the majority of 15-LOX is distributed in the cytoplasm, a small amount can associate with organelle membranes causing them to become permeabilized (van Leyen et al. 1998). Peroxisomes can be targeted by this permeabilization. Indeed, an immune-electron microscopy study demonstrated that 15-LOX is localized onto the surface of peroxisomes in rat hepatocytes (Yokota et al. 2001). This finding suggested 15-LOX plays a role in peroxisomal degradation.

3.1.3 Autophagy

The involvement of autophagy (degradation after transfer into the lysosome) in the degradation of peroxisomes was clearly demonstrated by electron microscopy studies of rat hepatocytes obtained after the animal was fed on a DEHP-containing diet and subsequently transferred to a regular diet (i.e. without DEHP) (Yokota 1993). Six days after the rat was put on a regular diet, the hepatocytes exhibited a staining pattern for catalase activity in the lysosomal compartment reminiscent of peroxisomes. This observation strongly suggested transfer of the whole organelle into the lysosome. Moreover, when hepatocytes were obtained from a rat injected with leupeptin, which is an inhibitor of lysosomal proteases, a characteristic double membrane structure surrounding the peroxisomes was observed (Fig. 3.1). These structures, termed autophagosomes, were derived from newly-synthesized

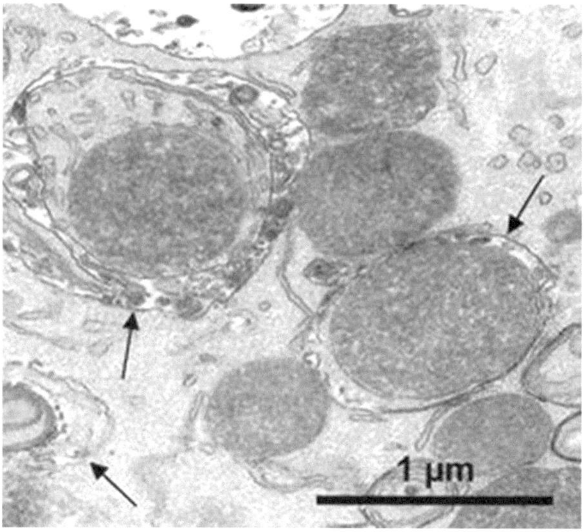

Fig. 3.1 Electron microscopy of rat hepatocytes exhibiting autophagosome formation. The cells were obtained from a rat treated with DHEP and subsequently injected with leupeptin. The arrows indicate autophagosomes. [This image is a reprint from an article in Yokota and Fahimi (2009)]

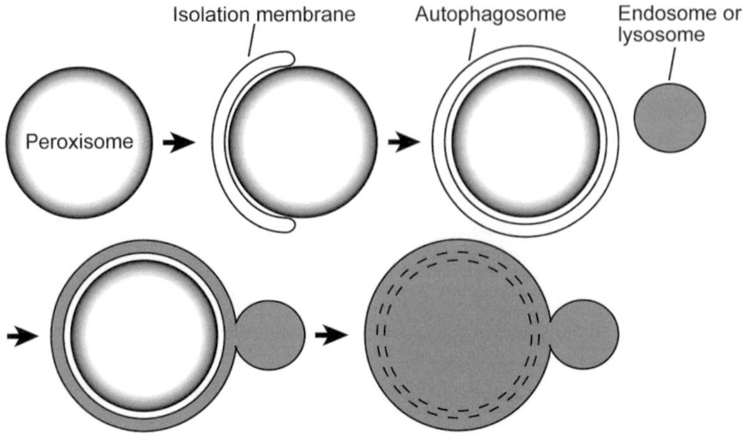

Fig. 3.2 Schematic drawing of mammalian pexophagy processes. On the surface of peroxisome, the isolation membrane is formed and elongated to completely enwrap the organelle, ending in the formation of the autophagosome (pexophagosome). The outer membrane of this structure is fused with endosome or lysosome, leading to the lysis of the autophagosome inner membrane and the trapped peroxisome

double-membrane structures, called isolation membranes, which eventually elongated along the whole surface of the peroxisome and completed the enclosure by membrane fusion (Fig. 3.2). The sequestration steps are so rapid that it was not possible to detect intermediate structures without pharmacologically inhibiting the final degradation process of the trapped peroxisomes. Biochemical determinations of enzyme activities in the cell lysates from rat hepatocytes were consistent with the morphological observations. Specifically, the decrease in the level of acyl-CoA oxidase, a peroxisomal enzyme, was blocked by an inhibitor of autophagy, 3-methyladenine (3-MA, described below) (Luiken et al. 1992).

Further detailed morphological analyses conducted by microscopic detection of lysosomal proteins demonstrated that the autophagosomes sequestering peroxisomes were initially negative for lysosomal marker proteins, but then acquired lysosomal composite proteins (Yokota et al. 1993). Based on these findings, a scheme for the degradation of peroxisomes through autophagy was established, which was consistent with the morphological changes observed during macroautophagy (Fig. 3.2).

3.2 Molecular Machinery of Macroautophagy

*3.2.1 Identification of Autophagy-Related (*ATG*) Gene Products*

Autophagy is a general term for the degradation of cytoplasmic components in lysosomes. Overall autophagy encompasses different intracellular traffic systems where cytoplasmic constituents are delivered to the lysosome. Among the various systems, one accompanying the formation of autophagosome is known as macroautophagy.

Historically, identification of gene products responsible for macroautophagy was initiated by research with the budding yeast *Saccharomyces cerevisiae* (Tsukada and Ohsumi 1993). Cytoplasmic materials transferred into the vacuole of *S. cerevisiae* (equivalent to the mammalian lysosome) through macroautophagy were readily detected by light microscopy in 'sac' structures within the vacuole under starvation conditions when the vacuolar hydrolase activities were inhibited. The first autophagy mutant was isolated based on visible and morphological screening of yeast mutants that lacked these sac-like structures inside the vacuole. The mutated gene giving rise to this autophagy mutant is now termed *ATG1*. Next, Ohsumi and his colleagues established that the mutant strain of *ATG1* was vulnerable to nitrogen starvation conditions in contrast to wild-type yeast that can survive a week under such conditions. The discovery of this phenotype led to the isolation of further 15 mutant types defective in macroautophagy.

The macroautophagy mutants regained autophagic activity and viability under nitrogen starvation conditions upon receipt of a normal version of the mutated gene from a yeast genomic library cloned in a plasmid. These experiments enabled isolation of macroautophagy-restored (complemented) yeast cells. Subsequent recovery of the corresponding plasmids from the isolated cells facilitated identification of the cloned genes. Using this strategy, multiple genes responsible for macroautophagy in *S. cerevisiae* were identified. Similar genetic approaches were performed in other yeast species, and a unified nomenclature of such genes as the *ATG* genes required for macroautophagic processes was established in 2003 (Klionsky et al. 2003). Subsequent studies with mammalian cells demonstrated that most of the identified genes are conserved in the experimental system. Moreover, developmental and pathological aspects of autophagy have been extensively studied (Mizushima and Komatsu 2011). These striking findings led Yoshinori Ohsumi to be awarded the Nobel Prize for physiology or medicine in 2016.

3.2.2 Functional Groups of ATG *Gene Products*

After identification of multiple *ATG* genes, functional details of the gene products were revealed by investigations of interactions between the factors and by microscopic observations of the factors. Taken together, these studies demonstrate two major points. Firstly, the *ATG* gene products comprise a set of 'core' factors that contribute to biogenesis of the autophagosome, while other factors act to 'fine-tune' membrane biogenesis (e.g. guidance of membrane biogenesis to surround target components) (Xie and Klionsky 2007; Suzuki et al. 2017). Secondly, the core factors can be categorized into six distinct groups as outlined in Table 3.1. These groups will be described in the following paragraphs, and briefly illustrated in Fig. 3.3.

Atg1 (Unc-51-like kinases ULK1 and 2 for mammalian cells) is a pivotal kinase for the initiation of autophagosome biogenesis and forms a complex with regulatory factors (i.e. others in this group) (Papinski and Kraft 2016). Alteration in nutrient availability or hormonal signaling changes the interactions in the complex of factors or phosphorylation status of these factors, which modulates the kinase activity of Atg1. Atg9 is the only transmembrane protein among these core factors. Membrane vesicles containing Atg9 are thought to shuttle between the site of autophagosome biogenesis and other authentic organelles (Yamamoto et al. 2012; Imai et al. 2016).

Table 3.1 ATG proteins responsible for autophagosome assembly and their classification

Group	Yeast component	Mammalian component
Atg1/ULK complex	Atg1	ULK1/2
	Atg11 (selective autophagy)	
	Atg13	ATG13
	Atg17 (bulk autophagy)	
	Atg29 (bulk autophagy)	
	Atg31 (bulk autophagy)	
		FIP200
		ATG101
Atg9 vesicle	Atg9	ATG9A/B
PI3K complex	Vps34	VPS34
	Vps15	VPS15
	Vps30 (Atg6)	Beclin 1
	Atg14	ATG14L
Atg2-18 complex	Atg2	ATG2A/B
	Atg18	WIPI1-4
Atg12-conjugation machinery	Atg12	ATG12
	Atg7	ATG7
	Atg10	ATG10
	Atg5	ATG5
	Atg16	ATG16L1/2
Atg8-conjugation machinery	Atg8	LC3A/B/C, GABARAP, GABALAPL1-3
	Atg7	ATG7
	Atg3	ATG3
	Atg4	ATG4A/B/C/D

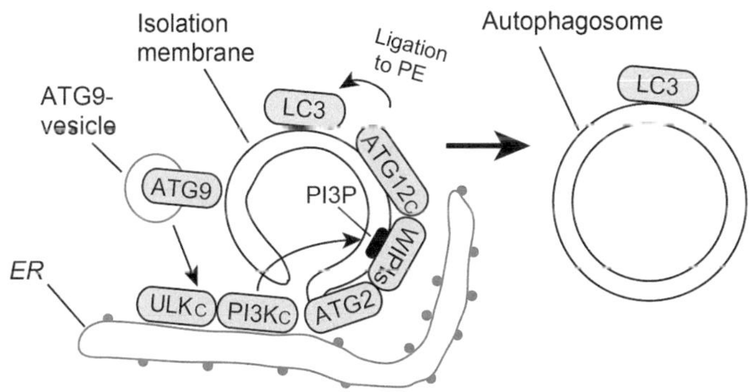

Fig. 3.3 Mammalian ATG proteins responsible for biogenesis of autophagosomes. The isolation membrane is formed in close vicinity to ER subdomain onto which ULK complex is recruited. The recruited ULK complex triggers relocation of ATG9-vesicles and induces the enzymatic activity of PI3K, leading to accumulation of PI3P on the isolation membrane. WIPI proteins that associate with ATG12 complex and/or ATG2 are then recruited to the membrane by recognizing the accumulated PI3P. The ATG12 complex functions in the ligation step of Atg8 to phosphatidylethanolamine on the isolation membrane, and ATG2 has a lipid transfer activity that may contribute to expansion of the isolation membrane. The leading edge of the isolation membrane fuses to form the autophagosome. ATG12c, ATG12 complex; LC3, light chain 3 protein, the orthologue of Atg8; PE, phosphatidylethanolamine; PI3Kc, phosphatidylinositol 3′-kinase complex; PI3P, phosphatidylinositol 3′-phosphate; ULKc, ULK (Unc-51-like kinase, the orthologue of Atg1) complex; WIPI, WD repeat protein interacting with phosphoinositides, the orthologue of Atg18

Phosphatidylinositol 3′-kinase (PI3K) Class III converts phosphatidylinositol, one of the phospholipids contained in the cell membrane, into phosphatidylinositol 3′-phosphate (PI3P). For this enzyme to act on autophagosome biogenesis, several Atg factors of this group associate with the enzyme (Kihara et al. 2001). The lipid product of PI3K Class III, PI3P, in turn recruits Atg proteins of another group, namely Atg2 and Atg18 (the mammalian orthologues termed WIPI, WD repeat protein interacting with phosphoinositides) proteins (Obara et al. 2008). The enzymatic activity of PI3K Class III is inhibited by 3-MA, a drug widely used to block autophagy (Seglen and Gordon 1982).

The remaining groups of core Atg factors function to covalently attach small proteins to other proteins or lipid molecules in a process reminiscent of ubiquitination. Ubiquitin is a highly conserved protein among eukaryotes comprising 76 amino-acid residues. The C-terminal glycine residue of ubiquitin forms an amide-bond with the ε-amino group of a lysine residue in the target protein. The ubiquitination reaction consists of three steps: activation, conjugation, and ligation. For autophagosome biogenesis, two small Atg proteins (Atg8 and Atg12) are attached to a phospholipid phosphatidylethanolamine (PE) and Atg5 through an amide and isopeptide conjugation, respectively (Mizushima et al. 1998; Ichimura et al. 2000). Atg7 acts on both of the two reactions as E1 enzyme for activation, and Atg3 and Atg10 function as E2 enzymes for conjugation (Ichimura et al. 2000; Shintani et al. 1999). The resulting Atg12-Atg5 complex mediates the ligation step of Atg8-PE attachment (Hanada et al. 2007). Overall, these reactions serve to modify lipids of the autophagosome membrane with Atg8 (LC3, light chain 3 protein in mammalian cells) (Kabeya et al. 2000).

3.3 Molecular Machinery of Peroxisome-Specific Autophagy (Pexophagy)

3.3.1 Insights from Yeast Studies

While autophagosomes formed during starvation-induced macroautophagy comprise the bulk of the cytoplasm, microscopic images of rat hepatocytes (Fig. 3.1) indicate that the autophagosomes enwrap the peroxisomes with a greater preference than the other intracellular components. Thus, this process is regarded as a type of selective autophagy toward the peroxisome, named pexophagy. Studies have been conducted to elucidate the molecular machinery of pexophagy using several yeast species. Of particular value were electron microscopy studies of methylotrophic yeasts (yeasts that can be grown on methanol as a solo carbon source), which selectively degrade peroxisomes under specific nutrient conditions (Veenhuis et al. 1983; Tuttle and Dunn Jr 1995).

Screening strategies for pexophagy factors in the methylotrophic yeasts are based on detection of the activity of a peroxisomal enzyme, alcohol oxidase (AO) (Titorenko et al. 1995; Mukaiyama et al. 2002). Peroxisomes containing AO are massively induced when these organisms are cultured on methanol; subsequent transfer of the organisms to another medium (glucose or ethanol medium) rapidly induces pexophagy, and AO is eventually degraded. Thus, after changing the medium the wild-type strain of methylotrophic yeast lost AO activity while pexophagy mutants retained the activity. Because AO activity in a single colony of methylotrophic yeast can be readily visualized by staining, it is possible to pick pexophagy mutants among a randomly-mutagenized population of cells.

The screening experiments with the methylotrophic yeast *Komagataella phaffii* (previously called *Pichia pastoris*) identified most of the core Atg factors as being required for pexophagy (Mukaiyama et al. 2002). This finding reflects the biogenesis of the autophagosome enwrapping the peroxisome (pexophagosome) and of an isolation-membrane-like structure termed MIPA (micropexophagy-specific membrane apparatus) (Mukaiyama et al. 2004), observed during macropexophagy and micropexophagy, respectively (Fig. 3.4). In addition, another study identified a key factor Atg30 that recruits the core Atg machinery onto the surface of the target peroxisomes (Farre et al. 2008). Atg30 interacts with peroxisomal membrane proteins Pex3 and Pex14 as well as the core Atg proteins (Atg11 and Atg8) to mediate localization of the core Atg protein complexes on the peroxisome (Fig. 3.5). Interestingly, interactions of Atg30 with Atg8 and Atg11 are primed by phosphorylation of serine residues within Atg30 (Farre et al. 2013). Thus, initiation of the pexophagosome formation is triggered by the phosphorylation of Atg30. The counterpart protein of

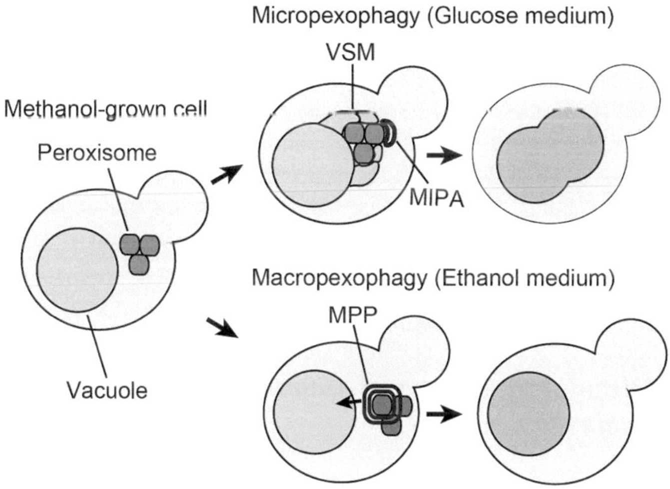

Fig. 3.4 Two types of pexophagy process in the methylotrophic yeast *Komagataella phaffii*. Peroxisomes are induced when the cells are cultured on methanol as a carbon source. When the cells are shifted to glucose medium, part of the vacuoles (equivalent to the lysosomes) extends toward and engulf the peroxisomes cluster by forming VSM (vacuolar sequestering membrane). An Atg8-positive double-membrane structure, termed MIPA (micropexophagy-specific membrane apparatus), is formed to fully enwrap peroxisomes from the cytoplasm. Then VSM and MIPA fuse to incorporate the peroxisomes into the vacuole. When the cells are transferred to ethanol medium, autophagosomal structures enwrapping a single peroxisome, termed macropexophagosome (MPP), is formed in a similar manner to pexophagy in mammalian cells

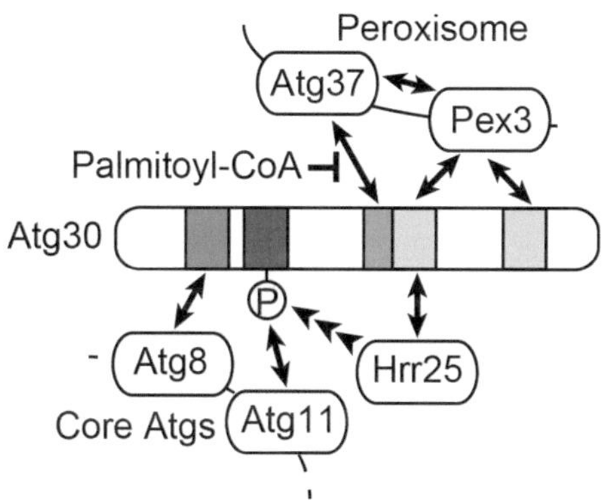

Fig. 3.5 Protein interactions intervened with the pexophagy adaptor protein, Atg30, in the yeast Komagataella phaffii. The arrows indicate the interactions. The arrowheads and P stand for phosphorylation. The inhibitory action of palmitoyl-CoA toward Atg37-Atg30 interaction is also shown

Atg30 in *K. phaffii* was found in the budding yeast *S. cerevisiae* and named Atg36 (Motley et al. 2012). Both Atg30 and Atg36 are phosphorylated by Hrr25 (Tanaka et al. 2014; Zientara-Rytter et al. 2018), a kinase belonging to the casein kinase 1 family that exerts diverse physiological functions (Schittek and Sinnberg 2014).

Another study on the interaction network centered on *K. phaffii* Atg30 identified a plausible link between pexophagy and peroxisome metabolism (Nazarko et al. 2014).

In this study, a peroxisomal membrane protein Atg37 was identified to be an interacting partner of both Pex3 and Atg30. Atg37 has an acyl-CoA-binding domain and is known to interact with palmitoyl-CoA. Noteworthy, the Atg37-Atg30 interaction competes with binding of Atg37 to palmitoyl-CoA, suggesting that palmitoyl-CoA exhibits an inhibitory effect toward the formation of the Atg complex, thereby preventing pexophagy (Fig. 3.5). This result seems to be reasonable because, under physiological conditions in the presence of an excess amount of palmitoyl-CoA, there is a requirement for enhanced peroxisomal β-oxidation activity to degrade the fatty acid. Thus, pexophagy should be inhibited under these growth conditions.

3.3.2 Molecular Machinery of Mammalian Pexophagy (1): Atg Proteins

The functional requirement of Atg proteins for the degradation of peroxisomes in mouse hepatocytes was demonstrated in a study using a conditional (liver-specific) knockout of ATG7 (Iwata et al. 2006). Degradation of DHEP-induced peroxisomes was severely impaired with concomitant loss of autophagosome detection in the knockout hepatocytes. Furthermore, decline in the amount of a pulse-labelled peroxisomal protein in a Chinese hamster ovary (CHO) cell line was also found to be inhibited by 3-MA treatment, which strongly suggests the contribution of autophagy for basal turnover of peroxisomes (Huybrechts et al. 2009).

3.3.3 Molecular Machinery of Mammalian Pexophagy (2): Ubiquitin

For induction of pexophagy, mammalian cells utilize ubiquitin as a ‘tag’. As noted earlier, the ubiquitination reaction yields covalently-attached ubiquitin to specific proteins, which is recognized by different proteins with ubiquitin-binding domains. In the case of pexophagy, proteins on the peroxisome surface are ubiquitinated and two adaptor proteins, neighbor of BRACA1 gene 1 (NBR1) and sequestosome 1 (SQSTM1 or p62), are known to bind to the ubiquitinated proteins (Kirkin et al. 2009). Both these proteins possess an LC3 (the orthologue of yeast Atg8)-interacting region (LIR), in addition to a ubiquitin-associated (UBA) domain, one of the ubiquitin-binding domains (Fig. 3.6). The central region of LIRs contains the amino acid sequence WTHL (SQSTM1) or YIII (NBR1), characterized by hydrophobic residues in the first (W or Y) and the fourth (L or I) positions, which is common to various LC3-interacting proteins (Noda et al. 2010). These adaptor proteins contribute to the formation of autophagosomes along the surface of the peroxisome by interacting with LC3, one of the core Atg proteins, and ubiquitin on the peroxisomal protein. In CHO cells, LC3 conjugated to PE also interacts with Pex14, a component

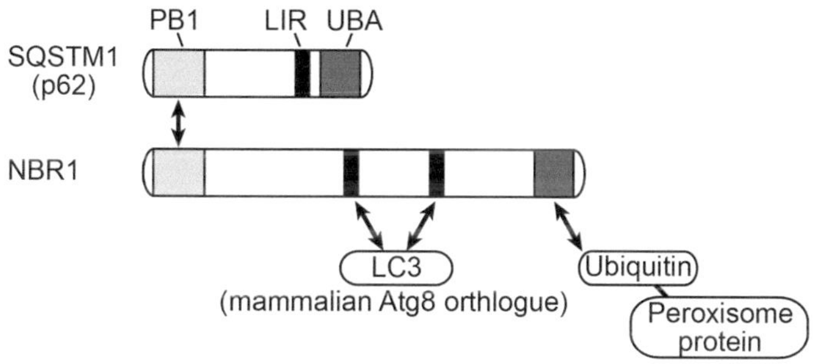

Fig. 3.6 Pexophagy adaptor proteins for mammalian pexophagy. The functional intra-molecular domains within the adaptor proteins, PB1, LIR, and UBA domains, are shown as the gray or black boxes

of peroxisome protein import machinery, during starvation-induced pexophagy (Hara-Kuge and Fujiki 2008; Jiang et al. 2015). SQSTM1 and NBR1 form self-oligomers or interact with each other through their N-terminal Phox and Bem1p (PB1) domains (Lamark et al. 2003).

When several peroxisomal membrane proteins are expressed in conjugation with ubiquitin, pexophagy dependent on SQSTM1 is induced (Kim et al. 2008). By contrast, overexpression of PEX3 induces NBR1-dependent pexophagy in the CHO cell line (Yamashita et al. 2014), and overexpression of NBR1 itself induces pexophagy in HeLa cells (Deosaran et al. 2013). These findings highlight the functional importance of NBR1. SOSTM1 also plays a role in clustering the peroxisomes (Yamashita et al. 2014), which appears to facilitate pexophagy.

Peroxisomal proteins known to be subject to ubiquitination during pexophagy include PEX5, the shuttling factor of the peroxisomal import machinery, and an ATP binding cassette (ABC) transporter PMP70 (ABCD3). When PEX5 was fused to GFP at its C-terminus, most of the protein was found to be located on the peroxisomal surface in a ubiquitinated state, with the concomitant induction of pexophagy (Nordgren et al. 2015). Ubiquitination of Pex5 and PMP70 in the HeLa cell line is induced in response to amino-acid starvation, which depends on one of the peroxisome assembly factors Pex2, a peroxisomal ubiquitin ligase (Sargent et al. 2016). Induction of Pex2 expression and pexophagy was observed in hepatocytes isolated from mice fed an amino-acid restricted diet (Sargent et al. 2016). The ubiquitination of peroxisomal proteins is reversed by an enzymatic action of USP30 to remove the ubiquitin moiety, which is thought to serve as a regulatory mechanism of pexophagy (Marcassa et al. 2018; Riccio et al. 2019). It remains to be elucidated whether other peroxisomal proteins are ubiquitinated and whether other ubiquitin ligases function in pexophagy under various physiological conditions. These issues are discussed in the next chapter.

3.4 Medical and Physiological Aspects of Mammalian Pexophagy

3.4.1 Peroxisomal Biogenesis Disorders and Pexophagy

Peroxisomal biogenesis disorders, such as Zellweger syndrome, are characterized by a severe decrease in the number of functional peroxisomes due to mutations in the Peroxin (PEX)-coding genes. One of the most common mutations associated with these disorders is found in PEX1-encoding gene, resulting in a substitution of glycine at position 843 of the protein product for aspartic acid (G843D). PEX1, forming a complex with PEX6 and PEX26, acts as ATPase associated with diverse cellular activities (AAA ATPase) to export PEX5 from the peroxisome, and thereby supports an efficient recycling of PEX5 for multiple rounds of shuttling between the cytoplasm and the peroxisome (Fig. 3.7). It had been proposed that the decrease in the number of peroxisomes resulting from the PEX1^{G843D} mutation is derived from accumulation of PEX5 in the peroxisome, which depletes the cytoplasmic pool of PEX5 eventually impairing peroxisomal protein transport. However, recent studies suggest the involvement of pexophagy in peroxisome loss associated with this mutation (Nuttall et al. 2014; Law et al. 2017).

One important finding from these studies was that the decrease in the number of peroxisomes by knockdown or knockout of PEX1 was reversed by inhibition of autophagy. In a study with the HeLa cell line, suppression of the core Atg protein (ATG12) or NBR1 expression led to recovery in peroxisome numbers (Law et al. 2017). Of note, a separate study with the budding yeast also gave similar results, which suggested loss of PEX1 gave rise to overrun of peroxisome protein degradation that was cancelled by simultaneous knockout of a core Atg protein (Atg1) or the key pexophagy factor Atg36 (Nuttall et al. 2014). Because enhanced ubiquitination of PEX5 was detected in the HeLa cell line where PEX1 was knocked down,

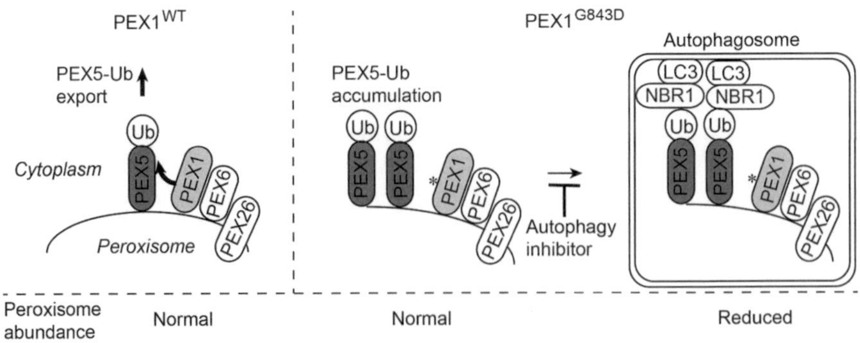

Fig. 3.7 Model scheme of pathological processes arising from PEX1 G843D mutation. The bold arrows indicate the action of PEX1 to export PEX5 to the cytoplasm, while the asterisk denotes mutation in PEX1. Ub ubiquitin

accumulation of ubiquitinated PEX5 on the peroxisome is thought to enhance the level of pexophagy (Fig. 3.7).

In *pex1* mutant fibroblasts with the G843D mutation, very-long-chain fatty acids accumulate due to impaired β-oxidation activity as a direct outcome of peroxisome deficiency. Intriguingly, treatment of these mutant cells with chloroquine, an FDA-approved chemical that inhibits autophagy, reduces the levels of very-long-chain fatty acids to those comparable with normal cells (Law et al. 2017). This finding suggests autophagy inhibitors may be promising drugs for alleviating symptoms of the disease arising from the PEX1 G843D mutation. Moreover, inhibition of autophagy not only partly recovers the lipid profile, but also rectifies the depletion in the number of peroxisomes of the mutant cells.

3.4.2 Oxygen, ROS and Pexophagy

Peroxisome metabolism contains various oxidases, and inevitably yields reactive oxygen species (ROS) as byproducts in the reactions. Therefore, the number of organelles should be tightly regulated in response to changes in the levels of oxygen or ROS. Specifically, when oxygen supply is limited, the organelle metabolism should be restricted accordingly. Likewise, when the ROS level from a population of peroxisomes is high enough to damage the organelle, the peroxisome portion should be promptly removed before it imposes detrimental effects on the cell. Pexophagy contributes to oxygen-related regulation of peroxisome abundance under varying physiological and pathological situations.

The sensing machinery of hypoxia (low concentration of oxygen), termed hypoxia-inducible factors (HIF), is composed of α and β subunits. The HIF α subunits (HIF-1α and HIF-2α) are degraded in the presence of normal concentrations of oxygen as a result of ubiquitination by a ligase complex containing von Hippel Lindau (VHL) tumor suppressor protein followed by targeting to the proteasome, the main protease complex responsible for rapid protein degradation. During hypoxia the HIFα subunits are stabilized and function as transcriptional factors together with HIF-1β (Keith et al. 2011). A recent study demonstrated that loss of the VHL protein causes HIF-2α-dependent enhancement of pexophagy, a decrease in the number of peroxisomes and a similar lipid profile to that seen in peroxisomal biogenesis disorders (Walter et al. 2014). In line with the fact that VHL protein is frequently mutated in carcinoma cells, accumulation of HIF-2α and a decrease in peroxisome numbers is often observed in renal cell carcinoma tissue samples.

Catalase is the main enzyme responsible for eliminating hydrogen peroxide generated within the peroxisome. When this enzyme is knocked down or its enzymatic activity is pharmacologically inhibited in a liver-derived HepG2 cell line under serum starved conditions, pexophagy is enhanced along with ubiquitination of PEX5 and recruitment of NBR1 to the peroxisomes (Lee et al. 2018). The HepG2 cell line also induces pexophagy in response to treatment with hydrogen peroxide, a representative ROS (Zhang et al. 2015). Similar induction of pexophagy by hydrogen peroxide depends on ataxia-telangiectasia mutated (ATM) kinase in another human cell line (HEK293) (Zhang et al. 2015). This kinase, which is also known to respond to DNA damage in the nucleus (Iijima et al. 2008), is recruited to the peroxisome and phosphorylates PEX5 in response to hydrogen peroxide treatment. The phosphorylation of PEX5 is required for its efficient ubiquitination.

3.5 Future Perspectives

Most of the studies that unveiled the molecular machinery of mammalian pexophagy employed artificial manipulations of the cellular status, such as overexpression or knockdown of certain proteins and inhibition of peroxisomal enzymes (e.g. catalase). Future studies will aim to analyze the molecular mechanisms of pexophagy under more physiological and pathological conditions. From a clinical perspective, inhibition of the overrun of pexophagy appears to be a promising strategy for alleviating at least one type of peroxisome biogenesis disorder. A similar approach will be sought to treat other diseases that are characterized by a severe reduction in peroxisome abundance.

References

Aksam EB et al (2007) A peroxisomal lon protease and peroxisome degradation by autophagy play key roles in vitality of Hansenula polymorpha cells. Autophagy 3:96–105

Colakoglu M, Tuncer S, Banerjee S (2018) Emerging cellular functions of the lipid metabolizing enzyme 15-lipoxygenase-1. Cell Prolif 51:e12472

Deosaran E et al (2013) NBR1 acts as an autophagy receptor for peroxisomes. J Cell Sci 126:939–952

Farre JC et al (2008) PpAtg30 tags peroxisomes for turnover by selective autophagy. Dev Cell 14:365–376

Farre JC et al (2013) Phosphorylation of mitophagy and pexophagy receptors coordinates their interaction with Atg8 and Atg11. EMBO Rep 14:441–449

Hanada T et al (2007) The Atg12-Atg5 conjugate has a novel E3-like activity for protein lipidation in autophagy. J Biol Chem 282:37298–37302

Hara-Kuge S, Fujiki Y (2008) The peroxin Pex14p is involved in LC3-dependent degradation of mammalian peroxisomes. Exp Cell Res 314:3531–3541

Huybrechts SJ et al (2009) Peroxisome dynamics in cultured mammalian cells. Traffic 10:1722–1733

Ichimura Y et al (2000) A ubiquitin-like system mediates protein lipidation. Nature 408:488–492

Iijima K et al (2008) Dancing on damaged chromatin: functions of ATM and the RAD50/MRE11/NBS1 complex in cellular responses to DNA damage. J Radiat Res 49:451–464

Imai K et al (2016) Atg9A trafficking through the recycling endosomes is required for autophagosome formation. J Cell Sci 129:3781

Iwata J et al (2006) Excess peroxisomes are degraded by autophagic machinery in mammals. J Biol Chem 281:4035–4041

Jiang L et al (2015) Peroxin Pex14p is the key component for coordinated autophagic degradation of mammalian peroxisomes by direct binding to LC3-II. Genes Cells 20:36–49

Kabeya Y et al (2000) LC3, a mammalian homologue of yeast Apg8p, is localized in autophagosome membranes after processing. EMBO J 19:5720–5728
Keith B, Johnson RS, Simon MC (2011) HIF1alpha and HIF2alpha: sibling rivalry in hypoxic tumour growth and progression. Nat Rev Cancer 12:9–22
Kihara A et al (2001) Two distinct Vps34 phosphatidylinositol 3-kinase complexes function in autophagy and carboxypeptidase Y sorting in Saccharomyces cerevisiae. J Cell Biol 152:519–530
Kikuchi M et al (2004) Proteomic analysis of rat liver peroxisome: presence of peroxisome-specific isozyme of Lon protease. J Biol Chem 279:421–428
Kim PK et al (2008) Ubiquitin signals autophagic degradation of cytosolic proteins and peroxisomes. Proc Natl Acad Sci U S A 105:20567–20574
Kirkin V et al (2009) A role for ubiquitin in selective autophagy. Mol Cell 34:259–269
Klionsky DJ et al (2003) A unified nomenclature for yeast autophagy-related genes. Dev Cell 5:539–545
Lamark T et al (2003) Interaction codes within the family of mammalian Phox and Bem1p domain-containing proteins. J Biol Chem 278:34568–34581
Law KB et al (2017) The peroxisomal AAA ATPase complex prevents pexophagy and development of peroxisome biogenesis disorders. Autophagy 13:868–884
Lee JN et al (2018) Catalase inhibition induces pexophagy through ROS accumulation. Biochem Biophys Res Commun 501:696–702
Luiken JJ et al (1992) Autophagic degradation of peroxisomes in isolated rat hepatocytes. FEBS Lett 304:93–97
Marcassa E et al (2018) Dual role of USP30 in controlling basal pexophagy and mitophagy. EMBO Rep 19:e45595
Mizushima N, Komatsu M (2011) Autophagy: renovation of cells and tissues. Cell 147:728–741
Mizushima N et al (1998) A protein conjugation system essential for autophagy. Nature 395:395–398
Motley AM, Nuttall JM, Hettema EH (2012) Pex3-anchored Atg36 tags peroxisomes for degradation in Saccharomyces cerevisiae. EMBO J 31:2852–2868
Mukaiyama H et al (2002) Paz2 and 13 other PAZ gene products regulate vacuolar engulfment of peroxisomes during micropexophagy. Genes Cells 7:75–90
Mukaiyama H et al (2004) Modification of a ubiquitin-like protein Paz2 conducted micropexophagy through formation of a novel membrane structure. Mol Biol Cell 15:58–70
Nazarko TY et al (2014) Peroxisomal Atg37 binds Atg30 or palmitoyl-CoA to regulate phagophore formation during pexophagy. J Cell Biol 204:541–557
Noda NN, Ohsumi Y, Inagaki F (2010) Atg8-family interacting motif crucial for selective autophagy. FEBS Lett 584:1379–1385
Nordgren M et al (2015) Export-deficient monoubiquitinated PEX5 triggers peroxisome removal in SV40 large T antigen-transformed mouse embryonic fibroblasts. Autophagy 11:1326–1340
Nuttall JM, Motley AM, Hettema EH (2014) Deficiency of the exportomer components Pex1, Pex6, and Pex15 causes enhanced pexophagy in Saccharomyces cerevisiae. Autophagy 10:835–845
Obara K et al (2008) The Atg18-Atg2 complex is recruited to autophagic membranes via phosphatidylinositol 3-phosphate and exerts an essential function. J Biol Chem 283:23972–23980
Papinski D, Kraft C (2016) Regulation of autophagy by signaling through the Atg1/ULK1 complex. J Mol Biol 428:1725–1741
Pomatto LC, Raynes R, Davies KJ (2017) The peroxisomal Lon protease LonP2 in aging and disease: functions and comparisons with mitochondrial Lon protease LonP1. Biol Rev Camb Philos Soc 92:739–753
Reddy JK et al (1982) Hepatic and renal effects of peroxisome proliferators: biological implications. Ann N Y Acad Sci 386:81–110
Riccio V et al (2019) Deubiquitinating enzyme USP30 maintains basal peroxisome abundance by regulating pexophagy. J Cell Biol 218:798–807

Sargent G et al (2016) PEX2 is the E3 ubiquitin ligase required for pexophagy during starvation. J Cell Biol 214:677–690
Schittek B, Sinnberg T (2014) Biological functions of casein kinase 1 isoforms and putative roles in tumorigenesis. Mol Cancer 13:231
Seglen PO, Gordon PB (1982) 3-methyladenine: specific inhibitor of autophagic/lysosomal protein degradation in isolated rat hepatocytes. Proc Natl Acad Sci U S A 79:1889–1892
Shintani T et al (1999) Apg10p, a novel protein-conjugating enzyme essential for autophagy in yeast. EMBO J 18:5234–5241
Suzuki H et al (2017) Structural biology of the core autophagy machinery. Curr Opin Struct Biol 43:10–17
Tanaka C et al (2014) Hrr25 triggers selective autophagy-related pathways by phosphorylating receptor proteins. J Cell Biol 207:91–105
Titorenko VI et al (1995) Isolation and characterization of mutants impaired in the selective degradation of peroxisomes in the yeast Hansenula polymorpha. J Bacteriol 177:357–363
Tsukada M, Ohsumi Y (1993) Isolation and characterization of autophagy-defective mutants of Saccharomyces cerevisiae. FEBS Lett 333:169–174
Tuttle DL, Dunn WA Jr (1995) Divergent modes of autophagy in the methylotrophic yeast Pichia pastoris. J Cell Sci 108:25–35
van Leyen K et al (1998) A function for lipoxygenase in programmed organelle degradation. Nature 395:392–395
Veenhuis M et al (1983) Degradation and turnover of peroxisomes in the yeast Hansenula polymorpha induced by selective inactivation of peroxisomal enzymes. Arch Microbiol 134:193–203
Walter KM et al (2014) Hif-2alpha promotes degradation of mammalian peroxisomes by selective autophagy. Cell Metab 20:882–897
Xie Z, Klionsky DJ (2007) Autophagosome formation: core machinery and adaptations. Nat Cell Biol 9:1102–1109
Yamamoto H et al (2012) Atg9 vesicles are an important membrane source during early steps of autophagosome formation. J Cell Biol 198:219
Yamashita S et al (2014) The membrane peroxin PEX3 induces peroxisome-ubiquitination-linked pexophagy. Autophagy 10:1549–1564
Yokota S (1993) Formation of autophagosomes during degradation of excess peroxisomes induced by administration of dioctyl phthalate. Eur J Cell Biol 61:67–80
Yokota S, Dariush Fahimi H (2009) Degradation of excess peroxisomes in mammalian liver cells by autophagy and other mechanisms. Histochem Cell Biol 131:455–458
Yokota S et al (1993) Formation of autophagosomes during degradation of excess peroxisomes induced by di-(2-ethylhexyl)phthalate treatment. II. Immunocytochemical analysis of early and late autophagosomes. Eur J Cell Biol 62:372–383
Yokota S, Oda T, Fahimi HD (2001) The role of 15-lipoxygenase in disruption of the peroxisomal membrane and in programmed degradation of peroxisomes in normal rat liver. J Histochem Cytochem 49:613–622
Yokota S, Haraguchi CM, Oda T (2008) Induction of peroxisomal Lon protease in rat liver after di-(2-ethylhexyl)phthalate treatment. Histochem Cell Biol 129:73–83
Zhang J et al (2015) ATM functions at the peroxisome to induce pexophagy in response to ROS. Nat Cell Biol 17:1259–1269
Zientara-Rytter K et al (2018) Pex3 and Atg37 compete to regulate the interaction between the pexophagy receptor, Atg30, and the Hrr25 kinase. Autophagy 14:368–384

Chapter 4
The Function of the Peroxisome

Masashi Morita and Tsuneo Imanaka

Abstract Peroxisomes are organelles that are essential for maintaining cellular function in lipid metabolism, redox homeostasis and intracellular signalling. They are involved in the β-oxidation of various fatty acids, especially very long chain fatty acids, as well as the synthesis of ether-phospholipids and bile acids in mammals. Substrates for the β-oxidation are transported into peroxisomes by the ABC transporters ABCD1–3, and the metabolites produced by the β-oxidation cycle are released from peroxisomes through certain unique transporters and channels on the peroxisomal membrane. The initial steps of ether-phospholipid synthesis take place in peroxisomes, with one of steps being tightly controlled by negative feedback regulation. As peroxisomes have various enzymes that generate reactive oxygen species such as H_2O_2, their metabolic activity significantly impacts cellular redox balance, which is closely connected to the redox homeostasis in mitochondria. It has been shown that the physical contact sites between peroxisomes and lysosomes, the ER, mitochondria or lipid droplets are formed by tethered proteins and play an important role in the transporting or shuttling of metabolites between peroxisomes and other organelles. Furthermore, peroxisomes tightly collaborate with mitochondria in the antiviral defence system. In this chapter, we describe the current understanding of the peroxisomal metabolic pathways including lipid metabolism and redox metabolism, as well as the crosstalk that takes place between peroxisomes and other organelles.

Keywords ABC transporter · Crosstalk of peroxisomes with other organelles · Ether-phospholipid synthesis · Fatty acid β- and α-oxidation · Oxidative stress · Peroxisome · Reactive oxygen species · Solute transporter

---

M. Morita (✉)
Graduate School of Medicine and Pharmaceutical Sciences, University of Toyama, Toyama, Japan
e-mail: masa@pha.u-toyama.ac.jp

T. Imanaka (✉)
Faculty of Pharmaceutical Sciences, Hiroshima International University, Kure, Hiroshima, Japan

University of Toyama, Toyama, Japan
e-mail: imanaka@hirokoku-u.ac.jp

T. Imanaka, N. Shimozawa (eds.), *Peroxisomes: Biogenesis, Function, and Role in Human Disease*, https://doi.org/10.1007/978-981-15-1169-1_4

Abbreviations

ABC	ATP-binding cassette
ACOXs	Acyl-CoA oxidases
CNS	Central nervous system
DHCA	Dihydroxycholestanoic acid
ER	Endoplasmic reticulum
LCFA	Long chain fatty acid
LDs	Lipid droplets
MAVS	Mitochondrial antiviral-signaling protein
NBD	Nucleotide-binding domain
PMP	Peroxisomal membrane protein
PPAR	Peroxisome proliferator-activated receptor
PUFAs	Polyunsaturated fatty acids
ROS	Reactive oxygen species
THCA	Trihydroxycholestanoic acid
TMD	Transmembrane domain
VLCFA	Very long chain fatty acid
X-ALD	X-linked adrenoleukodystrophy

4.1 Outline of Peroxisome Function

Peroxisomes are comprised of a dense matrix surrounded by a single membrane. They are highly dynamic organelles and can rapidly change in size, abundance and protein content in response to nutritional and environmental alterations. They contain a unique set of enzymes include various H_2O_2-generating oxidases and the H_2O_2-degrading enzyme catalase across most species. Human peroxisomes contain over 50 different enzymes. Most of these are involved in the various metabolic pathways unique to this organelle and play an essential role in cellular functions (Waterham et al. 2016).

Among the metabolic processes in the peroxisomes in mammals, the β-oxidation of fatty acids is the major catabolic pathway. Very long chain fatty acid (VLCFA), long-chain dicarboxylic acids, 2- or 3-methyl branched-chain fatty acids and bile acid intermediates are selectively β-oxidized in peroxisomes. These fatty acids are mostly imported into peroxisomes as CoA esters via the ATP-binding cassette (ABC) transporters ABCD1–3, and metabolized by one or more cycles of the β-oxidation. The metabolites, medium or short chain acyl-CoA as well as acetyl-CoA, are converted to other forms such as acylcarnitine and acetate, and released into the cytosol for further metabolism (Wanders 2014). In the case of bile acids, the choloyl-CoA and deoxycholoyl-CoA produced in peroxisomes are converted to taurine- or glycine-conjugates and exported into the cytosol (Rembacz et al. 2010).

Peroxisomes also catalyze the synthesis of the ether-phospholipids, a family of phospholipids in which one of the acyl chains is joined to glycerol by an ether bond at the $sn-1$ position. The most abundant ether-phospholipids are plasmalogens with a vinyl ether bond at the $sn-1$ position. The ether-phospholipids occupy approximately 20% of the total phospholipid mass in humans (Vance 2015). They are an important membrane constituent in the heart, brain and white blood cells. The ether-phospholipids have antioxidant properties and protect the biomembrane and myelin sheath from reactive oxygen species (ROS).

In addition to lipid metabolism, peroxisomes play a role in several non-lipid metabolic pathways, including purine, polyamine, glyoxylate and D-amino acid metabolism. Among these, human peroxisomes play a crucial role in the detoxification of the glyoxylate produced in the liver. The conversion of glyoxylate to glycine is catalyzed by peroxisomal alanine-glyoxylate aminotransferase (AGT), an enzyme expressed predominantly in the liver. Peroxisomes are also involved in ROS metabolism. Human peroxisomes contain anti-oxidant enzymes such as catalase and superoxide dismutase (SOD). Peroxisomes are potentially important organelles in cellular redox metabolism (Fransen et al. 2012).

Recently, a new function of peroxisomes was reported that is based on communication with other organelles. Dixit et al. suggested that peroxisomes as well as mitochondria are important sites of antiviral signal transduction through a sensor of cytosolic virus, called mitochondrial antiviral-signaling protein (MAVS), that is also located on the peroxisomal membrane (Dixit et al. 2010). MAVS induces antiviral signalling from this organelle by the expression of several antiviral factors and serves critical subcellular hubs for promoting MAVS dependent antiviral immunity (Odendall et al. 2014). Furthermore, Chu et al. reported a novel peroxisomal function in cellular cholesterol transport (Chu et al. 2015). They demonstrated, based on genome-wide RNAi screening, that peroxisomes have a role in the transport of free cholesterol from lysosomes to the endoplasmic reticulum (ER) through lysosome-peroxisome membrane contact. Furthermore, phosphatidylinositol 4,5-bisphosphate ($PI(4,5)P_2$) on the peroxisomal membrane might regulate the transport of cholesterol via peroxisomes from lysosomes (Hu et al. 2018). It is thus highly likely that peroxisomes are multifunctional organelles that interact with other organelles. The crosstalk of peroxisomes with mitochondria, lysosomes and other organelles is important in various metabolic and signalling pathways.

The peroxisome proliferator-activated receptor (PPAR) was discovered in 1990 (Issemann and Green 1990). PPAR is a general transcriptional regulator of lipid homeostasis, and several hypolipidemic and antidiabetic agents have been shown to be effective ligands. PPARα is known to be able to control not only peroxisomal metabolism, but also lipid homeostasis (Vamecq et al. 2014). However, the physiological importance of PPARα has not yet been well elucidated in humans, although PPARα has been shown to have a variety of functions at the cellular as well as organism level using PPARα knockout mice.

4.2 Peroxisome and Lipid Metabolism

As mentioned above, peroxisomes are important for lipid metabolism. Here we describe in detail the β- and α-oxidation of various fatty acids in peroxisomes, including both how the substrates are imported and the metabolites are then exported. In addition, the synthesis of ether-phospholipids and its regulation are mentioned briefly.

4.2.1 Transport of Substrate into Peroxisomes

Fatty acids exist mainly as CoA esters in the cytosol. Therefore, fatty acid-CoA (acyl-CoA) is transported into peroxisomes by transporters before being oxidized by peroxisomal enzymes. In mammals, three ABC transporters belonging to subfamily D, i.e. ABCD1–3, are involved in the transport of these fatty acids on the peroxisomal membrane (Morita and Imanaka 2012). ABCD1 and ABCD2 have a high sequence homology and both are involved in the transport of saturated and unsaturated VLCFA-CoA into peroxisomes (Fig. 4.1). In contrast, ABCD3 is involved in the import of bile acid intermediates such as trihydroxycholestanoic acid (THCA), dihydroxycholestanoic acid (DHCA), pristanic acid and phytanic acid. The recent identification of an ABCD3 deficiency in a patient with markedly elevated DHCA and THCA levels in the plasma, along with studies in the *Pmp70*$^{-/-}$ mouse, have provided support for this conclusion (Ferdinandusse et al. 2015).

4.2.2 Fatty Acid β-Oxidation

In mammals, certain fatty acids are metabolized exclusively in peroxisomes. These include VLCFA (such as C22:0, C24:0 and C26:0), polyunsaturated VLCFA (C24:5 n − 6, C24:6 n − 3), long chain dicarboxylic acids, 2-methyl branched chain fatty acid (pristanic acid, 2,6,10,14-tetramethylpentadecanoic acid), 3-methyl branched chain fatty acid (phytanic acid, 3,7,11,14-tetramethylhexadecanoic acid) and bile acid intermediates (DHCA and THCA) (Fig. 4.1). Phytanic acid and pristanic acid are taken into the body via food intake, including plant-derived oil, ruminant animal fats and certain kinds of fish. These fatty acids are imported into peroxisomes via ABCD1–3, degraded by one or more cycles of β-oxidation and released into the cytosol for further metabolism.

The enzymes that catalyse the formation of DHCA and THCA are localized in different subcellular compartments, including the cytosol, ER, and mitochondria. Cholesterol is converted to 25R-THCA and 25R-DHCA, and these are activated to 25R-THCA-CoA and 25R-DHCA-CoA by bile acid-CoA ligase on the ER membrane, entering the peroxisome through ABCD3. Then 25R-THCA-CoA and

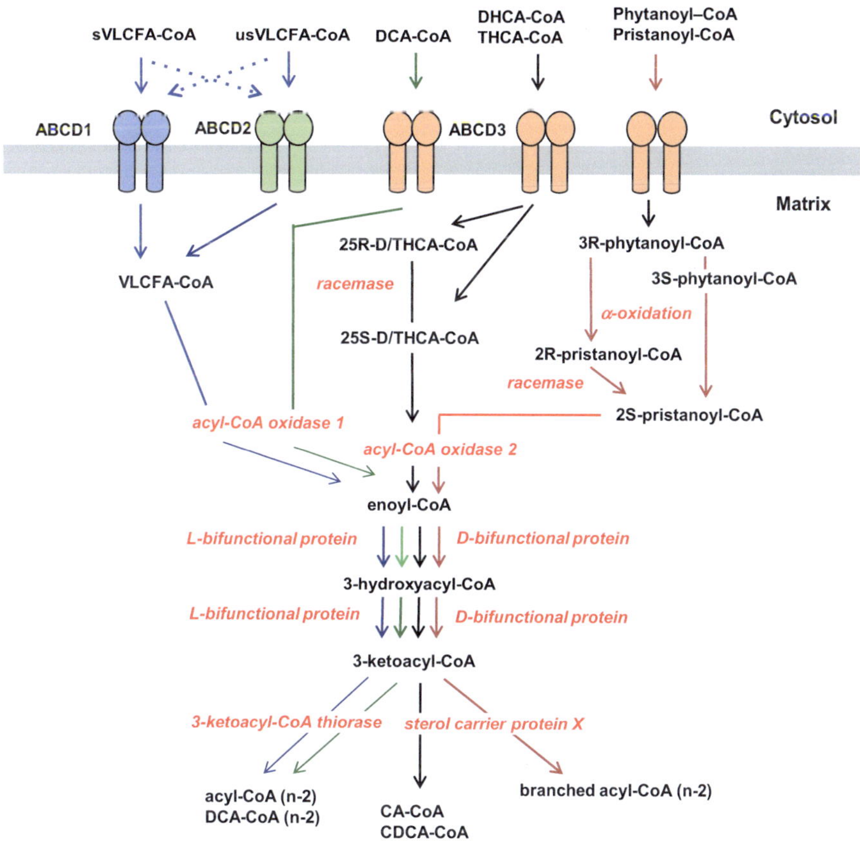

Fig. 4.1 Fatty acid oxidation pathway in the peroxisome. A variety of types of fatty acid-CoA is transported into peroxisomes via three ABC transporters. 3R- and 3S-phytanic acid-CoA are respectively metabolized to 2R- and 2S-pristanoyl-CoA. 2R-pristanoyl-CoA is converted by racemase to 2S-pristanoyl-CoA. 25R-DHCA-CoA and 25R-THCA-CoA are converted to 25S-DHCA-CoA and 25S-THCA-CoA by racemase, respectively. These substrates are then oxidized by fatty acid β-oxidation enzymes. sVLCFA, saturated very long chain fatty acid; usVLCFA, unsaturated very long chain fatty acid; CA, cholic acid; CDCA, chenodeoxycholic acid; DCA, dicarboxylic acid

25R-DHCA-CoA are converted to 25S-THCA-CoA and 25S-DHCA-CoA by 2-methyl-acyl-CoA racemase for β-oxidation, respectively.

Phytanic acid exists in 3R- and 3S-forms. Both are activated as CoA esters and transported into peroxisomes by ABCD3. As phytanic acid has a methyl group at the 3-position, it cannot be oxidized by the peroxisomal β-oxidation system. Therefore, 3R- and 3S-phytanic acid are first converted to 2R- and 2S-pristanic acid by α-oxidation in peroxisomes (see below). Subsequently, 2R-pristanoyl-CoA is converted to 2S-pristanoyl-CoA by 2-methyl-acyl-CoA racemase and 2S-pristanoyl-CoA enters the β-oxidation cycle (Fig. 4.1) (Wanders et al. 2011).

Like mitochondria, the β-oxidation that occurs in peroxisomes is conducted in 4 steps, dehydrogenation (oxidation), hydration, dehydrogenation again and thiolytic cleavage. In peroxisomes, FADH reacts with O_2 to produce H_2O_2, which is degraded by catalase into H_2O and O_2. The re-oxidation of FADH is coupled to the electron transport chain so as to produce ATP in mitochondria (Wanders et al. 2015).

The first step in β-oxidation is catalysed by acyl-CoA oxidases (ACOXs) that interact directly with molecular oxygen, generating H_2O_2. H_2O_2 is subsequently converted to molecular oxygen by catalase. There are at least three different ACOXs. ACOX1 preferentially degrades long and medium saturated and unsaturated straight chain fatty acids, whereas ACOX2 has a high affinity for 2-methyl branched fatty acids such as pristanoyl-CoA, DHCA-CoA and THCA-CoA. ACOX3 is also known to be involved in the degradation of 2-methyl branched fatty acids. Unlike the rat homolog, the human gene is expressed at very low levels in the liver and the physiological role of ACOX3 has yet to be elucidated.

In the second and third steps, bifunctional proteins possessing dual enoyl-CoA hydratase and 3-hydroxyacyl CoA dehydrogenase activities, conduct the process. They are an L-bifunctional (L-PBE) and D-bifunctional enzyme (D-PBE), or alternatively, multifunctional enzyme type 1 and type 2 (MEP1 and MEP2), respectively. It has been elucidated that D-PBE catalyses the hydration and subsequent dehydrogenation of enoyl-CoA esters of VLCFAs, pristanic acid, DHCA and THCA. In contrast, the physiological role of L-PBE is not well understood, but it might be involved in the degradation of long to medium chain dicarboxylic enoyl-CoA esters.

In the last step, the two thiolases 3-ketoacyl-CoA thiolase 1 (pTH1) and sterol carrier protein X (SCPX) (pTH2) are involved in the thiolytic cleavage of 3-ketoacyl-CoA to acetyl-CoA and acyl-CoA, shortened by two carbon atoms. pTH1 metabolizes only straight-chain fatty acids. The branched chain fatty acids and bile acid precursors are solely cleaved by pTH2. Once the fatty acid chains are shortened to medium-chain fatty acyl-CoA via peroxisomal β-oxidation, they are conjugated to carnitine or exist in part as free fatty acids, exit the peroxisomes, and undergo further β-oxidation in mitochondria. During β-oxidation, the NADH formed in peroxisomes is partially transported to mitochondria, re-oxidized to NAD^+, and returned back to peroxisomes. NADH is also re-oxidized in peroxisomes via lactate/pyruvate and the glycerophosphate shuttle (Antonenkov and Hiltunen 2006). Several different kinds of acyl-CoA thioesterases exist in peroxisomes in mammals. They might contribute to the recycling of CoA in the process of β-oxidation, and the free fatty acids that are produced are released into the cytosol (Hunt et al. 2014).

In contrast, the choloyl-CoA and deoxycholoyl-CoA produced in peroxisomes are converted to taurine- or glycine-conjugated cholic acid, or deoxycholic acid, by bile acid-CoA: amino acid *N*-acyltransferase and then exported into the cytosol (Rembacz et al. 2010).

4.2.3 *Fatty Acid α-Oxidation*

In contrast to the 2-methyl branched-chain fatty acids, the 3-methyl-branched chain fatty acids are not directly oxidized by the β-oxidation system because of the methyl-group at position 3. Such fatty acids first have the last carbon atom removed by α-oxidation to generate 2-methyl branched-chain fatty acids that can be β-oxidized. Phytanoyl-CoA is coverted to 2-hydroxyphytanoyl-CoA by phytanoyl-CoA 2-hydroxylase. Subsequently, 2-hydroxyphytanoyl-CoA is cleaved to pristanol and formyl-CoA by 2-hydroxyl-CoA lyase. Pristanol is then oxidized to pristanic acid by pristanol aldehyde dehydrogenase. Formyl-CoA is hydrolyzed to formic acid and CoA by acyl-CoA thioesterase. The pristanic acid activated as pristanoyl-CoA is oxidized by ACOX2 and enters the β-oxidation cycle (Fig. 4.2) (Wanders et al. 2011).

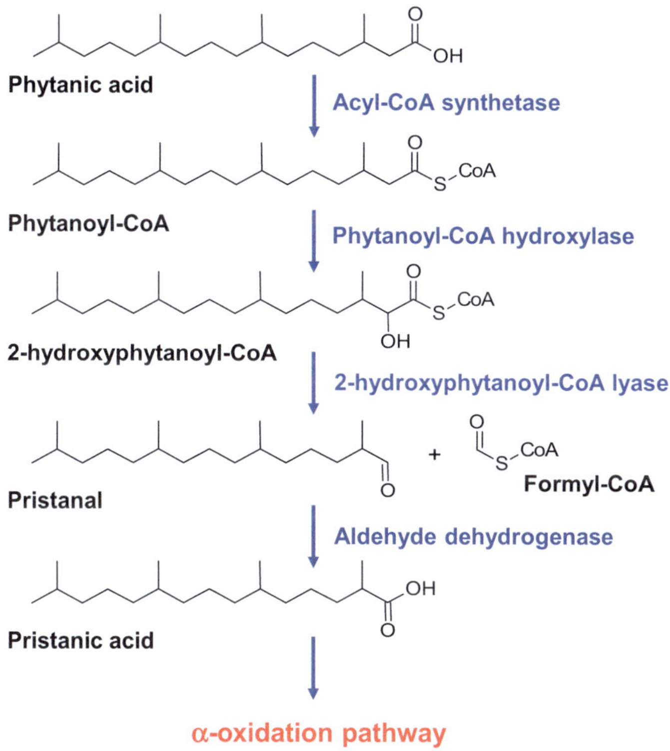

Fig. 4.2 α-oxidation of phytanic acid. The α-Oxidation pathway for phytanic acid. The pathway is necessary to remove the β-carbon blocked with a methyl group so that β-oxidation can be proceed

4.2.4 Ether-Phospholipid Synthesis

The ether-phospholipids are a special class of phospholipids with an ether bond at the *sn* − 1 position of the glycerol backbone. The most abundant ether-phospholipids are the plasmalogens, which have a vinyl ether bond at the *sn* − 1 position (Fig. 4.3a). The plasmalogens mostly have either ethanolamine or choline as the head group. They are a constituent of the membrane glycerophospholipids and are enriched in polyunsaturated fatty acids (PUFAs) at the *sn* − 2 position (Braverman and Moser 2012). The plasmalogens are largely found in the central nervous system (CNS) as major constituents of myelin phospholipids. As they contain PUFAs, they serve as antioxidants in the CNS. Plasmalogen biosynthesis deficiency is reported to result in a peroxisomal disease with abnormal myelin formation, suggesting that the plasmalogens play a pivotal role in the CNS (see Chap. 5) (Sztriha et al. 2000).

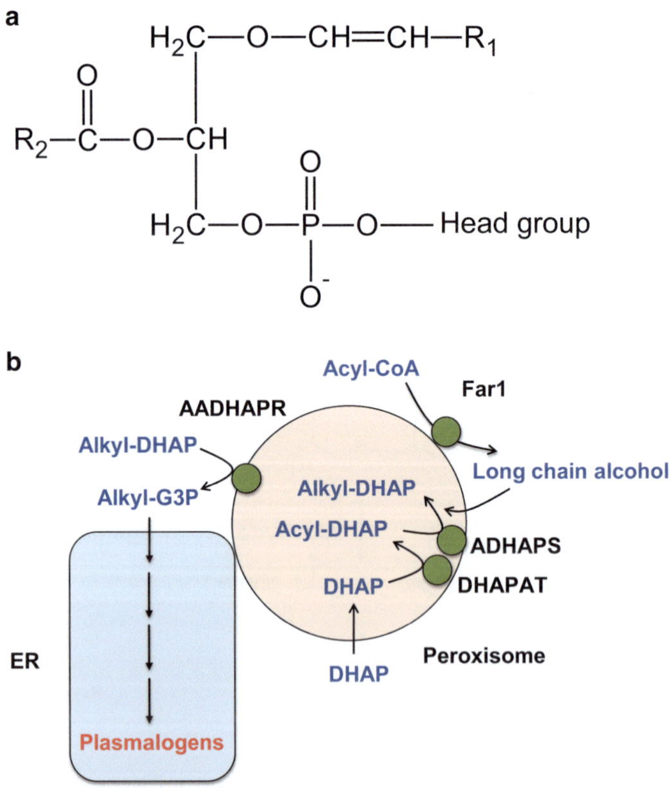

Fig. 4.3 Biogenesis of ether-phospholipids. (**a**) Structure of plasmalogen. (**b**) Pathway for plasmalogen biosynthesis. AADHAPR, acyl/alkyl-dihydroxyacetone phosphate dehydrogenase; ADHAPS, alkyldihydroxyacetonephosphate synthase; DHAPAT, dihydroxyacetone phosphate acyltransferase; Far1, fatty acyl-CoA reductase

Plasmalogen biosynthesis requires enzymes localized in both peroxisomes and the ER (Fig. 4.3b) (Nagan and Zoeller 2001). Plasmalogen synthesis is initiated in peroxisomes. Dihydroxyacetone phosphate (DHAP) is converted to acyl-DHAP by DHAP acyl transferase (DHAPAT). The ester linkage of acyl-DHAP is converted to an ether linkage by an exchange reaction with long chain alcohol by alkyldihydroxyacetonephosphate synthase (ADHAPS). Both DHAPAT and ADHAPS are intraperoxisomal enzymes, and they form a functional complex to increase efficiency. The fatty alcohol for the ADHAPS reaction is generated by the fatty acyl-CoA reductase (FAR1) that is a peroxisomal tail-anchored protein (Farr et al. 2016; Honsho et al. 2013). This step is suggested to be rate limiting in plasmalogen biosynthesis. The resulting precursor, alkyl-DHAP, moves to the outer leaflet of the peroxisomal membrane and alkyl-DHAP is converted to 1-O-alkyl-2-hydroxy-*sn*-glycerophosphatidic acid by acyl/alkyl-DHAP reductase (AADHAPR) and is transported to the ER (Dorninger et al. 2017). Subsequently, the plasmalogen synthesis completes in the ER by four additional modification steps before transport to the plasma membrane.

Plasmalogen biosynthesis is subjected to feedback regulation by the plasmalogen level. Honsho et al. reported that plasmalogen synthesis is regulated by the proteolytic stability of FAR1 on the peroxisomal membrane by a feedback mechanism in which the plasmalogen level on the inner plasma membrane leaflet is sensed (Honsho and Fujiki 2017). Deficiency of certain peroxisomal enzymes essential for plasmalogen synthesis, i.e. DHAPAT, ADHAPS, or FAR1, results in a fatal genetic disease, rhizomelic chondrodysplasia punctata (RCDP) (see Chap. 5).

4.3 Structure and Function of the ABC Transporter

Peroxisomes are involved in a variety of metabolic processes. Recently it has become clear that several transporters are involved in the transport of the resulting metabolites. One of the transporter families is the ABC transporter. The structure of the ABC transporters is highly conserved, and they catalyse ATP-dependent transmembrane transport of a variety of substrates into or out of cells so as to maintain cellular homeostasis. The human ABC transporter family is comprised of 48 members, if we exclude ABCC13, which is a pseudogene product lacking any functional ABC protein domain. The ABC transporters are divided into seven subfamilies, A–G, based on structural organization and amino acid homology (Dassa and Bouige 2001; Vasiliou et al. 2009). The human ABC proteins are involved in a number of important physiological processes, and dysfunction of the ABC transporters is known to result in severe disease.

4.3.1 ABCD Transporters

In mammals, four ABC transporters belonging to subfamily D, ABCD1–4, have been identified: the adrenoleukodystrophy protein (ALDP/ABCD1), ALDP-related protein (ALDRP/ABCD2), 70-kDa peroxisomal membrane protein (PMP70/ABCD3) and PMP70-related protein (P70R/ABCD4) (Kamijo et al. 1990; Mosser et al. 1993; Lombard-Platet et al. 1996; Holzinger et al. 1997a, b; Shani et al. 1997). ABCD3 is markedly induced by the administration of hypolipidemic agents in parallel with peroxisome proliferation and induction of peroxisomal β-oxidation enzymes (Kamijo et al. 1990).

The *ABCD1* gene was discovered by the positional cloning that was performed in an effort to identify the gene responsible for X-linked adrenoleukodystrophy (X-ALD) (Mosser et al. 1993). The deduced protein sequence showed significant sequence identity with ABCD3. There was considerable surprise at this unexpected result since it was suspected at that time that VLCFA-CoA synthetase was the gene responsible for X-ALD. The *ABCD2* gene was cloned from a cDNA library from the mouse monocyte/macrophage cell line P388D1 using PCR products encoding Walker A and a 12-mer conserved motif in the ABC transporters (Holzinger et al. 1997a). ABCD4 was identified by a homology search for ABCD1- and ABCD3-related sequences in a database of expressed sequence tags (ESTs) (Holzinger et al. 1997b; Shani et al. 1997).

ABCD1–3 exist in the peroxisomal membrane and are involved in the transport of various kinds of fatty acid-CoA. ABCD4 localizes to the lysosome and is suggested to be involved in the transport of vitamin B_{12} from the lysosome to the cytosol.

4.3.2 Structure of the ABCD Transporters

The ABCD transporters are half-size ABC transporters with one transmembrane domain (TMD) and one nucleotide-binding domain (NBD). From the hydropathy profile, the amino terminal half is hydrophobic, with six transmembrane segments, and the COOH-terminal half is hydrophilic, having the NBD (Fig. 4.4a). The TMD provides the passageway for the substrate across the membranes, while the NBD energizes the directional transport of these substrates by alternating cycles of ATP binding and hydrolysis. The NBD contains three conserved motifs: Walker A, Walker B and the ABC signature motif, which is situated upstream of the Walker B sequence (Dassa and Bouige 2001; Higgins 1992).

In early studies, we showed that the binding and hydrolysis of ATP induced conformational changes in ABCD3 close to the boundary between the TMD and NBD, as well as the helical domain between the Walker A and B motifs (Kashiwayama et al. 2002; Tanaka et al. 2002). On the other hand, Guimarães et al. showed with a protease-based assay that LCFA-CoA and VLCFA-CoA induced conformational

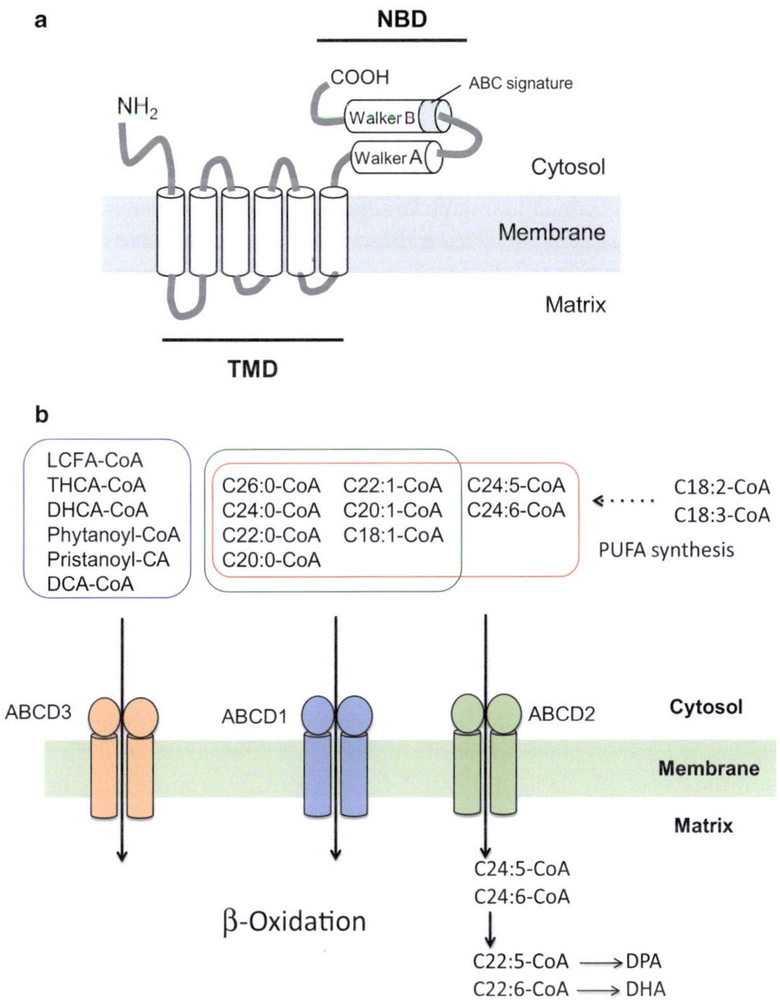

Fig. 4.4 Function of the ABC transporter. Hypothesized structure of the peroxisomal ABC transporter (**a**) and the possible function of ABC transporters on the peroxisomal membrane (**b**). The ABCD transporters are comprised of a half-size ABC transporter with one transmembrane domain (TMD) and one nucleotide-binding domain (NBD). Six transmembrane domains are located in the NH_2-terminal half of the transporter, and the Walker A, B and ATP-binding cassette signature sequence (the ABC signature) is located in the COOH-terminal half of the transporter. In mammals, ABCD1 and ABCD2 have overlapping substrate specificities toward saturated and monounsaturated VLCFA-CoAs. However, ABCD1 has a higher specificity for C24:0-CoA and C26:0-CoA than ABCD2. In contrast, ABCD2 has a higher specificity for C22:0-CoA and C20:0-CoA than ABCD1. ABCD2, but not ABCD1, has an affinity for polyunsaturated VLCFA-CoA. C24:5-CoA and C24:6-CoA synthesized in the ER are transported into peroxisomes by ABCD2 and converted to docosapentaenoic acid (DPA) and DHA, respectively. ABCD3 is thought to be involved in the transport of LCFA-CoA, branched chain acyl-CoA, THCA-CoA and DHCA-CoA

alterations in ABCD1, suggesting that the TMD is involved in the recognition of substrates (Guimaraes et al. 2005). These findings suggest that the NH_2-terminal TMD of ABCD1 is involved in the recognition of these substrates, and undergoes a conformational change on ATP binding to the COOH-terminal NBD of ABCD1.

How do the peroxisomal ABC transporters recognize their substrates and import them into the lumen of peroxisomes? This is one of the key outstanding questions at present. The X-ray crystallographic structure of the peroxisomal ABC transporters has yet to be obtained. However, a recent study on the structure of homomeric *Cyanidioschyzon merolae* P-glycoprotein (CmABCB1), a homolog of human P-glycoprotein (Mdr1), suggests a complex structure. This study revealed a two-fold structure at 1.9 Å and 3.0 Å resolution: an outward-facing conformation state with a bound nucleotide and an inward-facing conformation, respectively (Kodan et al. 2019). According to the model proposed, the inward facing conformation results in a large internal cavity open to both the cytoplasm and inner leaflet of the plasma membrane. This conformation represents the initial stage of the transport cycle and is suitable for substrate binding. Then, when substrate enters the internal substrate-binding pocket, ATP binds to the NBD, causing a large conformational change, thereby presenting the substrate and substrate-binding sites to the outer space. Based on these findings, the peroxisomal ABC transporters might have a similar catalytic mechanism of ATP binding and hydrolysis. On the other hand, the ABCD transporters possess acyl-CoA thioesterase activity (De Marcos Lousa et al. 2013; Okamoto et al. 2018), and it is suggested that this activity is required for the transport of VLCFA-CoA into peroxisomes. It is of considerable interest to determine the location of the active site of acyl-CoA thioesterase on the transporter and how the respective cavity in VLCFA-CoA or VLCFA is specifically organized.

In terms of the quaternary structures of the ABCD proteins, ABCD1, ABCD3 and ABCD4 are suggested to exist in the membrane mainly as a homodimer (Liu et al. 1999; Guimaraes et al. 2004; Hillebrand et al. 2007), although ABCD1 is able to form a complex with ABCD2 or ABCD3. Recently it was reported that ABCD1 and ABCD2 might assemble as a heterotetramer (Geillon et al. 2017).

4.3.3 Function of the ABCD Transporters

4.3.3.1 ABCD1

ABCD1 has been thought to transport VLCFA-CoA, as *ABCD1* mutation causes X-ALD, which is characterized by abnormal VLCFA tissue accumulation. Early studies showed that exogenous expression of ABCD1 in X-ALD skin fibroblasts restored VLCFA β-oxidation. Furthermore, it was demonstrated by expressing human ABCD1 in a *pxa1/pxa2* knockout yeast mutant that ABCD1 is involved in the transport of saturated, monounsaturated and polyunsaturated LCFA-CoA and VLCFA-CoA, such as such as C18:0-, C22:0-, C24:0-, C26:0-, C18:1- and C24:6-CoA, across the peroxisomal membrane (Fig. 4.4b) (van Roermund et al. 2011).

However, the mechanism of fatty acid transport by ABCD1 has yet to be well characterized. It is proposed that the CoA moiety of VLCFA-CoA is cleaved during the transport cycle since COMATOSE, a homolog of human ABCD1 in *A. thaliana*, itself possesses intrinsic acyl-CoA thioesterase activity (De Marcos Lousa et al. 2013). In addition, β-oxidation of oleic acid was analysed in peroxisomes in a medium including $[^{18}O]H_2O$ using yeast cells with a deficiency of thiolase but expressing carnitine palmitoyltransferase 2-SKL in peroxisomes (van Roermund et al. 2012). Quantitation of the amount of isotope-labeled carnitine suggests that acyl-CoA is hydrolyzed before it is metabolized. Another model suggests that esterified fatty acids are delivered directly to the peroxisomal matrix. This model depends on the observation that the β-oxidation of VLCFA-CoA in peroxisomes isolated from fibroblasts directly depends on ABCD1 without any need of additional re-esterification by an acyl-CoA synthetase (Wiesinger et al. 2013).

Recently we showed that purified ABCD1 reconstituted in proteoliposomes possesses acyl-CoA thioesterase activity (Okamoto et al. 2018). It is of interest to determine how acyl-CoA thioesterase activity is involved in the transport of VLCFA into peroxisomes. On the other hand, the activity of ABCD1 was shown to be very low in mammals compared with other acyl-CoA thioesterases found in peroxisomes, mitochondria, the ER and cytosol. In addition, the acyl chain covalently bound to ABCD1 was shown to be detectable under a condition in which such intermediates were not detectable in other acyl-CoA thioesterases. This suggests that acylation of ABCD1 occurs over a relatively long period of time. Therefore, acyl-CoA thioesterase might be required for regulating the transport of VLCFA through acylation of the ABCD1 protein.

The export of cholesterol from lysosomes is reduced by mutation of ABCD1 (Chu et al. 2015). It is interesting that there is crosstalk between peroxisomes and lysosomes in the course of cholesterol metabolism. In *Abcd1*-deficient mice brain and X-ALD fibroblasts, free cholesterol was shown to have accumulated in lysosomes, indicating that the peroxisomal ABCD1 protein is directly or indirectly involved in cholesterol trafficking.

4.3.3.2 ABCD2

ABCD2 shares functional redundancy with ABCD1 (Pujol et al. 2004). ABCD2 has an affinity for PUFAs such as C24:5-CoA and C24:6-CoA, although ABCD1 has a higher specificity for saturated VLCFA-CoA (Fig. 4.4b). This was shown by experiments using the yeast *pxa1/pxa2*Δ mutant, which expresses ABCD1 and/or ABCD2 (van Roermund et al. 2011). In addition, the abnormal fatty acid phenotype in *Abcd2*$^{-/-}$ mice is different from that in *Abcd1*$^{-/-}$ mice (Fourcade et al. 2009). *Abcd1*$^{-/-}$ mice display a higher accumulation of C24:0 and especially C26:0. In contrast, *Abcd2*$^{-/-}$ mice display different fatty acid abnormalities, especially monounsaturated and PUFAs. The β-oxidation activity of C24:6n − 3, an immediate precursor of docosahexaenoic acid (DHA), is reduced in the *Abcd2*$^{-/-}$ mouse brain. In addition, ABCD2 expression is highly sensitive to PUFAs, suggesting that ABCD2 is involved in the transport of PUFAs in DHA metabolism (van Roermund et al. 2011; Fourcade et al. 2009; Genin et al. 2011). Furthermore, the ABCD2 expression pattern is different from that of ABCD1 in tissues and cells in mice and humans. It is hypothesized that there are different benefits conferred on cells by the respective expression of ABCD1 and ABCD2.

ABCD2 appears to have a role in the metabolism of monounsaturated and polyunsaturated VLCFAs rather than saturated VLCFAs, and may be involved in the regulation of oxidative stress and synthesis of DHA (Fourcade et al. 2010). However, the actual function of ABCD2 *in vivo* is still unclear because the endogenous expression level of ABCD2 is quite low in human cells and there is no reported disease caused by mutation of the *ABCD2* gene.

4.3.3.3 ABCD3

ABCD3 is one of the most abundant PMPs, at least in the liver and kidney, and has been reported to be involved in the transport of various fatty acids that ABCD1 and ABCD2 cannot transport (Fig. 4.4b). We demonstrated that overexpression of ABCD3 stimulates C16:0 β-oxidation activity in CHO cells, but inhibits C24:0 β-oxidation activity (Imanaka et al. 1999). In addition, we showed that C24:0 β-oxidation activity was not decreased by the silencing of *ABCD3* in *ABCD1*-knockdown U87 cells (Morita et al. 2012). These data suggest that ABCD3 does not contribute to peroxisomal VLCFA β-oxidation. ABCD3 has been shown to have a role in the transport of more hydrophilic substrates compared to VLCFA. It has been reported in *Abcd3*$^{-/-}$ mice that the bile acid precursors THCA and DHCA, as well as pristanic acid, accumulate in plasma. There was a significant reduction of chenodeoxycholic acid and cholic acid conjugated with taurine or glycine shown in the liver, bile, and intestine. On the other hand, a remarkable increase of bile acid intermediates with C27 was observed (Ferdinandusse et al. 2015). These results suggest that ABCD3 is active in the transport of LCFA-CoA, THCA-CoA, DHCA-CoA, branched chain acyl-CoA and dicarboxylic acid-CoA (Fig. 4.4b). In the mouse liver, ABCD3 expression is induced by treatment with fibrates, such as ciprofibrate and fenofibrate, via PPARα. Recently, a patient with an ABCD3 defect reportedly exhibited hepatosplenomegaly and severe liver disease (Ferdinandusse et al. 2015). ABCD3 possesses intrinsic acyl-CoA thioesterase activity, as do ABCD1 and ABCD2 (Okamoto et al. 2018). Very recently peroxisomal fatty acid oxidation through ABCD3 was suggested to be important when mitochondrial fatty acid oxidation is impaired or overloaded (Violante et al. 2019).

4.3.3.4 ABCD4

It was a surprise when it was reported that mutation of *ABCD4* causes a lysosomal storage disease with vitamin B_{12} deficiency (Coelho et al. 2012). In humans, cobalamin in complex with transcobalamin in the blood is endocytosed into cells and delivered to lysosomes. The cobalamin in the lysosomes is then exported to the cytosol and in part to mitochondria, and converted into methylcobalamin (MeCbl) and adenosylcobalamin (AdoCbl), which are coenzymes of methionine synthase in the cytosol and methylmalonyl-CoA mutase in mitochondria, respectively (Coelho et al. 2008). A defect in ABCD4 results in lysosomal accumulation of cobalamin. In patient fibroblasts, the intracellular enzyme-bound cobalamin level is significantly lower than in control fibroblasts. Expression of wild-type ABCD4 in patient fibroblasts was shown to normalize the intracellular enzyme-bound cobalamin levels and remarkably increase the synthesis of MeCbl and AdoCbl. As mutations in ABCD4 and LMBD1 encoded by the limb region 1 domain containing 1 gene (*LMBDR1*) result in a quite similar phenotype (Rutsch et al. 2009), these two proteins function as a complex in the process of transporting cobalamin from lysosomes to the cytosol. Recently, we demonstrated that the ABCD4 dimer complexes with LMBD1 on the ER membrane and is transported to lysosomes by the lysosomal targeting signal of LMBD1 (Kawaguchi et al. 2016). The bacterial ABC transporter BtuCD is involved in the import of cobalamin into the cytoplasm across the inner membrane and possesses ATPase activity (Beedholm-Ebsen et al. 2010). ABCC1, a multidrug resistance-associated protein on the plasma membranes of mammalian cells, may be involved in the cobalamin export from cells (Borths et al. 2005). ABCC1 itself also possesses ATPase activity. Therefore, we postulate that ABCD4 comprises the transporter unit and that LMBD1 is an accessory protein. LMBD1 is a highly glycosylated protein and this might protect ABCD4 from proteases in lysosomes. Recently we showed that the ABCD4 reconstituted on liposomes exported cobalamin outside of the liposomes coupled to ATP hydrolysis (unpublished observation).

ABCD1–3 transport substrates from the cytosol to peroxisomes. In contrast, ABCD4 transports cobalamin from lysosomes to the cytosol. It is of special interest to determine the mechanisms underlying the difference in the direction of these transport activities.

4.4 Solute/Metabolite Traffic Across the Peroxisomal Membrane

The various fatty acids metabolized in peroxisomes are mainly transported into peroxisomes from the cytosol as CoA forms via ABCD1–3 (Figs. 4.1 and 4.5). They are then metabolized by β- and α-oxidation in the peroxisomes (Wanders 2014). However, as peroxisomal β-oxidation is not carried out completely in peroxisomes, the final products, a shortened fatty acyl-CoA (medium chain acyl-CoA with a carbon chain length of 6–8) as well as acetyl-CoA, formyl-CoA and propionyl-CoA, must be exported for further metabolism. In addition, the NADH formed in peroxisomes must be re-oxidized to NAD^+ to continue the β-oxidation process. On the other hand, in mammals VLCFA is partly activated to VLCFA-CoA in peroxisomes by VLCFA-CoA synthase. ATP and CoA must be transported to peroxisomes from the cytosol.

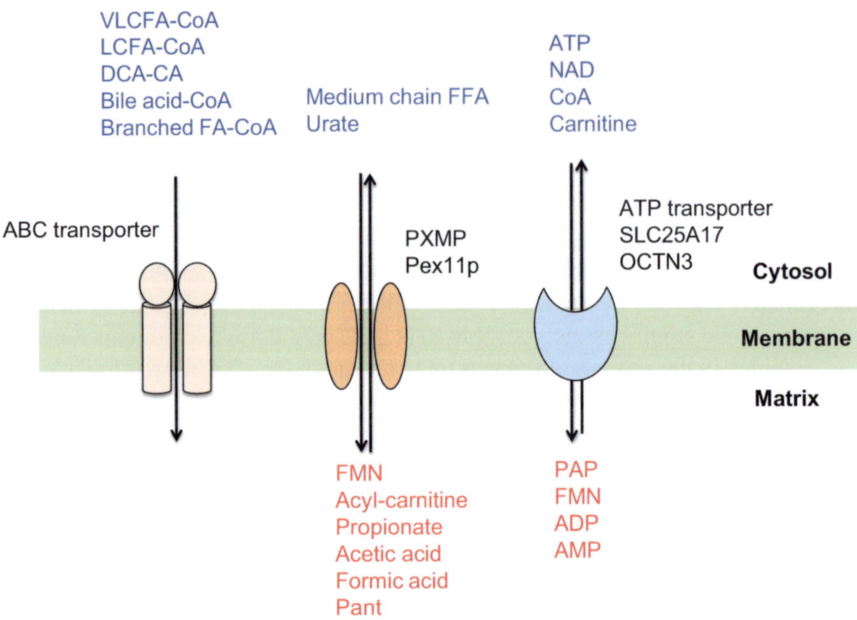

Fig. 4.5 Hypothetical model for the transport of substrate, metabolites and cofactors involved in lipid metabolism through peroxisomal membrane. The ABC transporters are involved in the transport of various types of fatty acid-CoA. The ATP transporters SLC25A17 and OCTN3 are involved in the transport of compounds that are required for fatty acid metabolism in peroxisomes and also the export of their metabolites. Non-selective channels such as PXMP2 are involved in the export of metabolites by fatty acid β-oxidation and degradation products from NADH and FAD. Pant, pantothenic acid; PAP, adenosine 3,5-diphosphate

The exchange of metabolites and cofactors through the peroxisomal membrane is important for adequate facilitation of the metabolic process in peroxisomes.

Early studies by de Duve et al. led to the concept that peroxisomes are freely permeable to compounds of low molecular weight based on evidence from equilibrium density gradient centrifugation in sucrose, and several of the peroxisomal enzymes failed to exhibit latency as a membrane barrier, unlike the lysosomal membrane (Beaufay et al. 1964; De Duve and Baudhuin 1966). Later Veldhoven et al. suggested that a PMP of 22- to 28-kDa is responsible for the nonspecific permeability of small solutes on the peroxisomal membrane (Van Veldhoven et al. 1987). Subsequently, Wanders' group at the University of Amsterdam predicted that a transporter and carrier must exist on the peroxisomal membrane from the evidence that the peroxisomal membrane in fibroblasts is not freely permeable to certain types of substrate (Wolvetang et al. 1990).

Around the same time, other important findings were reported. Disruption of the peroxisomal malate dehydrogenase gene (*MDH3*) in *Saccharomyces cerevisiae* was shown to cause impairment of the peroxisomal β-oxidation capacity, as measured in intact cells, whereas β-oxidation was normal in cell lysates. In addition, *mdh3*-

disrupted cells were unable to grow in oleate medium, whereas their growth in a glucose medium was normal. These data suggested that it is difficult for NADH to permeate the peroxisomal membrane, and it is oxidized to NAD^+ coupled to a conversion of oxaloacetate to malate by MDH (van Roermund et al. 1995). Furthermore, disruption of both the citrate synthase gene (*CIT2*) and carnitine acetyltransferase gene (*YCAT*) blocked the β-oxidation, but not in lysates. This suggests that the acetyl-CoA produced by β-oxidation does not permeate the peroxisomal membrane and it is released by the acetylcarnitine and maleate converted by the glyoxylate cycle. An ATP transporter, a mitochondrial inner membrane solute carrier superfamily was discovered in the peroxisome membrane obtained from different cell sources (Palmieri et al. 2001; Visser et al. 2002; Linka et al. 2008; Arai et al. 2008).

Based on these findings, the peroxisomal membrane is predicted to possess a non-selective channel and several transporters for selective compounds (Fig. 4.5). Hiltunen's group at the University of Oulu at Finland have made contributions in this area with their extensive studies (Antonenkov and Hiltunen 2006; Antonenkov et al. 2015; Mindthoff et al. 2016).

4.4.1 Non-selective Channels on the Peroxisomal Membrane

As described above, peroxisomal membrane protein 2 (PXMP2), also known as PMP22, is responsible for channel formation. Pxmp2 in rodents is one of the major integral membrane proteins in the liver peroxisomes of mammals (Hartl and Just 1987). It belongs to a small family of membrane proteins consisting of four members in mammals: Pxmp2, Mpv17 (Weiher et al. 1990), Mpv17-like protein (M-LP) (Iida et al. 2003) and the *FKS24* gene product. The yeast *S. cerevisiae* contains two proteins homologous to Pxmp2: Sym1p (Trott and Morano 2004) and the *YOR292c* gene product.

In an early study by Veldhoven et al., peroxisomes in the rat liver were predicted to possess pore-forming channel capacity with low molecular weight polypeptides such as Pxmp2 (PMP22) (Van Veldhoven et al. 1987). Subsequently, Hiltunen and his colleagues have taken up the challenge of detecting the channel forming activity of Pxmp2 using purified protein from mouse liver peroxisomes. Pxmp2 is a homotrimer that mainly exhibits channel-forming activity at three conductance levels: 0.45, 0.9, and 1.3 nS in 1.0 M, with KCl as the electrolyte. The channel is weakly cation-selective and in a stably open configuration for long periods of time. The estimated diameter of the channel is 1.4 nm. The size is estimated to be well over the dimensions of small solutes but it is close to the size of bulky metabolites. Further studies revealed that the Pxmp2 channel enables free diffusion across the membrane of compounds with molecular masses up to 300 Da. The permeation of larger molecules up to 500–600 Da is partially restricted, while larger metabolites do not enter the channel (Antonenkov and Hiltunen 2012; Rokka et al. 2009). Metabolites such as acylcarnitine, free fatty acid and acetoacetate all probably diffuse through the channel. In addition, a study on Pxmp2-deleted mice revealed a partial restriction of peroxisomal membrane permeability to solutes both *in vivo* and *in vitro* in the absence of Pxmp2. An inability of female Pxmp2$^{-/-}$ mice to nurse their pups was observed owing to an adverse effect on the development of the mammary gland epithelium (Vapola et al. 2014).

Mpv17 and M-LP are additional candidate peroxisomal membrane channels. Human MPV17 has been shown to form a membrane channel as Pxmp2 (Antonenkov et al. 2015). Initially these proteins were reported as being localized to peroxisomes. However, later they are reported to locate exclusively to the inner mitochondrial membrane (Spinazzola et al. 2006; Krick et al. 2008) and be involved in the maintenance of mitochondrial DNA (mtDNA) (El-Hattab et al. 2017; Iida et al. 2018). On the other hand, very recently Weiher et al. reported that MPV17 localizes to peroxisomes, endosomes and lysosomes, but not mitochondria in human U2OS cells. MPV17 might have a novel function on the peroxisomal membrane instead of mtDNA depletion (Weiher et al. 2016). Sym1p, a yeast homolog to Pxmp2, is localized to the inner mitochondrial membrane (Trott and Morano 2004). Recently, it was suggested that Sym1p might be involved in mitochondrial transport and is essential for maintenance of mtDNA (Dallabona et al. 2010).

Interestingly, Pex11p is a candidate cannel-forming protein on the peroxisomal membrane. Mindthoff et al. showed that yeast Pex11p has sequence similarity with TRPM cation-selective channels possess pore-forming activity (Mindthoff et al. 2016). Overexpression of Pex11p resulted in acceleration of fatty acids β-oxidation in intact cells, but not in the corresponding lysates. The results suggest that Pex11p as a non-selective channel responsible for the transfer of metabolites across the peroxisomal membrane.

Based on these studies, unique channel proteins exist on the peroxisomal membrane that are involved in the transport of the various metabolites produced in peroxisomes (Fig. 4.5).

4.4.2 The ATP and Cofactor Transporter

ATP is required for metabolism in peroxisomes. The molecular size of ATP (507.2 Da) and its strong negative charge predict that the transfer of ATP across the peroxisomal membrane is highly restricted. An ATP transporter must exist in the peroxisomal membrane as in the inner mitochondrial membrane.

The Pmp47 of *Candida boidinii* shows the highest similarity to a mitochondrial ATP/ADP translocase. Interestingly, the topology of Pmp47 in the peroxisomal membrane is in the opposite orientation as the mitochondrial ATP/ADP translocase. In yeast, medium-chain fatty acids such as laurate are activated in the peroxisomal matrix by Faa2p for β-oxidation. We demonstrated that the *pmp47Δ* strain could not grow on laureate and that the level of laureate oxidation in the permeabilized cells of the *pmp47Δ* strain was lower than that in the wild-type strain. These results suggest that Pmp47 may be involved in the transport of a small molecule (possibly ATP) required in the activation of laureate to its CoA form in peroxisomes (Nakagawa et al. 2000). Palmieri et al. subsequently found that Ant1p is essential as the sole carbon source for growth on medium-chain fatty acids in *S. cerevisiae* (Palmieri et al. 2001). Upon reconstitution of the overexpressed and purified Ant1p into liposomes, specific transport of ATP was demonstrated. The physiological role of Ant1p is to transport cytoplasmic ATP into the peroxisomal lumen in exchange for the AMP generated in the activation of fatty acids. To facilitate this process, PMP34 in mammalian peroxisomal membranes (Visser et al. 2002), and PNC1 and PNC2 in plant peroxisomal membranes (Linka et al. 2008; Arai et al. 2008) function as ATP transporters. They belong to a superfamily of mitochondrial solute carriers and accelerate the transfer of cytoplasmic ATP into the peroxisomal lumen in exchange for ADP and AMP, which provides a shift in the steady-state concentrations of the nucleotides inside the peroxisomes toward ATP (Fig. 4.5).

The cofactors CoA, FAD and NAD^+ that are essential for β-oxidation are synthesized outside the peroxisome and therefore must be imported. Solute carrier family 25 member 17 (SLC25A17), a member of the mitochondrial carrier superfamily, is located the peroxisomal membrane. Purified SLC25A17 reconstituted in liposomes displays transport properties and kinetic parameters that suggest SLC25A17 is a transporter of CoA, FAD, FMN and AMP, and to a lesser extent of NAD^+, adenosine 3, 5-diphosphate (PAP) and ADP (Agrimi et al. 2012). SLC25A17 is expressed to various degrees in all of the human tissues examined and is involved in the transport of free CoA, FAD and NAD^+ into peroxisomes in exchange for PAP, FMN and AMP generated in peroxisomes (Fig. 4.5). As mentioned above with regard to the re-oxidized NADH induced β-oxidation of fatty acids, a shuttle mechanism is used to re-oxidise NADH. In *S. cerevisiae*, peroxisomal NADH is re-oxidized to NAD^+ by malate dehydrogenase (Mdh3p) via the malate/oxaloacetate shuttle. The presence of multiple peroxisomal redox systems probably allows the cells to maintain the proper peroxisomal redox status when the metabolic conditions are altered.

It is known that the acetyl-CoA produced by the β-oxidation does not permeate the peroxisomal membrane. It may be released as acetylcarnitine or maleate converted by glyoxylate cycle. The exchange of acyl-CoA to acylcarnitine is important for keeping the acyl-CoA/CoA pool constant (Ferdinandusse et al. 1999). OCTN3 (SLC22A21) is an organic cation transporter family protein and human OCTN3 is reported to localize on the peroxisomal membrane. OCTN3 might contribute to the transport of carnitine from the cytosol to peroxisomes in the course of peroxisomal lipid metabolism (Lamhonwah et al. 2005) (Fig. 4.5).

4.4.3 *The Export of Metabolites by β-Oxidation and the Degraded Cofactors*

Peroxisomes contain various lipid metabolizing enzymes, including carnitine acetyltransferase, carnitine octanoyltransferase, acyl- or acetyl-CoA thioesterses, bile acid-CoA: amino acid *N*-acyltransferase (BAAT) and <u>nu</u>cleotide <u>di</u>phosphates linked to some moiety X (Nudix) hydrolases. These enzymes seem to contribute to the modification of metabolites produced by fatty acid β-oxidation as well as the degradation of NADH and NADPH.

Carnitine is a small and highly hydrophilic molecule that plays a key role in the transfer of acetyl or acyl moieties between different compartments in the cells. Carnitine acetyltransferase is active towards acetyl-CoA and propionyl-CoA (Farrell et al. 1984), and carnitine octanoyltransferase is specific for medium-chain acyl-CoAs (Ferdinandusse et al. 1999) in mammals. The acetyl- and acylcarnitines produced in peroxisomes might be released, since the size of these compounds are smaller than the corresponding CoA derivatives.

Acyl- and acetyl-CoA thioesterases comprise a large group of enzymes that catalyze the hydrolysis of acyl- or acetyl-CoA to either free fatty acids or acetic acid, and CoA. The thioesterases are localized to almost all cellular compartments, including peroxisomes. Mouse peroxisomes have been shown to contain six thioesterases. They have different substrate specificity and can hydrolyze various acyl-CoA, including long and medium-chain acyl-CoA, branched chain acyl-CoA, medium-chain dicarboxylic acyl-CoA, short-chain acyl-CoA, acetyl-CoA and propionyl-CoA. The thioesterases modulate the intraperoxisomal pools of CoA (Hunt et al. 2014, 2012). The thioesterases might also create conditions favourable for the transfer of at least some of the β-oxidation products out of peroxisomes via peroxisomal membrane channels (Fig. 4.5).

BAAT is involved in export of bile acids from peroxisomes by producing tauro- and glyco-bile acids by amidation (Issemann and Green 1990; He et al. 2003; Pellicoro et al. 2007; Visser et al. 2007). A significant portion of BAAT is known to be located in liver peroxisomes in mammals. However, it is not clear how these bile acids conjugates are transported across the peroxisomal membrane at present.

A portion of the cofactors such as NADH, NADPH and CoA are degraded in peroxisomes. Elimination of degraded cofactor molecules from peroxisomes also seems to contribute to the regulation of their steady-state concentration in the particles. The mechanism responsible for the removal of soluble cofactors from the peroxisomal lumen is quite unusual, combining the actions of membrane channels with matrix enzymes called Nudix hydrolases. The Nudix family consists of a large group of pyrophosphatases containing a highly conserved amino acid sequence, the Nudix box (McLennan 2006). The enzymes catalyze the hydrolysis of a pyrophosphate bond connecting nucleosides with various moieties, including components of the cofactor molecules. Several Nudix hydrolases are localized to peroxisomes. Mammalian NUDT7 and NUDT19 are CoA diphosphatases (Gasmi and McLennan 2001; Ofman et al. 2006), whereas NUDT12 cleaves NAD(P)H (Abdelraheim et al. 2003). Similarly, yeast peroxisomes contain two Nudix hydrolases, Pcd1p (Cartwright et al. 2000) and Npy1p (AbdelRaheim et al. 2001) that catalyse the hydrolysis of CoA and NAD(P)H, respectively. Peroxisomal Nudix hydrolases cleave pyrophosphate bonds connecting two 'bulky' cofactor molecules of nearly equal size. As a result, the size of the reaction products is approximately half that of the initial substrate. It is suggested that due to their size the products of cofactor cleavage might be freely released from peroxisomes via membrane channels (Fig. 4.5). Peroxisomal Nudix hydrolases show some preference for damaged cofactor molecules, indicating a role as 'housecleaning' enzymes. In addition, the mammalian Nudix hydrolases specific for CoA are active towards certain acyl-CoA derivatives. NUDT19 not only hydrolyses CoA but cleaves a wide range of acyl-CoA, including medium chain acyl-CoA and choloyl-CoA. Similarly, NUDT7 exhibits a high level of diphosphatase activity towards bile acid-CoA and lauroyl-CoA. The products of these reactions are acyl (bile acid)-phosphopantetheines and 3, 5-ADP. The compounds might be cleaved by thioesterases to free fatty acids or bile acids and 4-phosphopantetheine for release from peroxisomes (Reilly et al. 2008).

The system for producing metabolites in peroxisomes as well as the transporters and cannels on the peroxisome membrane act in coordination to release metabolites. Recently, it was reported that peroxisomes associate with other organelles through specific contact sites, including mitochondria. The contacts facilitate the exchange of metabolites between peroxisomes and these other organelles (see Sect. 4.6).

4.5 Peroxisomes and Oxidative Stress

The cellular redox state is closely related to a variety of important functions, including cell survival, proliferation and differentiation. Disruption of cellular redox homeostasis induces "oxidative stress", which refers to elevated intracellular levels of reactive oxygen and/or nitrogen species (ROS/RNS). When the level of ROS/RNS exceeds the capacity of the antioxidant defence and repair mechanisms, the ROS/RNS cause oxidative damage to various biomolecules, with deleterious effects on cellular function, resulting in various diseases, including neurodegenerative and age-related diseases. Recently, peroxisomes as well as mitochondria have come to be recognized as key organelles in the maintenance of cellular ROS metabolism. The role of peroxisomes in cellular redox homeostasis has been well described in two recent reviews (Fransen et al. 2012; Walker et al. 2018). In this section, we focus on the role of peroxisomes in redox regulation in connection to mitochondria.

4.5.1 Oxidative Stress and the Antioxidant Defence Systems

Mitochondria are organelles that generate a high amount of ROS in association with the production of ATP via oxidative phosphorylation. Peroxisomes also have an important role in the production and degradation of ROS/RNS, and act as modulators of cellular oxidative balance (Antonenkov et al. 2010). It has become clear that the interplay of peroxisomes with mitochondria is important for the regulation of the redox signalling pathway.

Peroxisomes contain various enzymes that generate ROS/RNS, such as hydrogen peroxide (H_2O_2), superoxide ($O_2^{\bullet-}$), or nitric oxide (NO•) (Fig. 4.6). The most abundant ROS-producing enzymes are flavin-containing oxidases, including ACOXs, urate oxidase, D-amino acid oxidase, polyamine oxidase and xanthine oxidase, which reduce O_2 to H_2O_2 (Demarquoy and Le Borgne 2015). Oxidation of fatty acids is a pivotal function of peroxisomes and the ACOXs are the most significant source of peroxisome-derived H_2O_2. It has been reported that peroxisomes contribute approximately 35% of the H_2O_2 generated in the cell (Schrader and Fahimi 2006). In contrast, the inducible form of nitric oxide synthase (iNOS) to generate NO• is partly localized peroxisomes after inflammatory cytokine stimulation (Loughran et al. 2013). Peroxisomes produce hydroxyl radical •OH and peroxynitrite $ONOO^-$ as secondary by-products from H_2O_2, $O_2^{\bullet-}$ and NO• (Fransen et al. 2012). Although a low level of H_2O_2 is an important signalling molecule to mediate beneficial metabolic responses, a high level of ROS leads to the oxidative damage of biomolecules, which in turn results in profound cellular dysfunction and disease.

To protect peroxisomal enzymes and their membranes from oxidative stress, peroxisomes have multiple enzymatic and non-enzymatic antioxidant defence systems that scavenge H_2O_2 and free radicals (Fig. 4.6). The antioxidant enzymes include catalase, Cu/Zn superoxide dismutase (Cu/Zn SOD), peroxiredoxin 5, epoxide hydrolase, glutathione peroxidase and epoxide hydrolase 2 (Fransen et al. 2012; Lismont et al. 2015; Bonekamp et al. 2009). Catalase is a heme-containing enzyme which scavenges H_2O_2 by converting it to water and oxygen (Glorieux and Calderon 2017). Cu/Zn SOD, also known as SOD1, converts $O_2^{\bullet-}$ to O_2 and H_2O_2 (Dixon and Stockwell 2014), which are subsequently decomposed by catalase and glutathione peroxidase in peroxisomes. Peroxiredoxin 5 reduces H_2O_2 to H_2O and also converts the alkyl hydroperoxides (ROOH) to their respective alcohols, and $ONOO^-$ to NO_2^- (Knoops et al. 2011; Walbrecq et al. 2015). Peroxisomal peroxidase coverts H_2O_2 into H_2O through the oxidation of two reduced glutathione (GSH) molecules to form a disulfide-bridged dimeric GSSG form. The GSSG form is again reduced to two GSH molecules by glutathione reductase. Epoxide hydrolase 2 catalyses the conversion of epoxides to the corresponding dihydrodiols (Decker et al. 2009; Morisseau and Hammock 2013). These anti-oxidizing enzymes exist not only in peroxisomes but also various subcellular compartments, including the cytosol, mitochondria and nucleus.

On the other hand, a role for the non-enzymatic antioxidants such as GSH and vitamin C in the regulation of the peroxisomal redox state is also reported (Lismont et al. 2015). In addition, plasmalogens are reported to delay the oxidative degradation of PUFAs (Engelmann et al. 1994; Reiss et al. 1997; Hahnel et al. 1999) because they function as antioxidants with their viny-ether bond serving as a trap for free radicals (Wallner and Schmitz 2011). PUFAs and plasmalogens are enriched in the CNS and have important roles in the brain.

As peroxisomal enzymes are subjected to oxidative modification and inactivation by ROS/RNS, the enzymes are eliminated from the organelle through selective proteolysis, as occurs in mitochondria as well (Pinti et al. 2015). The ATP-stimulated Lon protease known as LonP2, which we first identified by proteomic analysis of

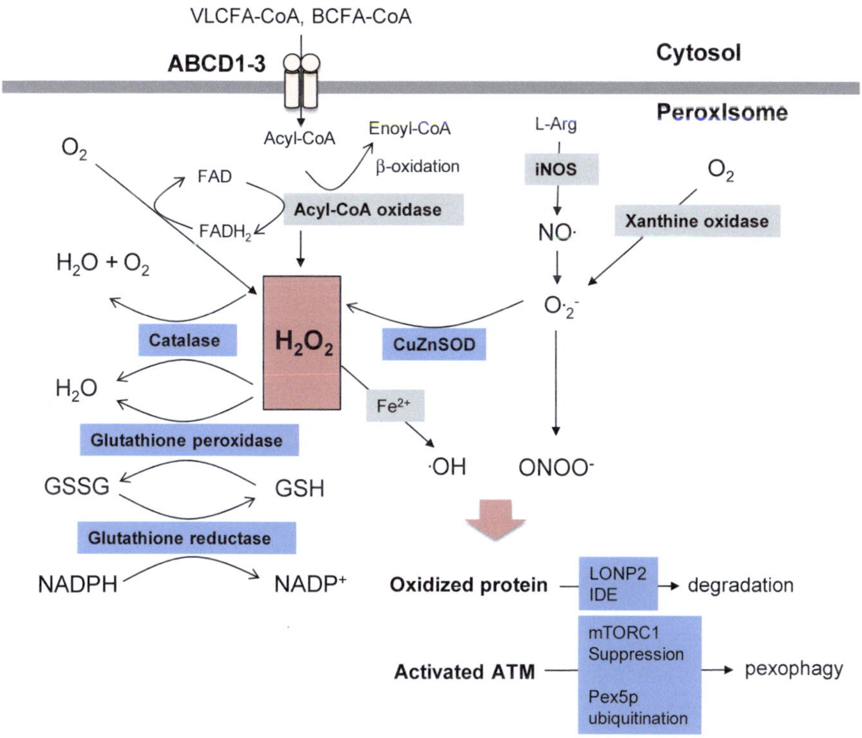

Fig. 4.6 Peroxisomal ROS/RNS generation and degradation. VLCFA or branched chain fatty acids (BCFA) are transported into peroxisomes via ABCD1–3 and subjected to the β-oxidation cycle, including acyl-CoA oxidase activity, which is a significant source of peroxisome-derived H_2O_2. Inducible nitric oxide synthase (iNOS) generates NO• by the oxidation of L-arginine (L-Arg). Peroxisomes also produce the hydroxyl radical•OH and peroxynitrite $ONOO^-$ as secondary by-products from H_2O_2, $O_2^{\bullet -}$ and NO•. Catalase scavenges H_2O_2 by converting it to H_2O and O_2. Cu/Zn SOD converts $O_2^{\bullet -}$ to O_2 and H_2O_2. Glutathione peroxidase coverts H_2O_2 into H_2O through the oxidation of two reduced glutathione molecules (GSH) to form a disulfide-bridged dimeric GSSG form. The GSSG form is again reduced to two GSH molecules by glutathione reductase. The oxidized enzymes in the peroxisomes are eliminated by degradation via selective proteolysis by LONP2 or insulin-degrading enzyme (IDE). ATM is activated by peroxisomal ROS, and the activated ATM signals to TSC to repress mTORC1 signaling, which leads to pexophagy. Activated ATM also leads to Pex5p ubiquitination and pexophagy

purified rat liver peroxisomes (Kikuchi et al. 2004), is involved in the removal of oxidatively damaged and misfolded proteins (Pomatto et al. 2017) (Fig. 4.6). In addition, we previously reported that insulin-degrading enzyme (IDE) isolated from rat liver peroxisomes is able to degrade oxidized proteins, suggesting that IDE is involved in maintaining peroxisomal function by eliminating oxidized proteins (Morita et al. 2000).

4.5.2 Impact of Peroxisomal ROS Production on Mitochondrial Function

It has become evident that a disturbance in peroxisomal ROS production directly and indirectly impacts the mitochondrial redox balance (Walton and Pizzitelli 2012; Wang et al. 2013). In *Pex* gene knockout mice that exhibit a peroxisomal import defect and lack of peroxisomes, mitochondria display structural abnormalities, defective respiratory chain complex and a marked increase in manganese superoxide dismutase (MnSOD) (Baumgart et al. 2001). Recently, Salpietro et al. reported the role of secondary mitochondrial dysfunction in Zellweger syndrome (Salpietro et al. 2015). These results indicate that peroxisome dysfunction leads to an increased production of ROS in mitochondria. This increased oxidative stress is probably due to the defective peroxisomal antioxidant mechanisms as well as the accumulation of lipid intermediates such as phytanic acid (Reiser et al. 2006), pristanic acid (Ronicke et al. 2009), and saturated VLCFAs (Baarine et al. 2012), which are substrates for peroxisomal fatty acid oxidation (Lismont et al. 2015) (Figs. 4.1 and 4.7). In *Abcd1*-deficient mice, which are known to be a model for adrenomyeloneuropathy (AMN), the saturated VLCFAs that accumulated in spinal cord were reported to induce oxidative damage, a process which might underlie the axonal degeneration observed in X-ALD patients (Fourcade et al. 2009). In X-ALD patient fibroblasts, an excess of C26:0 induces mtDNA oxidation and impairs oxidative phosphorylation, which in turn triggers mitochondrial ROS production (Lopez-Erauskin et al. 2013). It seems likely that the saturated VLCFAs introduce changes in the lipid composition of mitochondrial membranes that lead to ROS generation (Lismont et al. 2015). Although the precise mechanisms underlying the mitochondrial dysfunction caused by peroxisomal deficiency remain to be elucidated, functional peroxisomes play an important role in maintaining mitochondrial redox homeostasis.

M-LP, homologous to Pxmp2, was initially reported to localize to peroxisomes and is implicated in the regulation of peroxisomal ROS metabolism (Iida et al. 2006). However, M-LP was later found in the inner mitochondrial membrane and may be involved in the maintenance of mtDNA by protecting proteins essential for mtDNA stability (Iida et al. 2018). It is therefore of interest to elucidate how the channel-forming proteins such as Pxmp2 or M-LP are involved in ROS metabolism in peroxisomes and mitochondria. It is reported that a decrease in peroxisomal catalase activity by pharmacological treatment triggers elevated oxidative stress and reduced mitochondrial enzymatic activity (Walton and Pizzitelli 2012), suggesting that the catalase activity in peroxisomes exerts an impact on mitochondrial function.

These results show that peroxisomal ROS metabolism is tightly interconnected to mitochondrial redox homeostasis, although the precise mechanisms remain to be determined.

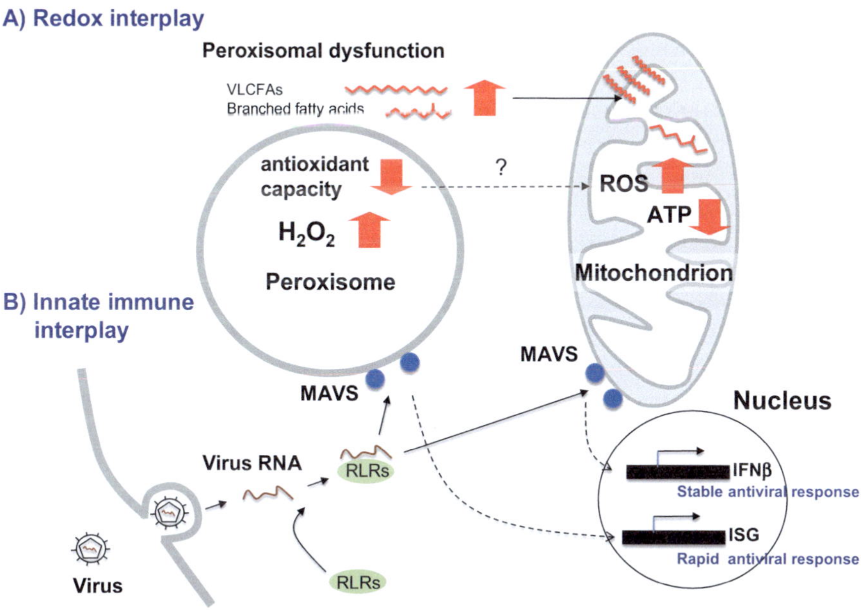

Fig. 4.7 Interplay between peroxisomes and mitochondria in redox homeostasis and the innate immune response. (**a**) Increased oxidative stress, presumablely due to defective peroxisomal antioxidant mechanisms may induce the impairment of mitochondrial oxidative phosphorylation and thereby bring about a reduction of the ATP level. The accumulation of lipid intermediates such as branched chain fatty acids (BCFA) and saturated VLCFAs due to the failure of peroxisomal β-oxidation could trigger mitochondrial ROS production. The mitochondrial dysfunction associated with decreased peroxisomal function is suggested to underlie the pathophysiological processes in various diseases, including the axonal degeneration observed in X-ALD patients. (**b**) Viral RNA recognized by retinoic acid-inducible gene I (RIG-I) like receptors (RLRs) binds to an adaptor protein called mitochondrial antiviral-signalling (MAVS), a tail anchored protein on the mitochondrial and peroxisomal membranes. Upon binding to MAVS, the mitochondrial MAVS leads to the expression of interferon-stimulated genes (ISGs) and sustains stable antiviral response. In contrast, the peroxisomal MAVS directly induces ISG expression independently of type I IFNs and thereby induces a rapid and transient expression of anti-viral factors

4.5.3 Oxidative Stress-Related Signaling Pathways

Peroxisomes have a potential role as regulators of oxidative stress-related signaling pathways by producing ROS/RNS. As both cysteine thiols and lipids are among the most prominent targets of ROS/RNS (Trachootham et al. 2008; Hekimi et al. 2011), many redox signals are conveyed through cysteine oxidation and lipid peroxidation. In contrast to an excess amount of H_2O_2, a low amount of H_2O_2 elicits desirable biological reactions in which H_2O_2 oxidizes proteins by converting the thiol groups of reactive cysteine to sulfenic acids or by forming disulfide bounds. It is now clear that H_2O_2 is important as a signaling molecule that plays a major role in the redox-dependent mechanisms of adaptive homeostasis.

It has been reported that damaged peroxisomes produce ROS, which in turn can signal for their own degradation (pexophagy) (Chen et al. 2009). In mammals, the serine/threonine kinase mTOR is involved in the regulation of autophagy. Mammalian target of rapamycin complex 1 (mTORC1), the kinase component of the TORC1 complex, is a negative regulator of autophagy. It does this by inhibiting ULK1, a key kinase involved in autophagy initiation, by phosphorylation. Recently, it was demonstrated that pexophagy requires ubiquitination of Pex5p (Fig. 4.6). It has become clear that the tumor suppressors ATM (ataxia telangiectasia mutated) and TSC2 (tuberous sclerosis complex 2), which regulate mTORC1 signalling, are localized to peroxisomes and Pex5p-dependent translocation of ATM to peroxisomes is involved in the activation of pexophagy (Tripathi and Walker 2016). In peroxisomes, ROS activate ATM by forming an intermolecular covalent linkage between two cysteine residues. Tripathi and Walke showed that when ATM is activated by peroxisomal ROS, ATM signals to the TSC to repress mTORC1 signaling. As mTORC1 is a repressor of autophagy, the mTORC1 repression induces autophagic flux in cells. In addition, the activated ATM phosphorylates Pex5p, which leads to Pex5p ubiquitination at Lys209 and eventually pexophagy. These results suggest that excess ROS production in peroxisomes is a signal of peroxisomal dysfunction that can trigger pexophagy.

On the other hand, the cellular oxidative balance is reported to be closely connected to the subcellular distribution of catalase, the most abundant peroxisomal H_2O_2-degrading enzyme. Catalase in the cytosol is involved in cellular defence to prevent oxidative damage of extra-peroxisomal origin. Recently, it has been reported that Pex5p is a redox-sensitive protein and functions as a stress sensor, retaining catalase in the cytosol under oxidative conditions (Apanasets et al. 2014; Walton et al. 2017). Catalase has a non-canonical PTS1 and is captured by Pex5p in the cytosol and imported to peroxisomes. After the Pex5p-catalase complex docks with the peroxisomal membrane, monoubiquitinylation of Pex5p at a conserved cysteine residue (Cys-11) is facilitated to dislocate it back into the cytosol. Under an oxidative condition, the oxidation of the cysteine residue in Pex5p blocks such recycling and the functional Pex5p in the cytosol is decreased. As a result, the catalase is retained in the cytosol and thereby can scavenge cellular H_2O_2. These results indicate that Pex5p regulates the cellular localization of catalase as an oxidative stress sensor and is involved in an antioxidant defence system.

Another feedback mechanism was revealed by Murakami et al. (Murakami et al. 2013). They found that the Valosin-containing protein (VCP) that belongs to the AAA class of ATPases is able to sense H_2O_2 level through redox changes in a cysteine group and regulate it by controlling the localization of catalase, although the control mechanism has yet to be elucidated. Under an oxidative condition, VCP ATPase is inactivated by oxidation, which in turn retains catalase in the cytoplasm, leading to reduced cellular H_2O_2 levels. On the other hand, catalase release from peroxisomes is regulated by the pro-apoptotic BCL-2 effector protein BAK, which contributes to the permeability of the peroxisomal membrane (Fujiki et al. 2017; Hosoi et al. 2017). Taken together, cellular catalase localization is surprisingly controlled by oxidative stress, which provides a defence mechanism against high levels of cellular H_2O_2.

4.5.4 *Oxidative Stress and Neurodegenerative Diseases*

Recently, it has been suggested that the oxidative stress caused by dysregulated peroxisomal ROS metabolism is associated with not only peroxisomal disorders but also aging-related disorders, such as neurodegenerative diseases (Fransen et al. 2012; Deori et al. 2018), because brain is highly vulnerable to oxidative stress and biomolecules damaged by oxidative stress accelerate the aging process (Cipolla and Lodhi 2017). Furthermore, most patients with peroxisomal disorders, including X-ALD and Zellweger syndrome, exhibit neurological defects, such as neurodevelopmental abnormalities, demyelination with neuroinflammation, and loss of axonal integrity (Berger et al. 2016). In X-ALD patients, a significant increase of plasma lipid peroxidation and glutathione peroxidase activity occurs in erythrocytes, and a deficient capacity to handle excess ROS has been observed (Vargas et al. 2004). Petrillo et al. found that both total glutathione and reduced glutathione were decreased in the lymphocytes of X-ALD patients, this was associated with high levels of oxidation of all glutathione forms (Petrillo et al. 2013). These results suggest that oxidative stress may be involved in the pathophysiology of X-ALD. It has been made apparent that the redox balance perturbation that occurs due to the increase of dysfunctional peroxisomes contributes to the development of various neurodegenerative disorders, such as Alzheimer's disease (Fanelli et al. 2013; Kou et al. 2011; Nell et al. 2017; Giordano et al. 2014) and Parkinson's disease (Yakunin et al. 2010). Therefore, the restoration of a normal peroxisomal oxidative state might slow the progression of these neurodegenerative conditions (Cipolla and Lodhi 2017).

4.6 Crosstalk Between the Peroxisome and Other Organelles

Peroxisomes have been recognized as being multifunctional organelles involved in lipid metabolism, ROS metabolism, and also the innate immune response. To fulfil their functions, peroxisomes form metabolic networks with other organelles in connection with lipid synthesis in ER, lipid storage in lipid droplets (LDs), β-oxidation in mitochondria, and lipid hydrolysis in lysosomes. It has become evident that peroxisomes share signals and exchange metabolites with these organelles (Castro et al. 2018; Shai et al. 2016; Schrader et al. 2019). This closstalk frequently occurs through inter-organelle membrane contacts (Fig. 4.8). Recently, using multispectral imaging techniques, Valm et al. have reported interactions among peroxisomes, the ER, the Golgi apparatus, lysosomes, mitochondria and LDs, each of which has a characteristic distribution in three-dimensional space and a reproducible pattern of contacts among the six organelles (Valm et al. 2017). Furthermore, Shai et al. reported that systematic mapping of the contact sites revealed that the organelles were held in a state of organized dynamic interplay through membrane contact sites (Shai et al. 2018). Therefore, it has now become clear that the physical membrane contact that is formed by protein-protein and/or protein-lipid interactions is a common approach for communication between organelles. Thus, the direct shuttling of metabolites, lipids and ions may largely depend on membrane contacts between peroxisomes and other organelles. This pattern of crosstalk, including signal transduction pathways, vesicular trafficking and the contact sites, has been reviewed elsewhere (Castro et al. 2018; Shai et al. 2016, 2018; Islinger et al. 2018; Farre et al. 2019; Schrader et al. 2015). In this section, we focus on lipid metabolic interplay between peroxisomes and other subcellular organelles. In addition, antiviral systems that act in coordination with peroxisomes and mitochondria are also described.

4.6.1 Crosstalk Through Lipid Metabolites

Multiple contact sites between peroxisomes and other organelles have been identified, most of which likely contribute to the transfer of lipid metabolites (Valm et al. 2017). Yeast is a good model system for studying the function of the contact sites between peroxisomes and mitochondria on lipid metabolism, because the fatty acid β-oxidation takes place in peroxisomes, but not in mitochondria. In this section, we focus on the crosstalk that takes place through lipid metabolites between peroxisomes and other organelles in both mammals and yeast.

4.6.1.1 Mitochondria

It has now been made apparent that physical contact sites between peroxisomes and mitochondria are one possible pathway for the exchange of small molecules. Shai et al. discovered two novel yeast contact sites between peroxisomes and mitochondria that form through the tethering molecules Fzo1 and Pex34p (Shai et al. 2016) (Fig. 4.8). Fzo1 is a homolog of the human mitofusins that mediate mitochondria-mitochondria tethering, but which have also been suggested to mediate peroxisome-mitochondria tethering. Pex34p interacts with Pex11p and is a peroxisomal protein involved in peroxisomal growth and division that also has the capacity to tether the peroxisome-mitochondria contact. They found that peroxisomes in close proximity to mitochondria increased under a culture condition in which yeast were grown with oleate as the sole carbon source. As yeast fatty acid β-oxidation takes place exclusively in peroxisomes, the peroxisomal proximity to mitochondria might facilitate the transport of β-oxidation products (acetyl-CoA and its derivatives) to mitochondria. These results indicate that the physical membrane contact tethered by Pex34p is important for the transfer of metabolites from peroxi-

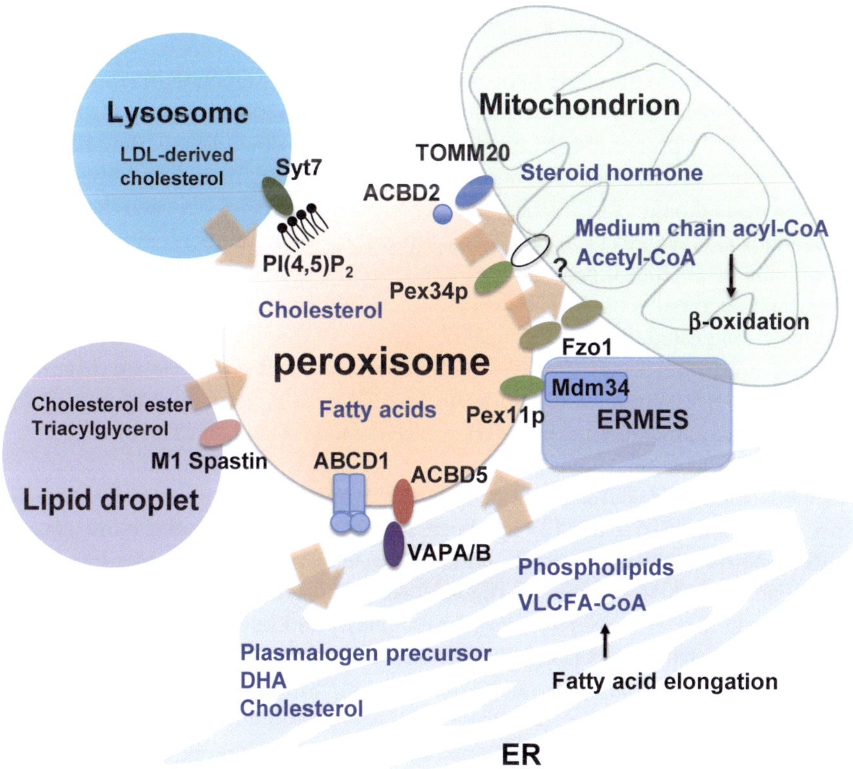

Fig. 4.8 Crosstalk between peroxisomes and other organelles through lipid metabolites. The β-Oxidation products in peroxisome, such as acetyl-CoA in yeast or medium chain acyl-CoA in mammals, are transferred to mitochondria for further oxidation, presumably through peroxisome-mitochondria contact sites. The contact sites via Pex34p or Fzo1 might facilitate the transport of acetyl-CoA and via ACBD2 assist in the exchange of metabolites for the purpose of steroid hormone synthesis. In addition, Pex11p tethers peroxisomes to mitochondria through direct interaction with Mdm34, a mitochondrial component of the ERMES complex. The peroxisome-ER contact site that is formed through the complex of VAPA/B and ACBD5 is involved in the transport of phospholipids, including the plasmalogen precursor. The contact sites may also assist the transfer of VLCFA-CoA to ABCD1 on peroxisomes. The peroxisome-lysosome contact site formed between synaptotagmin VII (Sty7) and phosphatidylinositol 4,5-bisphosphate (PI(4,5)P_2) at the peroxisomal membrane is involved in the transport of LDL-derived cholesterol from lysosomes to plasma membranes. The peroxisome–LD contact site functions in the β-oxidation of fatty acids liberated from LD to serve as an energy source

somes to mitochondria (Shai et al. 2018) although its mitochondrial counterpart has yet to be identified. In mammals, VLCFAs, branched chain fatty acids, dicarboxylic fatty acids and bile acid intermediates are degraded exclusively by the peroxisomal β- and α-oxidation pathways (Wanders 2014). However, as β-oxidation is not carried out to completion in peroxisomes, the final products, such as shortened fatty acyl-CoAs, must be shuttled into mitochondria for further oxidation. It is important to determine how the contact facilitates the transfer of peroxisomal β-oxidized products to mitochondria.

A possible clue to the tether between peroxisomes and mitochondria in yeast was previously suggested by the association of Pex11p with mitochondrial distribution and morphology protein 34 (Mdm34) (Mattiazzi Usaj et al. 2015) (Fig. 4.8). The ER-mitochondria encounter structure (ERMES) complex, including Mdm34 as a component, is thought to facilitate the exchange of molecules between the ER and mitochondria. As Pex11p physically interacts with mitochondrial Mdm34 (Kornmann et al. 2009), peroxisomes might form a contact site via specific mitochondrial subdomains of the ERMES complex (Cohen et al. 2014). Mattiazzi Usaj et al. have suggested that Pex11p may be a sensor of the metabolic state of the peroxisome matrix and acetyl-CoA a signalling molecule in this mechanism (Mattiazzi Usaj et al. 2015). These results indicate that ERMES is a major platform for the shuttling of lipid metabolites between mitochondria, ER and peroxisomes.

In contrast to yeast, direct contacts between peroxisomes and mitochondria in mammals remain obscure. However, acyl-CoA-binding domain (ACBD2)/enoyl-CoA delta isomerase 2 (ECI2) has been suggested to mediate the close contact between peroxisomes and mitochondria that occurs via a "peroxisome-like structure" (Fan et al. 2016). Fan et al. reported ACBD2/ECI2 has dual targeting signal (peroxisomal and mitochondrial targeting signals) and bind with both Pex5p and TOMM20, a translocator of outer mitochondrial membrane 20, which promote the contact between peroxisomes and mitochondria. Steroid hormone biosynthesis requires multiple biological processes with the rate-limiting step involving the import of cholesterol into mitochondria. They reported in Leydig cells, which produce steroid hormones, ACBD2/ECI2-mediated interaction between peroxisomes and mitochondria assist the exchange of metabolites between these two organelles for steroid hormone biosynthesis (Fan et al. 2016).

4.6.1.2 The ER

The ER produces several kinds of phospholipids and provides them to other compartments via vesicular or non-vesicular pathways. It has been suggested that the ER has the most abundant contact sites for peroxisomes among all organelles (Valm et al. 2017). Raychaudhuri and Prinz demonstrated the direct transfer of phospholipids from purified microsomes to purified peroxisomes in non-vesicular pathways (Raychaudhuri and Prinz 2008), indicating that lipid transfer from the ER to peroxisomes is mediated through ER-peroxisome contact. It has been demonstrated that the tethering of peroxisomes to the ER is necessary for both the regulation of peroxisome movement and membrane dynamics (Castro et al. 2018).

The transfer of lipids from the ER to peroxisomes is necessary for not only peroxisomal membrane expansion but also the synthesis and catabolism of lipids. The interplay of peroxisomes with the ER is needed to complete lipid metabolic reactions, including the biosynthesis of DHA, plasmalogen, bile acids and cholesterol

(Lodhi and Semenkovich 2014). The biosynthesis of DHA, C22:6 n − 3 is mainly conducted by the ER, but requires β-oxidation of the C24 precursor in peroxisomes. DHA is subsequently transferred back to the ER for incorporation in phospholipids. In plasmalogen synthesis, the intermediate metabolites must be transferred from peroxisomes to the ER (Cockcroft and Raghu 2018). In bile acid synthesis, the bile acids DHCA and THCA must undergo peroxisomal β-oxidation to generate bile acids. As for cholesterol, more than 30 enzymes that are related to cholesterol biosynthesis are distributed in subcellular compartments, such as the cytosol, ER and peroxisomes. Recently, peroxisomes were shown to mediate cholesterol trafficking of free cholesterol from lysosomes to the ER, as described below (Chu et al. 2015). Although it remains unclear how these metabolites are shuttled to each organelle, it seems likely that the shuttling of lipid metabolites between organelles preferentially occurs at contact sites.

The importance of vesicle-associated membrane protein-associated protein (VAP)–peroxisomal acyl-CoA binding protein 5 (ACBD5) mediated contact between ER and peroxisomes for lipid homeostasis has been demonstrated (Schuldiner and Zalckvar 2017) (Fig. 4.8). Costello et al. and Hua et al. independently identified a peroxisome-ER contact site in human cells through the complexing of VAP A/B with ACBD5 (Costello et al. 2017a, b; Hua et al. 2017). The VAPs are anchored to the ER membrane and implicated in the tethering various organelles, inter-organelle lipid exchange, and membrane trafficking. Peroxisomes are tethered to the ER through an interaction between ACBD5 with the ER-resident VAPs. It was demonstrated that loss of the ACBD5/VAP tethering complex alters the peroxisome-ER interaction and perturbs peroxisomal membrane expansion, suggesting that this tethering complex facilitates peroxisome transport from the ER to peroxisomes (Costello et al. 2017a). It has been reported that the loss of the ACBD5-VAP interaction by depleting either ACBD5 or VAPs reduce the plasmalogen level in cultured Hela cells (Hua et al. 2017), indicating that the interaction of ACBD5 and VAPs is required for plasmalogen precursor (alkyl-DHAP) shuttling and thereby efficient plasmalogen synthesis. A deficiency of ACBD5 impairs VLCFA β-oxidation in peroxisomes. In contrast, any involvement of ACBD5 in DHA synthesis has yet to be elucidated (Yagita et al. 2017).

4.6.1.3 Lysosomes

Recently, it has come to be understood that peroxisomes have a role in the transport of low density lipoprotein (LDL)-derived cholesterol from lysosomes to the plasma membrane (PM) (Chu et al. 2015). Cellular cholesterol is derived from the ER via *de novo* synthesis and from lysosomes via LDL receptor (LDLR)-mediated endocytosis of plasma LDL. LDL particles are endocytosed via the LDLR pathway and transported to lysosomes. The cholesterol esters in the LDL particles are hydrolyzed by acid lipase to free cholesterol, which is subsequently captured by the lysosomal NPC2 protein and delivered to NPC1, a lysosomal membrane protein (Sleat et al. 2004;

Pfeffer 2019). The free cholesterol in lysosomes is thought to be distributed to other cellular compartments through an NPC1-mediating pathway. As the cellular cholesterol level is tightly regulated at both the transcriptional and post-translational level, the intracellular cholesterol distribution pattern in each compartment has a significant impact on cholesterol homeostasis. It is known that less than 1% of the total cellular cholesterol is present in the ER membrane, while the PM is highly enriched in cholesterol (~60% to 80%). Therefore, the cholesterol synthesized in the ER and the LDL-derived cholesterol in lysosomes must be transported to the PM from the ER and/or lysosomes.

Chu et al. used a genome-wide shRNA screen to show that peroxisomes have an essential role in intracellular cholesterol transport through the formation of membrane contacts with lysosomes (Chu et al. 2015). When the *PEX1* or *PEX6* gene essential for peroxisomal biogenesis was reduced by gene silencing, the cholesterol transport from the lysosomes to the PM was decreased, suggesting that peroxisomes have some sort of functional connection with lysosomes during the transfer of LDL-derived cholesterol from lysosomes to the PM. They showed that LDL-derived cholesterol efficiently blocks sterol regulatory element binding protein (SREBP) processing and stimulates cholesterol esterification, but these effects were markedly diminished in the absence of lysosome-peroxisome membrane contact. In addition, it has been reported that the loss of ACBD5, which tethers peroxisomes to the ER, leads to an increase in the cellular cholesterol levels (Hua et al. 2017). This is presumably because the decrease in peroxisome-ER contact sites results in a reduced level of cholesterol in the ER, which induces SREBP processing and thereby an increase in the cellular cholesterol level. These results indicate that LDL-derived cholesterol in lysosomes is, at least in part, transported to ER via peroxisomes.

Furthermore, Song's group demonstrated that physical contacts are formed between lysosomal protein synaptotagmin VII and phosphatidylinositol 4,5-bisphosphate ($PI(4,5)P_2$) on the peroxisomal membrane (Fig. 4.8) (Luo et al. 2017). Hu et al. have reported that the level of $PI(4,5)P_2$ on the peroxisomal surface is important for regulating the transport of cholesterol via peroxisomes (Hu et al. 2018). It should be noted that efficient formation of peroxisomes and lysosome contact sites is required for ABCD1 as well as NPC1. Free cholesterol was shown to accumulate in lysosomes in both *abcd1*-deficient mouse brain and X-ALD fibroblasts, indicating that peroxisomal ABCD1 is directly or indirectly involved in cholesterol trafficking. The defects in cholesterol trafficking that occur due to the dysfunction of ABCD1 might be associated with the pathology of X-ALD. Although the possibility that peroxisomal dysfunction leads to secondary failure of lysosomes by impairing the turnover of glycolipids containing VLCFA cannot be excluded (Kleinecke et al. 2017), it is important to determine precisely how the ABCD1 protein is involved in cellular cholesterol trafficking and metabolism.

4.6.1.4 Lipid Droplets

Increasing evidence suggests that contact sites between lipid droplets (LDs) and other organelles play important roles in the trafficking of lipids and in the regulation of lipid metabolism (Valm et al. 2017). LDs are organelles that can store and mobilize fatty acids and other lipid species such as cholesterol ester and triacylglycerol, and serve as an energy source. It has come to be understood that the cellular distribution of LDs is in close proximity to peroxisomes as well as mitochondria (Schuldiner and Bohnert 2017; Henne et al. 2018). A deficiency of peroxisomal fatty acid metabolism and biogenesis leads to the enlargement of LDs and alterations in their number (Zhang et al. 2010), as observed in patients with Zellweger syndrome and X-ALD (Engelen et al. 2012). These results suggest that peroxisomes may interact with LDs by forming contact sites that facilitate the transport of fatty acids.

Furthermore, LDs reportedly contain several different β-oxidation enzymes (Binns et al. 2006; Barbosa et al. 2015). As yeast perform fatty acid β-oxidation only in peroxisomes, LD-peroxisome contacts seem to be essential for the fatty acid metabolism in LDs. These results suggest that the extensive physical contact between peroxisomes and LDs facilitate the direct transport of fatty acids. In monkey fibroblast cell line COS-7 cells, the LD-peroxisome contact sites are increased under starvation conditions (Valm et al. 2017). In this situation, the fatty acids liberated by lipase in LDs are thought to be transported to peroxisomes for β-oxidation, which ultimately generates ATP via the mitochondrial oxidative phosphorylation pathway. In contrast, the LD-peroxisome contact sites are decreased in response to excess fatty acids. These results suggest that the contact sites contribute to the transfer of fatty acids between LDs and peroxisomes.

Studies on the pathogenesis of the hereditary spastic paraplegias (HSPs), a group of inherited neurological disorders (Blackstone 2018), have provided insight into the proteins involved in the LD-peroxisome contact sites (Fig. 4.8). HSPs are neurologic disorders characterized by prominent lower-extremity spasticity, a condition resembling that of X-ALD. The mutations in the gene encoding Spastin that are the most common cause of autosomal dominant HSP lead to aberrant fatty acid metabolism in LDs (Papadopoulos et al. 2015). Recently, Chang et al. have reported a new role for peroxisomal ABCD1 in the fatty acid β-oxidation that occurs in lipid droplets (Chang et al. 2019). They showed that an isoform of Spastin on LDs named M1 Spastin plays a cooperative role with peroxisomal ABCD1 in the fatty acid trafficking from LDs to peroxisomes. M1 Spastin contains an integral membrane hairpin motif in the N-terminal region and forms an LD-peroxisome contact site by tethering to peroxisomal ABCD1. The extent of LD-peroxisome contacts induced by M1 Spastin was shown to be significantly reduced by a decrease in ABCD1. An immunoprecipitation experiment showed that M1 Spastin directly associates with ABCD1, although the ABCD1 domain that interacts with M1 Spastin has yet determined. These results indicate that ABCD1 is required for the formation of LD-peroxisomes contact sites. In contrast, M1 Spastin does not interact with

ACBD5, which tethers peroxisomes to the ER. Further studies are needed to properly elucidate the contribution of ABCD1 to the transfer of fatty acids from LDs.

4.6.2 Crosstalk During Antiviral Signalling

The mammalian innate immune system is constituted by several families of microbial detection receptors, i.e. pattern recognition receptors (PRPs) which are located in different subcellular compartments (Kagan 2012). Retinoic acid-inducible gene I (RIG-I) like receptors (RLRs) are one of PRPs which bind cytosolic viral RNA. Once viral RNA is recognized by RLR, it must bind to an adaptor protein called MAVS, a tail anchored protein on the mitochondrial membrane. Upon binding to MAVS, a series of signalling events are induced that produce pro-inflammatory cytokines and interferons (IFNs).

Dixit et al. first reported that not only mitochondria but also peroxisomes constitute an important site of antiviral signal transduction and cooperate with mitochondria to combat viral infections through activation of the RLR/MAVS signalling pathway (Fig. 4.7) (Dixit et al. 2010). MAVS localization to each organelle is important for antiviral signaling, because MAVS induces the expression of different antiviral factors when localized to either mitochondria or peroxisomes. Mitochondrial MAVS leads to the production of type I interferon, which leads to the expression of interferon-stimulated genes (ISGs) (Odendall et al. 2014). In contrast, peroxisomal MAVS directly induces ISG expression independently of type I IFNs. Therefore, peroxisomal MAVS induces the rapid and transient expression of anti-viral factors, whereas mitochondrial MAVS sustains the stable antiviral response. These results indicate that peroxisomal MAVS in cooperation with mitochondrial MAVS have a role in antiviral defence.

4.7 Concluding Remarks

Since the discovery of the fatty acid β-oxidation system in mammalian peroxisomes by Lazarow and de Duve at 1976, our understanding of the functions of peroxisomes has been deepened by a large number of important studies. It is now well understood that peroxisomes are indispensable for lipid metabolism, redox homeostasis and intracellular signalling. It is now recognized various fatty acids including bile acid intermediates are transported into peroxisomes through ABCD1–3, while the metabolites produced by the fatty acid β- and α-oxidation pathways are released outside of peroxisomes through various transporters and channels. However, the precise mechanism by which the ABCD transporters recognize and transport their substrates has yet to be elucidated. In addition, the actual form of the metabolites that can be released from peroxisomes is not well understood. A detailed analysis of the metabolites in peroxisomes is going to be needed, based on the knockout of enzymes such as peroxisomal acyl-CoA thioesterases and carnitine acyltransferases.

Increasing evidence has demonstrated that peroxisomes are involved in cellular redox homeostasis. An inherited metabolic defect of peroxisomes or reduced peroxisomal function due to aging can lead to increased levels of ROS, which eventually cause damage to mitochondria. Such damage may trigger mitochondrial ROS production, which in turn exacerbates the progression of age-related disorders and neurodegenerative diseases such as Alzheimer's disease, Parkinson's disease and X-ALD. However, further studies are required to understand how peroxisomal dysfunction specifically inflicts damage on mitochondria.

The progress of studies on the cross talk between and among organelles has been quite remarkable. The organelles within cells are much more tightly interconnected with each other than we had been previously imagined. The recent discovery of membrane contact sites between peroxisomes and other organelles has provided critically important insight into how metabolites shuttle between organelles. The disruption of contact sites between peroxisomes and other organelles leads to metabolic abnormalities, indicating the importance of these contacts. For example, a defect in the VLCFA-CoA binding protein ACBD5 on the peroxisomal membrane, which acts as a tether to form peroxisome-ER contact sites, results in decreased VLCFA β-oxidation. It is important to determine how the formation of membrane contact site is regulated and influences the efficiency of the metabolic processes that occur in peroxisomes. Furthermore, recent findings let us entertain the possibility of previously unexpected peroxisomal functions related to viral infection defence and intracellular cholesterol trafficking in coordination with mitochondria and lysosomes, respectively. In this regard, the further progress of peroxisome studies with the continued emergence of new technologies is greatly anticipated.

Acknowledgments The publication is supported in part by a Grant-in-Aid for Intractable Diseases from the Ministry of health, Labour and Welfare of Japan and from the Ministry of Education, Culture, Sports, Science and Technology of Japan. Pacific Edit reviewed the manuscript prior to submission.

References

AbdelRaheim SR, Cartwright JL, Gasmi L, McLennan AG (2001) The NADH diphosphatase encoded by the Saccharomyces cerevisiae NPY1 nudix hydrolase gene is located in peroxisomes. Arch Biochem Biophys 388(1):18–24. https://doi.org/10.1006/abbi.2000.2268

Abdelraheim SR, Spiller DG, McLennan AG (2003) Mammalian NADH diphosphatases of the Nudix family: cloning and characterization of the human peroxisomal NUDT12 protein. Biochem J 374(Pt 2):329–335. https://doi.org/10.1042/BJ20030441

Agrimi G, Russo A, Scarcia P, Palmieri F (2012) The human gene SLC25A17 encodes a peroxisomal transporter of coenzyme A, FAD and NAD+. Biochem J 443(1):241–247. https://doi.org/10.1042/BJ20111420

Antonenkov VD, Hiltunen JK (2006) Peroxisomal membrane permeability and solute transfer. Biochim Biophys Acta 1763(12):1697–1706. https://doi.org/10.1016/j.bbamcr.2006.08.044

Antonenkov VD, Hiltunen JK (2012) Transfer of metabolites across the peroxisomal membrane. Biochim Biophys Acta 1822(9):1374–1386. https://doi.org/10.1016/j.bbadis.2011.12.011
Antonenkov VD, Grunau S, Ohlmeier S, Hiltunen JK (2010) Peroxisomes are oxidative organelles. Antioxid Redox Signal 13(4):525–537. https://doi.org/10.1089/ars.2009.2996
Antonenkov VD, Isomursu A, Mennerich D, Vapola MH, Weiher H, Kietzmann T, Hiltunen JK (2015) The human mitochondrial DNA depletion syndrome gene MPV17 encodes a non-selective channel that modulates membrane potential. J Biol Chem 290(22):13840–13861. https://doi.org/10.1074/jbc.M114.608083
Apanasets O, Grou CP, Van Veldhoven PP, Brees C, Wang B, Nordgren M, Dodt G, Azevedo JE, Fransen M (2014) PEX5, the shuttling import receptor for peroxisomal matrix proteins, is a redox-sensitive protein. Traffic 15(1):94–103. https://doi.org/10.1111/tra.12129
Arai Y, Hayashi M, Nishimura M (2008) Proteomic identification and characterization of a novel peroxisomal adenine nucleotide transporter supplying ATP for fatty acid beta-oxidation in soybean and Arabidopsis. Plant Cell 20(12):3227–3240. https://doi.org/10.1105/tpc.108.062877
Baarine M, Andreoletti P, Athias A, Nury T, Zarrouk A, Ragot K, Vejux A, Riedinger JM, Kattan Z, Bessede G, Trompier D, Savary S, Cherkaoui-Malki M, Lizard G (2012) Evidence of oxidative stress in very long chain fatty acid--treated oligodendrocytes and potentialization of ROS production using RNA interference-directed knockdown of ABCD1 and ACOX1 peroxisomal proteins. Neuroscience 213:1–18. https://doi.org/10.1016/j.neuroscience.2012.03.058
Barbosa AD, Savage DB, Siniossoglou S (2015) Lipid droplet-organelle interactions: emerging roles in lipid metabolism. Curr Opin Cell Biol 35:91–97. https://doi.org/10.1016/j.ceb.2015.04.017
Baumgart E, Vanhorebeek I, Grabenbauer M, Borgers M, Declercq PE, Fahimi HD, Baes M (2001) Mitochondrial alterations caused by defective peroxisomal biogenesis in a mouse model for Zellweger syndrome (PEX5 knockout mouse). Am J Pathol 159(4):1477–1494. https://doi.org/10.1016/S0002-9440(10)62534-5
Beaufay H, Jacques P, Baudhuin P, Sellinger OZ, Berthet J, De Duve C (1964) Tissue fractionation studies. 18. Resolution of mitochondrial fractions from rat liver into three distinct populations of cytoplasmic particles by means of density equilibration in various gradients. Biochem J 92(1):184–205
Beedholm-Ebsen R, van de Wetering K, Hardlei T, Nexo E, Borst P, Moestrup SK (2010) Identification of multidrug resistance protein 1 (MRP1/ABCC1) as a molecular gate for cellular export of cobalamin. Blood 115(8):1632–1639. https://doi.org/10.1182/blood-2009-07-232587
Berger J, Dorninger F, Forss-Petter S, Kunze M (2016) Peroxisomes in brain development and function. Biochim Biophys Acta 1863(5):934–955. https://doi.org/10.1016/j.bbamcr.2015.12.005
Binns D, Januszewski T, Chen Y, Hill J, Markin VS, Zhao Y, Gilpin C, Chapman KD, Anderson RG, Goodman JM (2006) An intimate collaboration between peroxisomes and lipid bodies. J Cell Biol 173(5):719–731. https://doi.org/10.1083/jcb.200511125
Blackstone C (2018) Converging cellular themes for the hereditary spastic paraplegias. Curr Opin Neurobiol 51:139–146. https://doi.org/10.1016/j.conb.2018.04.025
Bonekamp NA, Volkl A, Fahimi HD, Schrader M (2009) Reactive oxygen species and peroxisomes: struggling for balance. Biofactors 35(4):346–355. https://doi.org/10.1002/biof.48
Borths EL, Poolman B, Hvorup RN, Locher KP, Rees DC (2005) In vitro functional characterization of BtuCD-F, the Escherichia coli ABC transporter for vitamin B12 uptake. Biochemistry 44(49):16301–16309. https://doi.org/10.1021/bi0513103
Braverman NE, Moser AB (2012) Functions of plasmalogen lipids in health and disease. Biochim Biophys Acta 1822(9):1442–1452. https://doi.org/10.1016/j.bbadis.2012.05.008
Cartwright JL, Gasmi L, Spiller DG, McLennan AG (2000) The Saccharomyces cerevisiae PCD1 gene encodes a peroxisomal nudix hydrolase active toward coenzyme A and its derivatives. J Biol Chem 275(42):32925–32930. https://doi.org/10.1074/jbc.M005015200
Castro IG, Schuldiner M, Zalckvar E (2018) Mind the organelle gap - peroxisome contact sites in disease. Trends Biochem Sci 43(3):199–210. https://doi.org/10.1016/j.tibs.2018.01.001

Chang CL, Weigel AV, Ioannou MS, Pasolli HA, Xu CS, Peale DR, Shtengel G, Freeman M, Hess HF, Blackstone C, Lippincott-Schwartz J (2019) Spastin tethers lipid droplets to peroxisomes and directs fatty acid trafficking through ESCRT-III. J Cell Biol 218(8):2583–2599 https://doi.org/10.1083/jcb.201902061

Chen Y, Azad MB, Gibson SB (2009) Superoxide is the major reactive oxygen species regulating autophagy. Cell Death Differ 16(7):1040–1052. https://doi.org/10.1038/cdd.2009.49

Chu BB, Liao YC, Qi W, Xie C, Du X, Wang J, Yang H, Miao HH, Li BL, Song BL (2015) Cholesterol transport through lysosome-peroxisome membrane contacts. Cell 161(2):291–306. https://doi.org/10.1016/j.cell.2015.02.019

Cipolla CM, Lodhi IJ (2017) Peroxisomal dysfunction in age-related diseases. Trends Endocrinol Metab 28(4):297–308. https://doi.org/10.1016/j.tem.2016.12.003

Cockcroft S, Raghu P (2018) Phospholipid transport protein function at organelle contact sites. Curr Opin Cell Biol 53:52–60. https://doi.org/10.1016/j.ceb.2018.04.011

Coelho D, Suormala T, Stucki M, Lerner-Ellis JP, Rosenblatt DS, Newbold RF, Baumgartner MR, Fowler B (2008) Gene identification for the cblD defect of vitamin B12 metabolism. N Engl J Med 358(14):1454–1464. https://doi.org/10.1056/NEJMoa072200

Coelho D, Kim JC, Miousse IR, Fung S, du Moulin M, Buers I, Suormala T, Burda P, Frapolli M, Stucki M, Nurnberg P, Thiele H, Robenek H, Hohne W, Longo N, Pasquali M, Mengel E, Watkins D, Shoubridge EA, Majewski J, Rosenblatt DS, Fowler B, Rutsch F, Baumgartner MR (2012) Mutations in ABCD4 cause a new inborn error of vitamin B12 metabolism. Nat Genet 44(10):1152–1155. https://doi.org/10.1038/ng.2386

Cohen Y, Klug YA, Dimitrov L, Erez Z, Chuartzman SG, Elinger D, Yofe I, Soliman K, Gartner J, Thoms S, Schekman R, Elbaz-Alon Y, Zalckvar E, Schuldiner M (2014) Peroxisomes are juxtaposed to strategic sites on mitochondria. Mol Biosyst 10(7):1742–1748. https://doi.org/10.1039/c4mb00001c

Costello JL, Castro IG, Hacker C, Schrader TA, Metz J, Zeuschner D, Azadi AS, Godinho LF, Costina V, Findeisen P, Manner A, Islinger M, Schrader M (2017a) ACBD5 and VAPB mediate membrane associations between peroxisomes and the ER. J Cell Biol 216(2):331–342. https://doi.org/10.1083/jcb.201607055

Costello JL, Castro IG, Schrader TA, Islinger M, Schrader M (2017b) Peroxisomal ACBD4 interacts with VAPB and promotes ER-peroxisome associations. Cell Cycle 16(11):1039–1045. https://doi.org/10.1080/15384101.2017.1314422

Dallabona C, Marsano RM, Arzuffi P, Ghezzi D, Mancini P, Zeviani M, Ferrero I, Donnini C (2010) Sym1, the yeast ortholog of the MPV17 human disease protein, is a stress-induced bioenergetic and morphogenetic mitochondrial modulator. Hum Mol Genet 19(6):1098–1107. https://doi.org/10.1093/hmg/ddp581

Dassa E, Bouige P (2001) The ABC of ABCS: a phylogenetic and functional classification of ABC systems in living organisms. Res Microbiol 152(3-4):211–229

De Duve C, Baudhuin P (1966) Peroxisomes (microbodies and related particles). Physiol Rev 46(2):323–357. https://doi.org/10.1152/physrev.1966.46.2.323

De Marcos Lousa C, van Roermund CW, Postis VL, Dietrich D, Kerr ID, Wanders RJ, Baldwin SA, Baker A, Theodoulou FL (2013) Intrinsic acyl-CoA thioesterase activity of a peroxisomal ATP binding cassette transporter is required for transport and metabolism of fatty acids. Proc Natl Acad Sci U S A 110(4):1279–1284. https://doi.org/10.1073/pnas.1218034110

Decker M, Arand M, Cronin A (2009) Mammalian epoxide hydrolases in xenobiotic metabolism and signalling. Arch Toxicol 83(4):297–318. https://doi.org/10.1007/s00204-009-0416-0

Demarquoy J, Le Borgne F (2015) Crosstalk between mitochondria and peroxisomes. World J Biol Chem 6(4):301–309. https://doi.org/10.4331/wjbc.v6.i4.301

Deori NM, Kale A, Maurya PK, Nagotu S (2018) Peroxisomes: role in cellular ageing and age related disorders. Biogerontology 19(5):303–324. https://doi.org/10.1007/s10522-018-9761-9

Dixit E, Boulant S, Zhang Y, Lee AS, Odendall C, Shum B, Hacohen N, Chen ZJ, Whelan SP, Fransen M, Nibert ML, Superti-Furga G, Kagan JC (2010) Peroxisomes are signaling platforms for antiviral innate immunity. Cell 141(4):668–681. https://doi.org/10.1016/j.cell.2010.04.018

Dixon SJ, Stockwell BR (2014) The role of iron and reactive oxygen species in cell death. Nat Chem Biol 10(1):9–17. https://doi.org/10.1038/nchembio.1416

Dorninger F, Forss-Petter S, Berger J (2017) From peroxisomal disorders to common neurodegenerative diseases - the role of ether phospholipids in the nervous system. FEBS Lett 591(18):2761–2788. https://doi.org/10.1002/1873-3468.12788

El-Hattab AW, Craigen WJ, Scaglia F (2017) Mitochondrial DNA maintenance defects. Biochim Biophys Acta Mol Basis Dis 1863(6):1539–1555. https://doi.org/10.1016/j.bbadis.2017.02.017

Engelen M, Kemp S, de Visser M, van Geel BM, Wanders RJ, Aubourg P, Poll-The BT (2012) X-linked adrenoleukodystrophy (X-ALD): clinical presentation and guidelines for diagnosis, follow-up and management. Orphanet J Rare Dis 7:51. https://doi.org/10.1186/1750-1172-7-51

Engelmann B, Brautigam C, Thiery J (1994) Plasmalogen phospholipids as potential protectors against lipid peroxidation of low density lipoproteins. Biochem Biophys Res Commun 204(3):1235–1242. https://doi.org/10.1006/bbrc.1994.2595

Fan J, Li X, Issop L, Culty M, Papadopoulos V (2016) ACBD2/ECI2-mediated peroxisome-mitochondria interactions in leydig cell steroid biosynthesis. Mol Endocrinol 30(7):763–782. https://doi.org/10.1210/me.2016-1008

Fanelli F, Sepe S, D'Amelio M, Bernardi C, Cristiano L, Cimini A, Cecconi F, Ceru MP, Moreno S (2013) Age-dependent roles of peroxisomes in the hippocampus of a transgenic mouse model of Alzheimer's disease. Mol Neurodegen 8:8. https://doi.org/10.1186/1750-1326-8-8

Farr RL, Lismont C, Terlecky SR, Fransen M (2016) Peroxisome biogenesis in mammalian cells: the impact of genes and environment. Biochim Biophys Acta 1863(5):1049–1060. https://doi.org/10.1016/j.bbamcr.2015.08.011

Farre JC, Mahalingam SS, Proietto M, Subramani S (2019) Peroxisome biogenesis, membrane contact sites, and quality control. EMBO Rep 20(1):e46864. https://doi.org/10.15252/embr.201846864

Farrell SO, Fiol CJ, Reddy JK, Bieber LL (1984) Properties of purified carnitine acyltransferases of mouse liver peroxisomes. J Biol Chem 259(21):13089–13095

Ferdinandusse S, Mulders J, IJlst IJ, Denis S, Dacremont G, Waterham HR, Wanders RJ (1999) Molecular cloning and expression of human carnitine octanoyltransferase: evidence for its role in the peroxisomal beta-oxidation of branched-chain fatty acids. Biochem Biophys Res Commun 263(1):213–218. https://doi.org/10.1006/bbrc.1999.1340

Ferdinandusse S, Jimenez-Sanchez G, Koster J, Denis S, Van Roermund CW, Silva-Zolezzi I, Moser AB, Visser WF, Gulluoglu M, Durmaz O, Demirkol M, Waterham HR, Gokcay G, Wanders RJ, Valle D (2015) A novel bile acid biosynthesis defect due to a deficiency of peroxisomal ABCD3. Hum Mol Genet 24(2):361–370. https://doi.org/10.1093/hmg/ddu448

Fourcade S, Ruiz M, Camps C, Schluter A, Houten SM, Mooyer PA, Pampols T, Dacremont G, Wanders RJ, Giros M, Pujol A (2009) A key role for the peroxisomal ABCD2 transporter in fatty acid homeostasis. Am J Physiol Endocrinol Metab 296(1):E211–E221. https://doi.org/10.1152/ajpendo.90736.2008

Fourcade S, Ruiz M, Guilera C, Hahnen E, Brichta L, Naudi A, Portero-Otin M, Dacremont G, Cartier N, Wanders R, Kemp S, Mandel JL, Wirth B, Pamplona R, Aubourg P, Pujol A (2010) Valproic acid induces antioxidant effects in X-linked adrenoleukodystrophy. Hum Mol Genet 19(10):2005–2014. https://doi.org/10.1093/hmg/ddq082

Fransen M, Nordgren M, Wang B, Apanasets O (2012) Role of peroxisomes in ROS/RNS-metabolism: implications for human disease. Biochim Biophys Acta 1822(9):1363–1373. https://doi.org/10.1016/j.bbadis.2011.12.001

Fujiki Y, Miyata N, Mukai S, Okumoto K, Cheng EH (2017) BAK regulates catalase release from peroxisomes. Mol Cell Oncol 4(3):e1306610. https://doi.org/10.1080/23723556.2017.1306610

Gasmi L, McLennan AG (2001) The mouse Nudt7 gene encodes a peroxisomal nudix hydrolase specific for coenzyme A and its derivatives. Biochem J 357(Pt 1):33–38

Geillon F, Gondcaille C, Raas Q, Dias AMM, Pecqueur D, Truntzer C, Lucchi G, Ducoroy P, Falson P, Savary S, Trompier D (2017) Peroxisomal ATP-binding cassette transporters form mainly tetramers. J Biol Chem 292(17):6965–6977. https://doi.org/10.1074/jbc.M116.772806

Genin EC, Geillon F, Gondcaille C, Athias A, Gambert P, Trompier D, Savary S (2011) Substrate specificity overlap and interaction between adrenoleukodystrophy protein (ALDP/ABCD1) and adrenoleukodystrophy-related protein (ALDRP/ABCD2). J Biol Chem 286(10):8075–8084. https://doi.org/10.1074/jbc.M110.211912

Giordano CR, Terlecky LJ, Bollig-Fischer A, Walton PA, Terlecky SR (2014) Amyloid-beta neuroprotection mediated by a targeted antioxidant. Sci Rep 4:4983. https://doi.org/10.1038/srep04983

Glorieux C, Calderon PB (2017) Catalase, a remarkable enzyme: targeting the oldest antioxidant enzyme to find a new cancer treatment approach. Biol Chem 398(10):1095–1108. https://doi.org/10.1515/hsz-2017-0131

Guimaraes CP, Domingues P, Aubourg P, Fouquet F, Pujol A, Jimenez-Sanchez G, Sa-Miranda C, Azevedo JE (2004) Mouse liver PMP70 and ALDP: homomeric interactions prevail in vivo. Biochim Biophys Acta 1689(3):235–243. https://doi.org/10.1016/j.bbadis.2004.04.001

Guimaraes CP, Sa-Miranda C, Azevedo JE (2005) Probing substrate-induced conformational alterations in adrenoleukodystrophy protein by proteolysis. J Hum Genet 50(2):99–105. https://doi.org/10.1007/s10038-004-0226-4

Hahnel D, Beyer K, Engelmann B (1999) Inhibition of peroxyl radical-mediated lipid oxidation by plasmalogen phospholipids and alpha-tocopherol. Free Radic Biol Med 27(9-10):1087–1094

Hartl FU, Just WW (1987) Integral membrane polypeptides of rat liver peroxisomes: topology and response to different metabolic states. Arch Biochem Biophys 255(1):109–119

He D, Barnes S, Falany CN (2003) Rat liver bile acid CoA:amino acid N-acyltransferase: expression, characterization, and peroxisomal localization. J Lipid Res 44(12):2242–2249. https://doi.org/10.1194/jlr.M300128-JLR200

Hekimi S, Lapointe J, Wen Y (2011) Taking a "good" look at free radicals in the aging process. Trends Cell Biol 21(10):569–576. https://doi.org/10.1016/j.tcb.2011.06.008

Henne WM, Reese ML, Goodman JM (2018) The assembly of lipid droplets and their roles in challenged cells. EMBO J 37(12):e98947. https://doi.org/10.15252/embj.201898947

Higgins CF (1992) ABC transporters: from microorganisms to man. Annu Rev Cell Biol 8:67–113. https://doi.org/10.1146/annurev.cb.08.110192.000435

Hillebrand M, Verrier SE, Ohlenbusch A, Schafer A, Soling HD, Wouters FS, Gartner J (2007) Live cell FRET microscopy: homo- and heterodimerization of two human peroxisomal ABC transporters, the adrenoleukodystrophy protein (ALDP, ABCD1) and PMP70 (ABCD3). J Biol Chem 282(37):26997–27005. https://doi.org/10.1074/jbc.M702122200

Holzinger A, Kammerer S, Berger J, Roscher AA (1997a) cDNA cloning and mRNA expression of the human adrenoleukodystrophy related protein (ALDRP), a peroxisomal ABC transporter. Biochem Biophys Res Commun 239(1):261–264. https://doi.org/10.1006/bbrc.1997.7391

Holzinger A, Kammerer S, Roscher AA (1997b) Primary structure of human PMP69, a putative peroxisomal ABC-transporter. Biochem Biophys Res Commun 237(1):152–157. https://doi.org/10.1006/bbrc.1997.7102

Honsho M, Fujiki Y (2017) Plasmalogen homeostasis - regulation of plasmalogen biosynthesis and its physiological consequence in mammals. FEBS Lett 591(18):2720–2729. https://doi.org/10.1002/1873-3468.12743

Honsho M, Asaoku S, Fukumoto K, Fujiki Y (2013) Topogenesis and homeostasis of fatty acyl-CoA reductase 1. J Biol Chem 288(48):34588–34598. https://doi.org/10.1074/jbc.M113.498345

Hosoi KI, Miyata N, Mukai S, Furuki S, Okumoto K, Cheng EH, Fujiki Y (2017) The VDAC2-BAK axis regulates peroxisomal membrane permeability. J Cell Biol 216(3):709–722. https://doi.org/10.1083/jcb.201605002

Hu A, Zhao XT, Tu H, Xiao T, Fu T, Wang Y, Liu Y, Shi XJ, Luo J, Song BL (2018) PIP4K2A regulates intracellular cholesterol transport through modulating PI(4,5)P2 homeostasis. J Lipid Res 59(3):507–514. https://doi.org/10.1194/jlr.M082149

Hua R, Cheng D, Coyaud E, Freeman S, Di Pietro E, Wang Y, Vissa A, Yip CM, Fairn GD, Braverman N, Brumell JH, Trimble WS, Raught B, Kim PK (2017) VAPs and ACBD5 tether peroxisomes to the ER for peroxisome maintenance and lipid homeostasis. J Cell Biol 216(2):367–377. https://doi.org/10.1083/jcb.201608128

Hunt MC, Siponen MI, Alexson SE (2012) The emerging role of acyl-CoA thioesterases and acyltransferases in regulating peroxisomal lipid metabolism. Biochim Biophys Acta 1822(9):1397–1410. https://doi.org/10.1016/j.bbadis.2012.03.009

Hunt MC, Tillander V, Alexson SE (2014) Regulation of peroxisomal lipid metabolism: the role of acyl-CoA and coenzyme A metabolizing enzymes. Biochimie 98:45–55. https://doi.org/10.1016/j.biochi.2013.12.018

Iida R, Yasuda T, Tsubota E, Takatsuka H, Masuyama M, Matsuki T, Kishi K (2003) M-LP, Mpv17-like protein, has a peroxisomal membrane targeting signal comprising a transmembrane domain and a positively charged loop and up-regulates expression of the manganese superoxide dismutase gene. J Biol Chem 278(8):6301–6306. https://doi.org/10.1074/jbc.M210886200

Iida R, Yasuda T, Tsubota E, Takatsuka H, Matsuki T, Kishi K (2006) Human Mpv17-like protein is localized in peroxisomes and regulates expression of antioxidant enzymes. Biochem Biophys Res Commun 344(3):948–954. https://doi.org/10.1016/j.bbrc.2006.04.008

Iida R, Ueki M, Yasuda T (2018) Knockout of Mpv17-like protein (M-LPH) gene in human hepatoma cells results in impairment of mtDNA integrity through reduction of TFAM, OGG1, and LIG3 at the protein levels. Oxid Med Cell Longev 2018:6956414. https://doi.org/10.1155/2018/6956414

Imanaka T, Aihara K, Takano T, Yamashita A, Sato R, Suzuki Y, Yokota S, Osumi T (1999) Characterization of the 70-kDa peroxisomal membrane protein, an ATP binding cassette transporter. J Biol Chem 274(17):11968–11976

Islinger M, Voelkl A, Fahimi HD, Schrader M (2018) The peroxisome: an update on mysteries 2.0. Histochem Cell Biol 150(5):443–471. https://doi.org/10.1007/s00418-018-1722-5

Issemann I, Green S (1990) Activation of a member of the steroid hormone receptor superfamily by peroxisome proliferators. Nature 347(6294):645–650. https://doi.org/10.1038/347645a0

Kagan JC (2012) Signaling organelles of the innate immune system. Cell 151(6):1168–1178. https://doi.org/10.1016/j.cell.2012.11.011

Kamijo K, Taketani S, Yokota S, Osumi T, Hashimoto T (1990) The 70-kDa peroxisomal membrane protein is a member of the Mdr (P-glycoprotein)-related ATP-binding protein superfamily. J Biol Chem 265(8):4534–4540

Kashiwayama Y, Morita M, Kamijo K, Imanaka T (2002) Nucleotide-induced conformational changes of PMP70, an ATP binding cassette transporter on rat liver peroxisomal membranes. Biochem Biophys Res Commun 291(5):1245–1251. https://doi.org/10.1006/bbrc.2002.6588

Kawaguchi K, Okamoto T, Morita M, Imanaka T (2016) Translocation of the ABC transporter ABCD4 from the endoplasmic reticulum to lysosomes requires the escort protein LMBD1. Sci Rep 6:30183. https://doi.org/10.1038/srep30183

Kikuchi M, Hatano N, Yokota S, Shimozawa N, Imanaka T, Taniguchi H (2004) Proteomic analysis of rat liver peroxisome: presence of peroxisome-specific isozyme of Lon protease. J Biol Chem 279(1):421–428. https://doi.org/10.1074/jbc.M305623200

Kleinecke S, Richert S, de Hoz L, Brugger B, Kungl T, Asadollahi E, Quintes S, Blanz J, McGonigal R, Naseri K, Sereda MW, Sachsenheimer T, Luchtenborg C, Mobius W, Willison H, Baes M, Nave KA, Kassmann CM (2017) Peroxisomal dysfunctions cause lysosomal storage and axonal Kv1 channel redistribution in peripheral neuropathy. Elife 6:23332. https://doi.org/10.7554/eLife.23332

Knoops B, Goemaere J, Van der Eecken V, Declercq JP (2011) Peroxiredoxin 5: structure, mechanism, and function of the mammalian atypical 2-Cys peroxiredoxin. Antioxid Redox Signal 15(3):817–829. https://doi.org/10.1089/ars.2010.3584

Kodan A, Yamaguchi T, Nakatsu T, Matsuoka K, Kimura Y, Ueda K, Kato H (2019) Inward- and outward-facing X-ray crystal structures of homodimeric P-glycoprotein CmABCB1. Nat Commun 10(1):88. https://doi.org/10.1038/s41467-018-08007-x

Kornmann B, Currie E, Collins SR, Schuldiner M, Nunnari J, Weissman JS, Walter P (2009) An ER-mitochondria tethering complex revealed by a synthetic biology screen. Science 325(5939):477–481. https://doi.org/10.1126/science.1175088

Kou J, Kovacs GG, Hoftberger R, Kulik W, Brodde A, Forss-Petter S, Honigschnabl S, Gleiss A, Brugger B, Wanders R, Just W, Budka H, Jungwirth S, Fischer P, Berger J (2011) Peroxisomal alterations in Alzheimer's disease. Acta Neuropathol 122(3):271–283. https://doi.org/10.1007/s00401-011-0836-9

Krick S, Shi S, Ju W, Faul C, Tsai SY, Mundel P, Bottinger EP (2008) Mpv17l protects against mitochondrial oxidative stress and apoptosis by activation of Omi/HtrA2 protease. Proc Natl Acad Sci U S A 105(37):14106–14111. https://doi.org/10.1073/pnas.0801146105

Lamhonwah AM, Ackerley CA, Tilups A, Edwards VD, Wanders RJ, Tein I (2005) OCTN3 is a mammalian peroxisomal membrane carnitine transporter. Biochem Biophys Res Commun 338(4):1966–1972. https://doi.org/10.1016/j.bbrc.2005.10.170

Linka N, Theodoulou FL, Haslam RP, Linka M, Napier JA, Neuhaus HE, Weber AP (2008) Peroxisomal ATP import is essential for seedling development in Arabidopsis thaliana. Plant Cell 20(12):3241–3257. https://doi.org/10.1105/tpc.108.062042

Lismont C, Nordgren M, Van Veldhoven PP, Fransen M (2015) Redox interplay between mitochondria and peroxisomes. Front Cell Dev Biol 3:35. https://doi.org/10.3389/fcell.2015.00035

Liu LX, Janvier K, Berteaux-Lecellier V, Cartier N, Benarous R, Aubourg P (1999) Homo- and heterodimerization of peroxisomal ATP-binding cassette half-transporters. J Biol Chem 274(46):32738–32743

Lodhi IJ, Semenkovich CF (2014) Peroxisomes: a nexus for lipid metabolism and cellular signaling. Cell Metab 19(3):380–392. https://doi.org/10.1016/j.cmet.2014.01.002

Lombard-Platet G, Savary S, Sarde CO, Mandel JL, Chimini G (1996) A close relative of the adrenoleukodystrophy (ALD) gene codes for a peroxisomal protein with a specific expression pattern. Proc Natl Acad Sci U S A 93(3):1265–1269

Lopez-Erauskin J, Galino J, Ruiz M, Cuezva JM, Fabregat I, Cacabelos D, Boada J, Martinez J, Ferrer I, Pamplona R, Villarroya F, Portero-Otin M, Fourcade S, Pujol A (2013) Impaired mitochondrial oxidative phosphorylation in the peroxisomal disease X-linked adrenoleukodystrophy. Hum Mol Genet 22(16):3296–3305. https://doi.org/10.1093/hmg/ddt186

Loughran PA, Stolz DB, Barrick SR, Wheeler DS, Friedman PA, Rachubinski RA, Watkins SC, Billiar TR (2013) PEX7 and EBP50 target iNOS to the peroxisome in hepatocytes. Nitric Oxide 31:9–19. https://doi.org/10.1016/j.niox.2013.02.084

Luo J, Liao YC, Xiao J, Song BL (2017) Measurement of cholesterol transfer from lysosome to peroxisome using an in vitro reconstitution assay. Methods Mol Biol 1583:141–161. https://doi.org/10.1007/978-1-4939-6875-6_11

Mattiazzi Usaj M, Brloznik M, Kaferle P, Zitnik M, Wolinski H, Leitner F, Kohlwein SD, Zupan B, Petrovic U (2015) Genome-wide localization study of yeast Pex11 identifies peroxisome-mitochondria interactions through the ERMES complex. J Mol Biol 427(11):2072–2087. https://doi.org/10.1016/j.jmb.2015.03.004

McLennan AG (2006) The Nudix hydrolase superfamily. Cell Mol Life Sci 63(2):123–143. https://doi.org/10.1007/s00018-005-5386-7

Mindthoff S, Grunau S, Steinfort LL, Girzalsky W, Hiltunen JK, Erdmann R, Antonenkov VD (2016) Peroxisomal Pex11 is a pore-forming protein homologous to TRPM channels. Biochim Biophys Acta 1863(2):271–283. https://doi.org/10.1016/j.bbamcr.2015.11.013

Morisseau C, Hammock BD (2013) Impact of soluble epoxide hydrolase and epoxyeicosanoids on human health. Annu Rev Pharmacol Toxicol 53:37–58. https://doi.org/10.1146/annurev-pharmtox-011112-140244

Morita M, Imanaka T (2012) Peroxisomal ABC transporters: structure, function and role in disease. Biochim Biophys Acta 1822(9):1387–1396. https://doi.org/10.1016/j.bbadis.2012.02.009

Morita M, Kurochkin IV, Motojima K, Goto S, Takano T, Okamura S, Sato R, Yokota S, Imanaka T (2000) Insulin-degrading enzyme exists inside of rat liver peroxisomes and degrades oxidized proteins. Cell Struct Funct 25(5):309–315

Morita M, Shinbo S, Asahi A, Imanaka T (2012) Very long chain fatty acid beta-oxidation in astrocytes: contribution of the ABCD1-dependent and -independent pathways. Biol Pharm Bull 35(11):1972–1979

Mosser J, Douar AM, Sarde CO, Kioschis P, Feil R, Moser H, Poustka AM, Mandel JL, Aubourg P (1993) Putative X-linked adrenoleukodystrophy gene shares unexpected homology with ABC transporters. Nature 361(6414):726–730. https://doi.org/10.1038/361726a0

Murakami K, Ichinohe Y, Koike M, Sasaoka N, Iemura S, Natsume T, Kakizuka A (2013) VCP Is an integral component of a novel feedback mechanism that controls intracellular localization of catalase and H2O2 Levels. PLoS One 8(2):e56012. https://doi.org/10.1371/journal.pone.0056012

Nagan N, Zoeller RA (2001) Plasmalogens: biosynthesis and functions. Prog Lipid Res 40(3):199–229

Nakagawa T, Imanaka T, Morita M, Ishiguro K, Yurimoto H, Yamashita A, Kato N, Sakai Y (2000) Peroxisomal membrane protein Pmp47 is essential in the metabolism of middle-chain fatty acid in yeast peroxisomes and is associated with peroxisome proliferation. J Biol Chem 275(5):3455–3461

Nell HJ, Au JL, Giordano CR, Terlecky SR, Walton PA, Whitehead SN, Cechetto DF (2017) Targeted antioxidant, catalase-SKL, reduces beta-amyloid toxicity in the rat brain. Brain Pathol 27(1):86–94. https://doi.org/10.1111/bpa.12368

Odendall C, Dixit E, Stavru F, Bierne H, Franz KM, Durbin AF, Boulant S, Gehrke L, Cossart P, Kagan JC (2014) Diverse intracellular pathogens activate type III interferon expression from peroxisomes. Nat Immunol 15(8):717–726. https://doi.org/10.1038/ni.2915

Ofman R, Speijer D, Leen R, Wanders RJ (2006) Proteomic analysis of mouse kidney peroxisomes: identification of RP2p as a peroxisomal nudix hydrolase with acyl-CoA diphosphatase activity. Biochem J 393(Pt 2):537–543. https://doi.org/10.1042/BJ20050893

Okamoto T, Kawaguchi K, Watanabe S, Agustina R, Ikejima T, Ikeda K, Nakano M, Morita M, Imanaka T (2018) Characterization of human ATP-binding cassette protein subfamily D reconstituted into proteoliposomes. Biochem Biophys Res Commun 496(4):1122–1127. https://doi.org/10.1016/j.bbrc.2018.01.153

Palmieri L, Rottensteiner H, Girzalsky W, Scarcia P, Palmieri F, Erdmann R (2001) Identification and functional reconstitution of the yeast peroxisomal adenine nucleotide transporter. EMBO J 20(18):5049–5059. https://doi.org/10.1093/emboj/20.18.5049

Papadopoulos C, Orso G, Mancuso G, Herholz M, Gumeni S, Tadepalle N, Jungst C, Tzschichholz A, Schauss A, Honing S, Trifunovic A, Daga A, Rugarli EI (2015) Spastin binds to lipid droplets and affects lipid metabolism. PLoS Genet 11(4):e1005149. https://doi.org/10.1371/journal.pgen.1005149

Pellicoro A, van den Heuvel FA, Geuken M, Moshage H, Jansen PL, Faber KN (2007) Human and rat bile acid-CoA:amino acid N-acyltransferase are liver-specific peroxisomal enzymes: implications for intracellular bile salt transport. Hepatology 45(2):340–348. https://doi.org/10.1002/hep.21528

Petrillo S, Piemonte F, Pastore A, Tozzi G, Aiello C, Pujol A, Cappa M, Bertini E (2013) Glutathione imbalance in patients with X-linked adrenoleukodystrophy. Mol Genet Metab 109(4):366–370. https://doi.org/10.1016/j.ymgme.2013.05.009

Pfeffer SR (2019) NPC intracellular cholesterol transporter 1 (NPC1)-mediated cholesterol export from lysosomes. J Biol Chem 294(5):1706–1709. https://doi.org/10.1074/jbc.TM118.004165

Pinti M, Gibellini L, Liu Y, Xu S, Lu B, Cossarizza A (2015) Mitochondrial Lon protease at the crossroads of oxidative stress, ageing and cancer. Cell Mol Life Sci 72(24):4807–4824. https://doi.org/10.1007/s00018-015-2039-3

Pomatto LC, Raynes R, Davies KJ (2017) The peroxisomal Lon protease LonP2 in aging and disease: functions and comparisons with mitochondrial Lon protease LonP1. Biol Rev Camb Philos Soc 92(2):739–753. https://doi.org/10.1111/brv.12253

Pujol A, Ferrer I, Camps C, Metzger E, Hindelang C, Callizot N, Ruiz M, Pampols T, Giros M, Mandel JL (2004) Functional overlap between ABCD1 (ALD) and ABCD2 (ALDR) transporters: a therapeutic target for X-adrenoleukodystrophy. Hum Mol Genet 13(23):2997–3006. https://doi.org/10.1093/hmg/ddh323

Raychaudhuri S, Prinz WA (2008) Nonvesicular phospholipid transfer between peroxisomes and the endoplasmic reticulum. Proc Natl Acad Sci U S A 105(41):15785–15790. https://doi.org/10.1073/pnas.0808321105

Reilly SJ, Tillander V, Ofman R, Alexson SE, Hunt MC (2008) The nudix hydrolase 7 is an acyl-CoA diphosphatase involved in regulating peroxisomal coenzyme A homeostasis. J Biochem 144(5):655–663. https://doi.org/10.1093/jb/mvn114

Reiss D, Beyer K, Engelmann B (1997) Delayed oxidative degradation of polyunsaturated diacyl phospholipids in the presence of plasmalogen phospholipids in vitro. Biochem J 323(Pt 3):807–814

Reiser G, Schönfeld P, Kahlert S (2006) Mechanism of toxicity of the branched-chain fatty acid phytanic acid, a marker of Refsum disease, in astrocytes involves mitochondrial impairment. Int J Dev Neurosci 24(2–3):113–122.https://doi.org/10.1016/j.ijdevneu.2005.11.002

Rembacz KP, Woudenberg J, Hoekstra M, Jonkers EZ, van den Heuvel FA, Buist-Homan M, Woudenberg-Vrenken TE, Rohacova J, Marin ML, Miranda MA, Moshage H, Stellaard F, Faber KN (2010) Unconjugated bile salts shuttle through hepatocyte peroxisomes for taurine conjugation. Hepatology 52(6):2167–2176. https://doi.org/10.1002/hep.23954

van Roermund CW, Elgersma Y, Singh N, Wanders RJ, Tabak HF (1995) The membrane of peroxisomes in Saccharomyces cerevisiae is impermeable to NAD(H) and acetyl-CoA under in vivo conditions. EMBO J 14(14):3480–3486

van Roermund CW, Visser WF, Ijlst L, Waterham HR, Wanders RJ (2011) Differential substrate specificities of human ABCD1 and ABCD2 in peroxisomal fatty acid beta-oxidation. Biochim Biophys Acta 1811(3):148–152. https://doi.org/10.1016/j.bbalip.2010.11.010

van Roermund CW, Ijlst L, Majczak W, Waterham HR, Folkerts H, Wanders RJ, Hellingwerf KJ (2012) Peroxisomal fatty acid uptake mechanism in Saccharomyces cerevisiae. J Biol Chem 287(24):20144–20153. https://doi.org/10.1074/jbc.M111.332833

Rokka A, Antonenkov VD, Soininen R, Immonen HL, Pirila PL, Bergmann U, Sormunen RT, Weckstrom M, Benz R, Hiltunen JK (2009) Pxmp2 is a channel-forming protein in Mammalian peroxisomal membrane. PLoS One 4(4):e5090. https://doi.org/10.1371/journal.pone.0005090

Ronicke S, Kruska N, Kahlert S, Reiser G (2009) The influence of the branched-chain fatty acids pristanic acid and Refsum disease-associated phytanic acid on mitochondrial functions and calcium regulation of hippocampal neurons, astrocytes, and oligodendrocytes. Neurobiol Dis 36(2):401–410. https://doi.org/10.1016/j.nbd.2009.08.005

Rutsch F, Gailus S, Miousse IR, Suormala T, Sagne C, Toliat MR, Nurnberg G, Wittkampf T, Buers I, Sharifi A, Stucki M, Becker C, Baumgartner M, Robenek H, Marquardt T, Hohne W, Gasnier B, Rosenblatt DS, Fowler B, Nurnberg P (2009) Identification of a putative lysosomal cobalamin exporter altered in the cblF defect of vitamin B12 metabolism. Nat Genet 41(2):234–239. https://doi.org/10.1038/ng.294

Salpietro V, Phadke R, Saggar A, Hargreaves IP, Yates R, Fokoloros C, Mankad K, Hertecant J, Ruggieri M, McCormick D, Kinali M (2015) Zellweger syndrome and secondary mitochondrial myopathy. Eur J Pediatr 174(4):557–563. https://doi.org/10.1007/s00431-014-2431-2

Schrader M, Fahimi HD (2006) Peroxisomes and oxidative stress. Biochim Biophys Acta 1763(12):1755–1766. https://doi.org/10.1016/j.bbamcr.2006.09.006

Schrader M, Costello J, Godinho LF, Islinger M (2015) Peroxisome-mitochondria interplay and disease. J Inherit Metab Dis 38(4):681–702. https://doi.org/10.1007/s10545-015-9819-7

Schrader M, Kamoshita M, Islinger M (2019) Organelle interplay-peroxisome interactions in health and disease. J Inherit Metab Dis. https://doi.org/10.1002/jimd.12083

Schuldiner M, Bohnert M (2017) A different kind of love - lipid droplet contact sites. Biochim Biophys Acta Mol Cell Biol Lip 1862(10 Pt B):1188–1196. https://doi.org/10.1016/j.bbalip.2017.06.005

Schuldiner M, Zalckvar E (2017) Incredibly close-a newly identified peroxisome-ER contact site in humans. J Cell Biol 216(2):287–289. https://doi.org/10.1083/jcb.201701072

Shai N, Schuldiner M, Zalckvar E (2016) No peroxisome is an island - peroxisome contact sites. Biochim Biophys Acta 1863(5):1061–1069. https://doi.org/10.1016/j.bbamcr.2015.09.016

Shai N, Yifrach E, van Roermund CWT, Cohen N, Bibi C, IJlst IJ, Cavellini L, Meurisse J, Schuster R, Zada L, Mari MC, Reggiori FM, Hughes AL, Escobar-Henriques M, Cohen MM, Waterham HR, Wanders RJA, Schuldiner M, Zalckvar E (2018) Systematic mapping of contact sites reveals tethers and a function for the peroxisome-mitochondria contact. Nat Commun 9(1):1761. https://doi.org/10.1038/s41467-018-03957-8

Shani N, Jimenez-Sanchez G, Steel G, Dean M, Valle D (1997) Identification of a fourth half ABC transporter in the human peroxisomal membrane. Hum Mol Genet 6(11):1925–1931

Sleat DE, Wiseman JA, El-Banna M, Price SM, Verot L, Shen MM, Tint GS, Vanier MT, Walkley SU, Lobel P (2004) Genetic evidence for nonredundant functional cooperativity between NPC1 and NPC2 in lipid transport. Proc Natl Acad Sci U S A 101(16):5886–5891. https://doi.org/10.1073/pnas.0308456101

Spinazzola A, Viscomi C, Fernandez-Vizarra E, Carrara F, D'Adamo P, Calvo S, Marsano RM, Donnini C, Weiher H, Strisciuglio P, Parini R, Sarzi E, Chan A, DiMauro S, Rotig A, Gasparini P, Ferrero I, Mootha VK, Tiranti V, Zeviani M (2006) MPV17 encodes an inner mitochondrial membrane protein and is mutated in infantile hepatic mitochondrial DNA depletion. Nat Genet 38(5):570–575. https://doi.org/10.1038/ng1765

Sztriha L, Al-Gazali LI, Wanders RJ, Ofman R, Nork M, Lestringant GG (2000) Abnormal myelin formation in rhizomelic chondrodysplasia punctata type 2 (DHAPAT-deficiency). Dev Med Child Neurol 42(7):492–495

Tanaka AR, Tanabe K, Morita M, Kurisu M, Kasiwayama Y, Matsuo M, Kioka N, Amachi T, Imanaka T, Ueda K (2002) ATP binding/hydrolysis by and phosphorylation of peroxisomal ATP-binding cassette proteins PMP70 (ABCD3) and adrenoleukodystrophy protein (ABCD1). J Biol Chem 277(42):40142–40147. https://doi.org/10.1074/jbc.M205079200

Trachootham D, Lu W, Ogasawara MA, Nilsa RD, Huang P (2008) Redox regulation of cell survival. Antioxid Redox Signal 10(8):1343–1374. https://doi.org/10.1089/ars.2007.1957

Tripathi DN, Walker CL (2016) The peroxisome as a cell signaling organelle. Curr Opin Cell Biol 39:109–112. https://doi.org/10.1016/j.ceb.2016.02.017

Trott A, Morano KA (2004) SYM1 is the stress-induced Saccharomyces cerevisiae ortholog of the mammalian kidney disease gene Mpv17 and is required for ethanol metabolism and tolerance during heat shock. Eukaryot Cell 3(3):620–631. https://doi.org/10.1128/EC.3.3.620-631.2004

Valm AM, Cohen S, Legant WR, Melunis J, Hershberg U, Wait E, Cohen AR, Davidson MW, Betzig E, Lippincott-Schwartz J (2017) Applying systems-level spectral imaging and analysis to reveal the organelle interactome. Nature 546(7656):162–167. https://doi.org/10.1038/nature22369

Vamecq J, Cherkaoui-Malki M, Andreoletti P, Latruffe N (2014) The human peroxisome in health and disease: the story of an oddity becoming a vital organelle. Biochimie 98:4–15. https://doi.org/10.1016/j.biochi.2013.09.019

Van Veldhoven PP, Just WW, Mannaerts GP (1987) Permeability of the peroxisomal membrane to cofactors of beta-oxidation. Evidence for the presence of a pore-forming protein. J Biol Chem 262(9):4310–4318

Vance JE (2015) Phospholipid synthesis and transport in mammalian cells. Traffic 16(1):1–18. https://doi.org/10.1111/tra.12230

Vapola MH, Rokka A, Sormunen RT, Alhonen L, Schmitz W, Conzelmann E, Warri A, Grunau S, Antonenkov VD, Hiltunen JK (2014) Peroxisomal membrane channel Pxmp2 in the mammary fat pad is essential for stromal lipid homeostasis and for development of mammary gland epithelium in mice. Dev Biol 391(1):66–80. https://doi.org/10.1016/j.ydbio.2014.03.022

Vargas CR, Wajner M, Sirtori LR, Goulart L, Chiochetta M, Coelho D, Latini A, Llesuy S, Bello-Klein A, Giugliani R, Deon M, Mello CF (2004) Evidence that oxidative stress is increased in patients with X-linked adrenoleukodystrophy. Biochim Biophys Acta 1688(1):26–32

Vasiliou V, Vasiliou K, Nebert DW (2009) Human ATP-binding cassette (ABC) transporter family. Hum Genomics 3(3):281–290

Violante S, Achetib N, van Roermund CWT, Hagen J, Dodatko T, Vaz FM, Waterham HR, Chen H, Baes M, Yu C, Argmann CA, Houten SM (2019) Peroxisomes can oxidize medium- and long-chain fatty acids through a pathway involving ABCD3 and HSD17B4. FASEB J 33(3):4355–4364. https://doi.org/10.1096/fj.201801498R

Visser WF, van Roermund CW, Waterham HR, Wanders RJ (2002) Identification of human PMP34 as a peroxisomal ATP transporter. Biochem Biophys Res Commun 299(3):494–497

Visser WF, van Roermund CW, Ijlst L, Waterham HR, Wanders RJ (2007) Demonstration of bile acid transport across the mammalian peroxisomal membrane. Biochem Biophys Res Commun 357(2):335–340. https://doi.org/10.1016/j.bbrc.2007.03.083

Walbrecq G, Wang B, Becker S, Hannotiau A, Fransen M, Knoops B (2015) Antioxidant cytoprotection by peroxisomal peroxiredoxin-5. Free Radic Biol Med 84:215–226. https://doi.org/10.1016/j.freeradbiomed.2015.02.032

Walker CL, Pomatto LCD, Tripathi DN, Davies KJA (2018) Redox regulation of homeostasis and proteostasis in peroxisomes. Physiol Rev 98(1):89–115. https://doi.org/10.1152/physrev.00033.2016

Wallner S, Schmitz G (2011) Plasmalogens the neglected regulatory and scavenging lipid species. Chem Phys Lipids 164(6):573–589. https://doi.org/10.1016/j.chemphyslip.2011.06.008

Walton PA, Pizzitelli M (2012) Effects of peroxisomal catalase inhibition on mitochondrial function. Front Physiol 3:108. https://doi.org/10.3389/fphys.2012.00108

Walton PA, Brees C, Lismont C, Apanasets O, Fransen M (2017) The peroxisomal import receptor PEX5 functions as a stress sensor, retaining catalase in the cytosol in times of oxidative stress. Biochim Biophys Acta Mol Cell Res 1864(10):1833–1843. https://doi.org/10.1016/j.bbamcr.2017.07.013

Wanders RJ (2014) Metabolic functions of peroxisomes in health and disease. Biochimie 98:36–44. https://doi.org/10.1016/j.biochi.2013.08.022

Wanders RJ, Komen J, Ferdinandusse S (2011) Phytanic acid metabolism in health and disease. Biochim Biophys Acta 1811(9):498–507. https://doi.org/10.1016/j.bbalip.2011.06.006

Wanders RJ, Waterham HR, Ferdinandusse S (2015) Metabolic interplay between peroxisomes and other subcellular organelles including mitochondria and the endoplasmic reticulum. Front Cell Dev Biol 3:83. https://doi.org/10.3389/fcell.2015.00083

Wang B, Van Veldhoven PP, Brees C, Rubio N, Nordgren M, Apanasets O, Kunze M, Baes M, Agostinis P, Fransen M (2013) Mitochondria are targets for peroxisome-derived oxidative stress in cultured mammalian cells. Free Radic Biol Med 65:882–894. https://doi.org/10.1016/j.freeradbiomed.2013.08.173

Waterham HR, Ferdinandusse S, Wanders RJ (2016) Human disorders of peroxisome metabolism and biogenesis. Biochim Biophys Acta 1863(5):922–933. https://doi.org/10.1016/j.bbamcr.2015.11.015

Weiher H, Noda T, Gray DA, Sharpe AH, Jaenisch R (1990) Transgenic mouse model of kidney disease: insertional inactivation of ubiquitously expressed gene leads to nephrotic syndrome. Cell 62(3):425–434

Weiher H, Pircher H, Jansen-Durr P, Hegenbarth S, Knolle P, Grunau S, Vapola M, Hiltunen JK, Zwacka RM, Schmelzer E, Reumann K, Will H (2016) A monoclonal antibody raised against bacterially expressed MPV17 sequences shows peroxisomal, endosomal and lysosomal localisation in U2OS cells. BMC Res Notes 9:128. https://doi.org/10.1186/s13104-016-1939-0

Wiesinger C, Kunze M, Regelsberger G, Forss-Petter S, Berger J (2013) Impaired very long-chain acyl-CoA beta-oxidation in human X-linked adrenoleukodystrophy fibroblasts is a direct consequence of ABCD1 transporter dysfunction. J Biol Chem 288(26):19269–19279. https://doi.org/10.1074/jbc.M112.445445

Wolvetang EJ, Tager JM, Wanders RJ (1990) Latency of the peroxisomal enzyme acyl-CoA:dihydroxyacetonephosphate acyltransferase in digitonin-permeabilized fibroblasts: the effect of ATP and ATPase inhibitors. Biochem Biophys Res Commun 170(3):1135–1143

Yagita Y, Shinohara K, Abe Y, Nakagawa K, Al-Owain M, Alkuraya FS, Fujiki Y (2017) Deficiency of a retinal dystrophy protein, acyl-CoA binding domain-containing 5 (ACBD5), impairs peroxisomal beta-oxidation of very-long-chain fatty acids. J Biol Chem 292(2):691–705. https://doi.org/10.1074/jbc.M116.760090

Yakunin E, Moser A, Loeb V, Saada A, Faust P, Crane DI, Baes M, Sharon R (2010) alpha-Synuclein abnormalities in mouse models of peroxisome biogenesis disorders. J Neurosci Res 88(4):866–876. https://doi.org/10.1002/jnr.22246

Zhang SO, Trimble R, Guo F, Mak HY (2010) Lipid droplets as ubiquitous fat storage organelles in C. elegans. BMC Cell Biol 11:96. https://doi.org/10.1186/1471-2121-11-96

Part II
Dysfunction of Peroxisome and Human Disease

Chapter 5
Peroxisomal Disorders

Nobuyuki Shimozawa

Abstract Peroxisomal disorders (PD) are genetic disorders caused by peroxisome dysfunction and are classified into two groups: genetic defects in peroxisome-localized proteins and genetic defects in peroxisomal biogenesis. The dawn of PD research came with the detailed analysis of the Zellweger syndrome, the prototype of PD. Even recently, new PD are still being identified by whole-exome sequencing analysis, which means that the concept of PD has been expanding. Furthermore, the role of peroxisome in cancer and age-related diseases has also been studied. In contrast, PD pathophysiology and treatment are not clarified yet completely and even in adrenoleukodystrophy, which is the most common PD, the prognosis of phenotype and disease in pre-symptomatic patients is a difficult task.

In this chapter, various types of PD based on patient clinical data will be described, which will be useful to researchers and clinicians. I hope that this chapter will be a valuable aid to many researchers and clinicians in a conjoint effort to overcome this intractable disease.

Keywords Peroxisome · β-oxidation · VLCFA · Plasmalogens · Phytanic acid · Zellweger syndrome · Adrenoleukodystrophy · Whole-exome sequencing

5.1 Introduction

Peroxisomes are single-membrane lined organelles present in all eukaryotic cells. They have many metabolic functions in humans, such as β-oxidation of saturated very long-chain fatty acids (VLCFA), unsaturated fatty acids, and bile acids; α-oxidation of phytanic acid; plasmalogen synthesis; hydrogen peroxide degradation; and glyoxylic acid detoxification (see Chap. 4).

N. Shimozawa (✉)
Division of Genomics Research, Life Science Research Center, Gifu University, Gifu, Japan

Department of Pediatrics, Gifu University School of Medicine, Gifu, Japan
e-mail: nshim@gifu-u.ac.jp

T. Imanaka, N. Shimozawa (eds.), *Peroxisomes: Biogenesis, Function, and Role in Human Disease*, https://doi.org/10.1007/978-981-15-1169-1_5

Peroxisomal disorders (PD) are a group of inherited metabolic diseases with genetically defective peroxisomal functions. Zellweger syndrome (ZS), a prototype of PD with generally impaired peroxisomal function, has greatly contributed to the discovery of other PD as well as to the understanding of their pathophysiology.

ZS was first reported as a 'familial syndrome of multiple congenital defects' (Bowen et al. 1964). A peroxisomal defect in the liver of ZS patients was reported (Goldfischer et al. 1973) however, at that time, the peroxisomal function had not been clarified, so not much attention was given to this finding. Later, the β-oxidation system was found not only in the mitochondria but also in the peroxisomes (Lazarow and De Duve 1976), and since then, research on the peroxisomal metabolic function in humans and in the pathology of ZS was accelerated. For some time, the genetic etiology of ZS was unknown, however, we succeeded in identifying the first gene responsible for ZS, the peroxisome assembly factor 1 (PAF1, called PEX2 later) (Shimozawa et al. 1992). Since then, 12 PEX genes responsible for ZS have been identified to date (Shimozawa et al. 2004). There are more than 30 kinds of PD caused by genes involved in faulty peroxisomal biogenesis and metabolic pathways. Furthermore, the concept of PD is still expanding by advances in basic research regarding peroxisomal function, progress of mass spectrometry for peroxisomal metabolite measurements in the patients, and emergence of the next generation sequencer (NGS).

PD are caused by mutations in genes that can be classified into two major groups (Table 5.1); including: (a) genes involved in the import of peroxisomal membrane and matrix proteins, as well as in peroxisomal proliferation and fission (Peroxisome biogenesis disorders: PBD) and (b) genes encoding enzymes that are located in the peroxisomes, where they exert their function (Single enzyme deficiencies: SED).

In this chapter, PBD, SED, and adrenoleukodystrophy (ALD)-the most frequent PD, and finally the role of peroxisomes in cancer and age-related diseases are described.

5.2 Peroxisome Biogenesis Disorders (PBD)

Peroxisomal matrix proteins can be imported into peroxisomes by the various PEX proteins via the peroxisome-targeting sequences, PTS1 and PTS2, after their synthesis on free ribosomes. PTS1 proteins bind to their receptor, a short isoform of PEX5 (PEX5S) in the cytosol, and PTS2 proteins bind to their receptors, PEX7 and the long isoform of PEX5 (PEX5L); these complexes are transferred to the peroxisome and are docked on PEX13 and 14, and only matrix proteins are imported into peroxisomes by PEX2, 10, and 12. On the other hand, the receptors are recycled in the cytosol by PEX1, 6 and 26, and recently the involvement of also TRIM37 in the recycle has been reported (see Sect. 5.2.3.2). In addition, the peroxisomal membrane proteins are recognized by PEX19 in the cytosol followed by docking on PEX3 whereas PEX16 is required for peroxisome membrane biogenesis and could play a role in the early stages of peroxisome assembly (Fig. 5.1) (Shimozawa 2011).

Table 5.1 Classification of Peroxisomal diseases (PD) (*Disease-causable genes*)

A. Peroxisome biogenesis disorders (PBD)	
Zellweger spectrum disorders (ZSD) (*PEX 1,2,3,5,6,10,12,13,14,16,19,26*)	
Zellweger syndrome (ZS)	
Neonatal adrenoleukodystrophy (NALD)	
Infantile Refsum disease (IRD)	
Rhizomelic chondrodysplasia punctata (RCDP) type 1 (*PEX 7*)	
	type 5 (*PEX 5-long isoform*)
Broad phenotypes of PEX gene defects (*PEX1,2,3,6,7,10,12,16*)	
Mulibrey nanism (*TRIM37*)	
PEX11β deficiency (*PEX11β*)	
Encephalopathy due to defective mitochondrial and peroxisomal fission 1 (EMPF1) (*DNM1L*)	
Charcot-Marie-Tooth disease Type 4A (*GDAP1*)	
B. Single enzyme deficiencies (SED)	
Impaired β-oxidation of fatty acids	
Adrenoleukodystrophy (ALD) (*ABCD1*)	
Acyl-CoA oxidase 1 (ACOX1) deficiency (*ACOX1*)	
D-bifunctional protein (DBP) deficiency (*HSD17B4*)	
Sterol carrier protein X (SCPX) deficiency (*SCP2*)	
2-Methylacyl-CoA racemase (AMACR) deficiency (*AMACR*)	
Acyl-CoA-binding domain-containing protein 5 (ACBD5) deficiency (*ACBD5*)	
Impaired bile acids synthesis	
Acyl-CoA oxidase 2 (ACOX2) deficiency (*ACOX2*)	
Peroxisomal membrane protein, 70KD (PMP70) deficiency (*ABCD3*)	
Bile acid-CoA: amino acid N-acyltransferase (BAAT) deficiency (*BAAT*)	
Impaired α-oxidation of fatty acids	
Refsum disease (phytanoyl-CoA hydroxylase deficiency) (*PHYH*)	
Impaired ether-phospholipid biosynthesis	
RCDP type 2 (dihydroxyacetone phosphate acyltransferase deficiency) (*GNPAT*)	
RCDP type 3 (alkyl-dihydroxyacetone phosphate synthase deficiency) (*AGPS*)	
RCDP type 4 (fatty acyl-CoA reductase 1 deficiency) (*FAR1*)	
Impaired hydrogen peroxide metabolism	
Acatalasemia, Hypocatalasemia (catalase deficiency) (*CAT*)	
Impaired glyoxylate metabolism	
Hyperoxaluria type 1 (alanine: glyoxylate aminotransferase deficiency) (*AGXT*)	
Glycolate oxidase 1 (GOX1) deficiency (*HAO1*)	

As generalized peroxisomal metabolic abnormalities occur similarly in patients with *PEX* gene defects, patients with different *PEX* gene defects tend to show common clinical phenotypes. Indeed, most patients with mutations in *PEX1, 2, 5, 6, 10, 12, 13, 14*, or *26* involved in both PTS1 and PTS2 protein import tend to manifest common clinical features of Zellweger spectrum disorders (ZSD) including ZS, neonatal adrenoleukodystrophy (NALD) and infantile Refsum diseases (IRD), and their severity depend on the mutational severity and residual function of these PEX

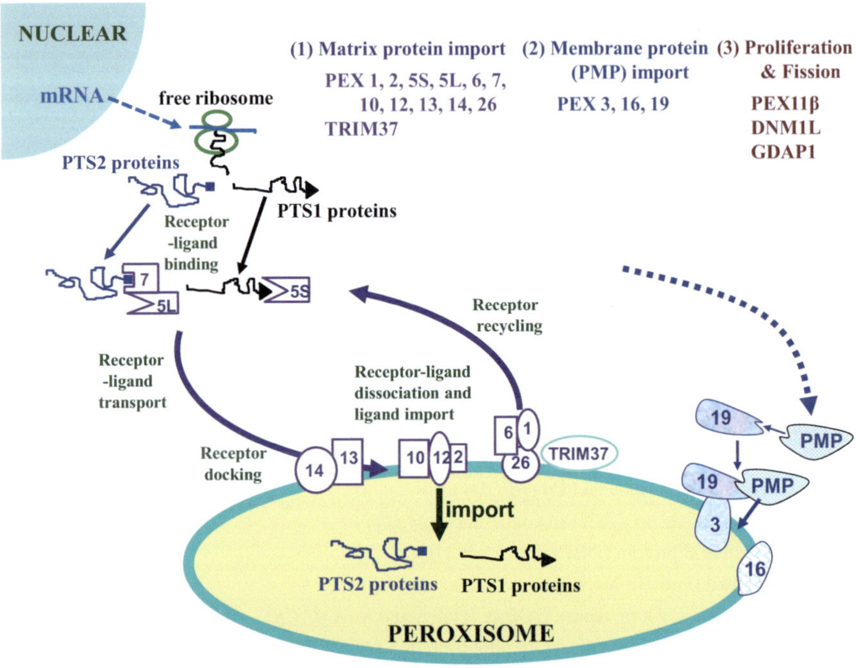

Fig. 5.1 Peroxisomal protein import and peroxisome biogenesis disorders

proteins. Most patients with defected PEX3, 16, and 19 involved in peroxisomal membrane protein import or synthesis tend to manifest phenotypes similar to the most severe ZS. On the other hand, most patients with mutated *PEX7* and *PEX5L*, which are involved only in PTS2 protein import, manifest a clinical type of rhizomelic chondrodysplasia punctate (RCDP).

With the recent expansion of whole-exome sequencing (WES), the abnormalities of these *PEX* genes have been identified in undiagnosed atypical patients, especially those with inherited neurological disorders, leading to the establishment of broad clinical phenotypes derived from *PEX* gene defects.

5.2.1 Zellweger Spectrum Disorders

ZSD are divided into different phenotypes according to clinical severity, ZS is the most severe phenotype, and NALD and IRD are milder variants, however, there is no clear distinction among these three phenotypes.

ZS is characterized by facial dysmorphism such as enlarged anterior fontanelles, high forehead, hypertelorism, broad nasal bridge, epicanthal folds, micrognathia, and malformed ears (Fig. 5.2a). Furthermore, severe hypotonia at birth (Fig. 5.2b, c), absent or weak sucking, hepatomegaly (Fig. 5.2c) with prolonged

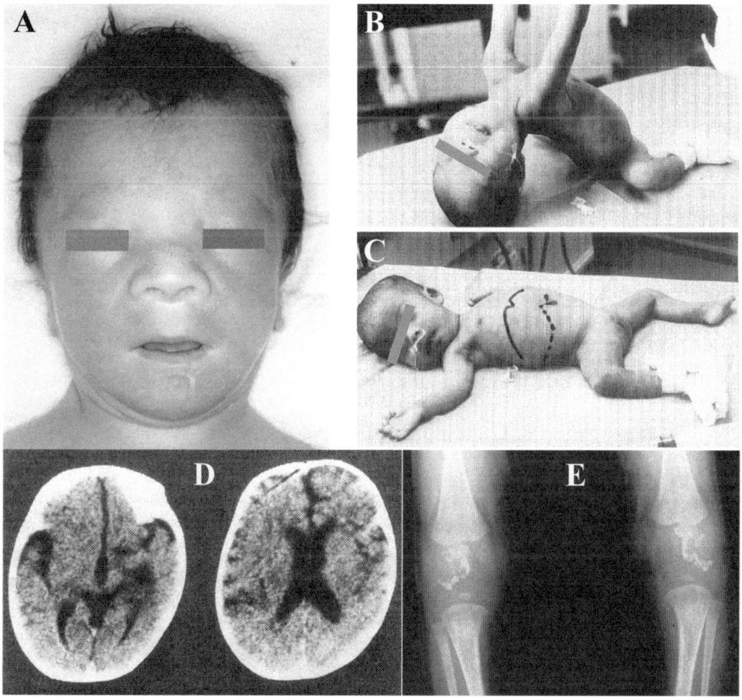

Fig. 5.2 Clinical features of Zellweger syndrome at neonatal age. (**a**) Facial features. Enlarged anterior fontanelles, high forehead, broad nasal bridge, and low-set ears. (**b**) Traction response. Severe hypotonia was seen. (**c**) Supine position. Severe hypotonia and hepatomegaly. (**d**) Brain CT. Enlarged lateral ventricles are seen. (**e**) X-ray photo of knee joints. Abnormal calcific stippling

jaundice and liver dysfunctions, renal cortical microcysts, ventricular enlargement in the brain (Fig. 5.2d), abnormal calcific stippling of multiple joints (Fig. 5.2e), cataracts and pigmentary retinopathy in the ophthalmic finding also appear. ZS patients typically show no developmental progress and die in early infancy.

In contrast, NALD patients have a less severe clinical phenotype than ZS, usually survive until the late infantile period, and exhibit mild facial dysmorphism (Fig. 5.3a) and no chondrodysplasia. Developmental regression and intractable seizures occur during the clinical course of NALD, and demyelination and progressive cortical atrophy in the brain become remarkable with age (Fig. 5.3b) as the survival time of patients is longer than that of ZS patients.

IRD, the mildest phenotype among the various ZSD, is very different from ZS. IRD patients manifest minimum facial dysmorphism, hearing impairment, retinal degeneration, and psychomotor retardation. Many patients with IRD develop up to walking alone (Fig. 5.4a) and acquiring meaningful words, therefore, IRD diagnosis in early childhood is difficult. In brain MRI findings, white matter degeneration appears first (Fig. 5.4b), and cerebral and spinocerebellar atrophies gradually become evident with increasing age (Fig. 5.4c, d); therefore, early diagnosis and

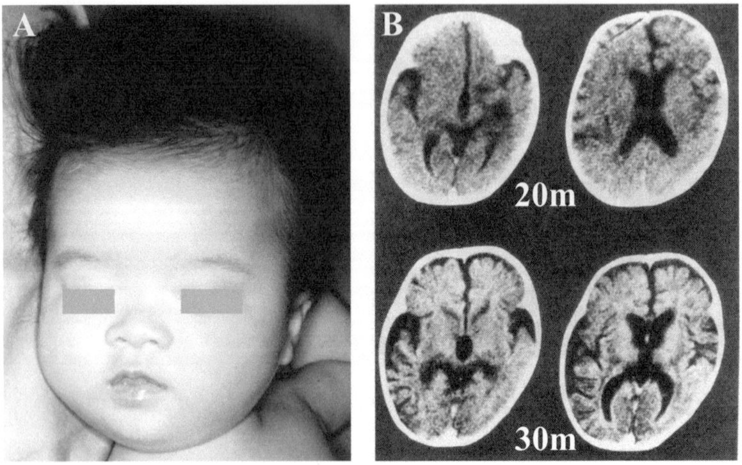

Fig. 5.3 Clinical features of neonatal adrenoleukodystrophy. (**a**) Facial features at neonatal. Mild or subtle facial dysmorphism. (**b**) Brain CT at the age of 20 months (upper row) and 30 months (lower row). Progressive cortical atrophy

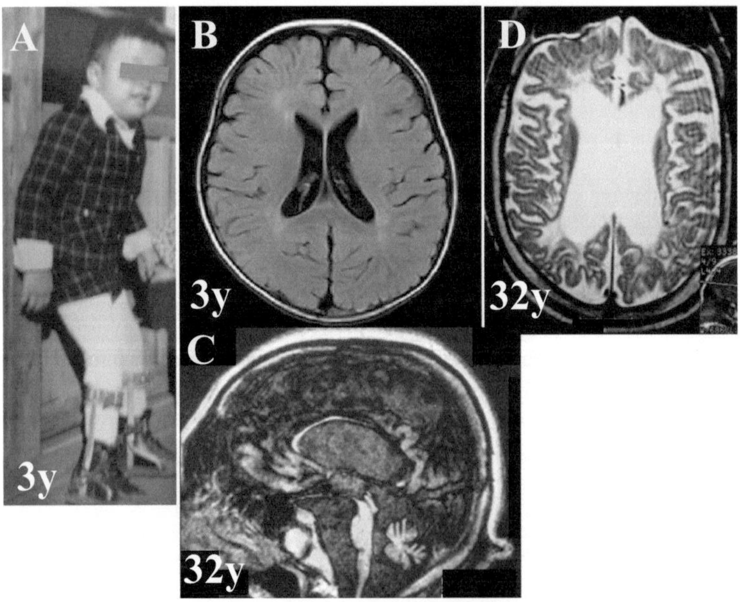

Fig. 5.4 Clinical features of infantile Refsum disease. (**a**) Standing alone at the age of 3 years. (**b**) Fluid-attenuated inversion recovery (FLAIR) image of brain MRI at the age of 3 years. High signal regions slightly in white matter of around the lateral ventricles (Matsunami et al. 2016). (**c**, **d**) Brain T2 weighted MRI at the age of 32 years. Severe atrophy of cerebellum and brain stem (**c**) and cortical atrophy and severe enlarged lateral ventricles (**d**) (Matsui et al. 2013)

subsequent treatment are important for a better prognosis. Many patients survive beyond the second decade of their life.

ZSD patients show generalized peroxisomal metabolic disturbances, such as accumulation of VLCFA, phytanic and pristanic acids, and intermediate metabolites of bile acids of di-/trihydroxycholestanoic acid (D/THCA) in the blood, as well as decreased levels of plasmalogens and docosahexaenoic acid (DHA), however, it should be noted that some biochemical parameters do not appear abnormal, especially in the mild phenotypes (see Table 7.1).

In ZSD therapy, although curative treatment is difficult, early liver transplantation may improve prognosis of patients with a mild ZSD phenotype. We performed a liver transplantation from a heterozygous parent to a 3-year-old patient diagnosed with IRD at 1 year of age, and observed that the increased VLCFA and phytanic acid concentrations in the patient's serum were improved (Matsunami et al. 2016), and no obvious symptoms worsening was noticed during the first 3 years after transplantation. Furthermore, Demaret et al. also speculated that liver transplantation performed before the onset of severe sensorineural defects in mild ZSD enables partial metabolic remission and improves long-term clinical outcome (Demaret et al. 2018). Dietary treatment with decreased phytanic acid intake may be effective for mild ZSD patients with elevated phytanic acid levels (Sá et al. 2016). Further treatment options for ZSD refer to the overview by Braverman et al. (2016). Recently, the Food and Drug Administration (FDA) in the U.S. has approved cholic acid for adjunctive treatment of ZSD patients with symptoms of liver disease, steatorrhea, or complications from decreased absorption of fat-soluble vitamins (https://www.fda.gov/Drugs/InformationOnDrugs/ucm446282.htm).

5.2.2 *Rhizomelic Chondrodysplasia Punctate Type 1 and 5*

Patients with rhizomelic chondrodysplasia punctate (RCDP) type 1 and type 5 caused by PEX7 and PEX5L defects respectively, exhibit limited peroxisomal metabolic abnormalities, including a deficiency in plasmalogen synthesis and

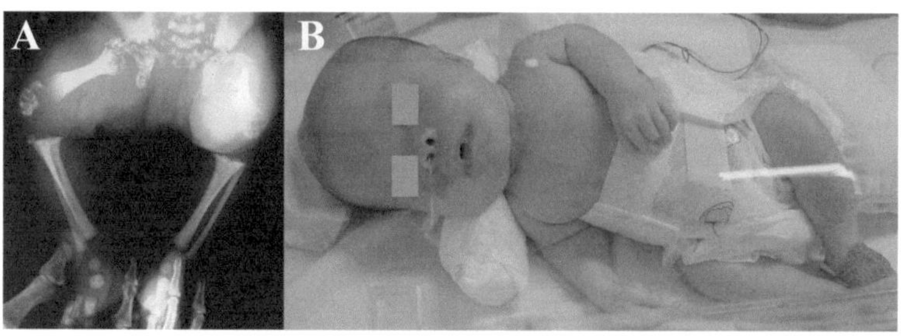

Fig. 5.5 Clinical features of rhizomelic chondrodysplasia punctata type 1 at neonatal. (**a**) X-ray photograph of knee joints. Abnormal calcific stippling. (**b**) Whole picture. The proximal extremity truncated short stature

α-oxidation of phytanic acid. Defective PEX7 and PEX5L are involved only in PTS2 protein import (Fig. 5.1) and result in the common clinical phenotype of RCDP. RCDP type 1 has been classified as a skeletal dysplasia characterized by the presence of calcific stippling in multiple joints (Fig. 5.5a); the patients show a disproportionally short stature with symmetric shortening of the proximal extremities, typical craniofacial dysmorphism resembling that of ZS (Fig. 5.5b), ichthyosis, cataract, failure to thrive, and severe mental retardation. Patients with RCDP type 1 show biochemical alterations, such as accumulation of phytanic acid and decrease in plasmalogens, whereas their pristanic acid and VLCFA levels are normal (see Table 7.1). Patients with a defect in *PEX7* display various clinical phenotypes, including severe typical RCDP and milder bone lesions with moderately decreased plasmalogen levels. Many patients die in the first 2 years of life, but some survive beyond the second decade of life. Furthermore, there are patients with a *PEX7* defect manifesting similar phenotypes to those with Refsum disease, who do not display bone lesions and their plasmalogen levels remain normal (see Sect. 5.2.3.1).

RCDP type 5 caused by a mutation in *PEX5L* was recently identified by WES and biochemical verification (Barøy et al. 2015). *PEX5* encodes two distinct isoforms, PEX5L and PEX5S, and previous patients with a *PEX5* defect manifested ZSD only, mainly owing to deficient import of PTS1 and PTS2 proteins. However, patients with *PEX5* mutations located in *PEX5L* specific exon 9, loose only PEX5L, a co-receptor of PTS2-proteins with PEX7, resulted in intact PEX5S, a receptor of PTS1-proteins (Fig. 5.1). The clinical and biochemical features of RCDP type 5 patients have similar manifestations to mild *PEX7* defect patients who manifest less pronounced skeletal abnormalities, milder growth delay and intellectual disability, and fewer biochemical disturbances.

5.2.3 *Broad Phenotypes of Known* PEX *Gene Defects and Newly Identified Disease-Causing Genes*

5.2.3.1 Broad Phenotypes of *PEX* Gene Defects

The phenotypic spectrum of *PEX7* mutations appears broad, including not only the severe phenotype of RCDP, but also relatively mild phenotypes. Some patients with the mild phenotypes display clinical symptoms similar to those of patients with Refsum disease, which are characterized by increased phytanic acid caused by a single enzyme deficiency [Phytanoyl-CoA hydroxylase (PHYH) deficiency (see Sect. 5.3.3)]. The biochemical abnormalities in Refsum patients due to *PEX7* defect show only increase in phytanic acid whereas the plasmalogen levels remain normal (Braverman et al. 2002; van den Brink et al. 2003).

There have been also reported extensive phenotypic heterogeneity among patients with mutation of several PEX genes which cause the ZSD. Indeed, there was a report in 2002 on patients with compound heterozygous mutations of PEX6 manifesting as Usher syndrome characterized by sensory hearing loss and retinitis pigmentosa

(Raas-Rothschild et al. 2002). Furthermore, since 2010, gene mutations in *PEX10* (Régal et al. 2010), *PEX16* (Ebberink et al. 2010), *PEX2* (Sevin et al. 2011), and *PEX6* (Tran et al. 2014) have been detected in undiagnosed patients with cerebellar ataxia and progressive leukodystrophy by combined biochemical analysis and Sanger sequencing, following detection of mild peroxisomal metabolite abnormalities in the patients. Among these cases, there was a 51-year-old man who presented with childhood onset and slowly progressive disease, caused by a mutated *PEX2*, with symptoms of ataxia, areflexia, nystagmus and strabismus (Mignarri et al. 2012).

After NGS became widely available, further disease-causing *PEX* mutations have been reported, mainly in patients with neurodegenerative diseases who had mutations in genes, such as *PEX1* (Ventura et al. 2016), *PEX3* (Bjørgo et al. 2017), *PEX10* (Renaud et al. 2016; Blomqvist et al. 2017; Yamashita et al. 2017), *PEX12* (Schabhüttl et al. 2014), and *PEX16* (Ohba et al. 2013; Bacino et al. 2015; Kumar et al. 2016), and in patients with Heimler syndrome who had mutations in *PEX1* and *PEX6* (Ratbi et al. 2015; Smith et al. 2016) (in detail, see Sect. 7.3).

Interestingly, it was reported that the allelic expression imbalance (AEI) induces mutant *PEX6* allele to cause ZSD, and the AEI of *PEX6* was correlated with heterozygosity of a frequent variant in the 3′ untranslated region (UTR) of the mutant allele (Falkenberg et al. 2017). The patients that carry this mutation presented multiple symptoms similar to the symptoms of patients with a mild phenotype of ZSD, characterized by neurological abnormalities, such as profound hypotonia, gait abnormalities, developmental delay, neuropathy, visual impairment, sensorineural hearing loss, and white matter abnormalities detected through brain MRI. The biochemical findings showed elevated VLCFA levels in the serum of patients and impaired peroxisomal biogenesis in their fibroblasts.

5.2.3.2 Mulibrey Nanism (TRIM37 Deficiency)

TRIM37, which was identified as a disease-causing gene of muscle–liver–brain–eye (Mulibrey) nanism, encodes a peroxisomal RING-B-box-coiled-coil protein; therefore, Mulibrey nanism is characterized by severe growth retardation of prenatal-onset, characteristic dysmorphic features, pericardial constriction, and hepatomegaly and was classified as a PD. TRIM37 was localized in the peroxisomes; however, peroxisomes of fibroblasts from the patients appeared normal by immunocytochemical methods, using both the peroxisomal matrix and membrane protein markers, suggesting normal peroxisomal biogenesis (Kallijärvi et al. 2002). Subsequent studies have revealed that TRIM37-mediated ubiquitination stabilizes PEX5 and promotes peroxisomal matrix protein import, which means that inactivation of TRIM37 leads to reduced PEX5 accumulation by inducing proteasomal degradation and compromising PEX5 functions in cargo binding and PTS protein import (Fig. 5.1); thereby Mulibrey nanism has been classified as a new PBD (Wang et al. 2017).

5.2.4 *Dysfunction of Peroxisomal Proliferation and Fission*

5.2.4.1 PEX11β Deficiency

A homozygous nonsense mutation of *PEX11β* involved in peroxisomal growth and division, was found in a 26-year-old male patient with congenital cataract, mild intellectual disability, progressive hearing loss, sensory nerve involvement, gastrointestinal problems, recurrent migraine-like episodes, and Chiari I malformation on MRI. Biochemical parameters of peroxisomal metabolites were normal, including VLCFA, phytanic and pristanic acids, bile acid intermediates and plasmalogens, whereas patient's fibroblasts did not contain PEX11β protein and exhibited a low number of enlarged and elongated peroxisomes (Ebberink et al. 2012) (see Table 7.1).

5.2.4.2 Encephalopathy Due to Defective Mitochondrial and Peroxisomal Fission 1 (EMPF1) (DNM1L Deficiency)

Waterham et al. reported a case of a newborn girl with microcephaly, abnormal brain development, optic atrophy and hypoplasia caused by heterozygous dominant negative mutations in dynamin 1-like (*DNM1L*) gene involved in the fission of both mitochondria and peroxisomes. The patient's biochemical profile showed persistent lactic acidemia and mildly elevated VLCFA. Peroxisomes in patient's fibroblasts were less in number, varied in size and were regularly arranged in rows (Waterham et al. 2007) (see Table 7.1). These patients developed epileptic encephalopathy with intractable seizures, followed by neurologic decline and died during childhood. Additionally, in several families with autosomal dominant optic-atrophy 5, heterozygous mutations in *DNM1L* have been identified (Gerber et al. 2017). Furthermore, Yoon et al. reported that in patients with autosomal recessive encephalopathy due to defective mitochondrial and peroxisomal fission resulting in early infantile death, complex heterozygous truncating mutations in *DNM1L* were identified (Yoon et al. 2016).

5.2.4.3 Charcot-Marie-Tooth Disease Type 4A (GDAP1 Deficiency)

Distinct mutations in the ganglioside-induced differentiation-associated protein 1 (*GDAP1*) gene, expressing a tail-anchored mitochondrial protein that induces mitochondrial fragmentation, were found in patients with Charcot-Marie-Tooth disease type 4A (Cuesta et al. 2002). The patients showed demyelinating peripheral neuropathy characterized by distal motor and sensory impairment resulting in gait difficulties and foot deformities. It was reported that GDAP1 was imported to peroxisomes by the import receptor Pex19, and regulated peroxisomal fission (Huber et al. 2013).

5.3 Single Enzyme Deficiencies (SED)

Single enzyme deficiencies (SED) exhibit clinical phenotypes caused by individual metabolic disturbances. Here, we describe the metabolic dysfunction in each PD, including β-oxidation of fatty acids, bile acids synthesis, α-oxidation of fatty acids, plasmalogen biosynthesis, hydrogen peroxide metabolism and glyoxylate metabolism (Table 5.1). Broad phenotypes and newly identified PD have also been reported in SED through WES.

5.3.1 Impaired β-Oxidation of Fatty Acids

5.3.1.1 Adrenoleukodystrophy (ALD)

ABCD1 was identified as the gene responsible for adrenoleukodystrophy (ALD) and encoding a peroxisomal membrane protein that may transport VLCFA-CoA through the peroxisomal membrane. Defect in this protein results in accumulation of VLCFA, however, various phenotypes of ALD are not correlated to genotypes and VLCFA values. The mechanism underlying the onset of cerebral ALD remains unknown. ALD is described in more detail later (see Sect. 5.4).

5.3.1.2 Acyl-CoA Oxidase 1 (ACOX1) Deficiency

ACOX1 catalyzes the initial step in peroxisomal fatty acid β-oxidation. Patients with ACOX1 deficiency have decreased muscle tone since the neonatal period, convulsions since infancy, hearing and vision disturbances, but absence of prominent facial dysmorphism (Fig. 5.6a). These symptoms gradually regress, whereas patients survive until childhood. The clinical findings resemble those of the mild ZSD phenotypes. In brain MRI, abnormal findings in cerebellar and cerebral white matter progress with age (Fig. 5.6b, c). As this enzyme oxidizes only the saturated fatty acids, VLCFA accumulation is the only biochemical abnormality that occurs. Peroxisomes appear larger than usual in the immunocytochemical analysis (Funato et al. 2006). Furthermore, adult patients with ACOX1 deficiency characterized by cerebellum and brain stem atrophy have been reported (Ferdinandusse et al. 2010). In therapy, although there is no curative treatment, a sibling comparison study on the effects of hematopoietic stem cell transplantation (HSCT) has been reported (Wang et al. 2014).

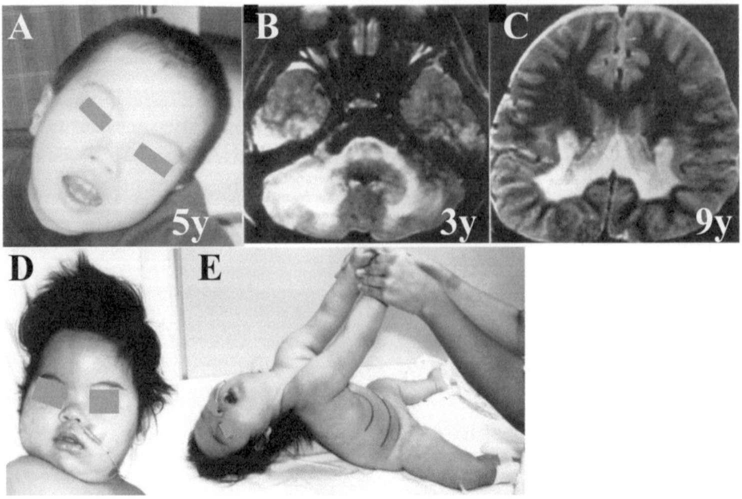

Fig. 5.6 Clinical features of acyl-CoA oxidase 1 (ACOX1) and D-bifunctional protein (DBP) deficiencies. (**a**) Facial feature of ACOX1 deficiency at the age of 5 years. (**b**) Brain T2 weighted MRI of ACOX1 deficiency at the age of 3 years. High signal regions in cerebellar white matter, peduncle and pons. (**c**) Brain T2 weighted MRI of ACOX1 deficiency at the age of 9 years. High signal regions in white matter around the occipital horn of lateral ventricles, subcortical white matter, and splenium of corpus callosum. (**d**) Facial features of patient with DBP deficiency. High forehead and broad nasal bridge. (**e**) Traction response of patient with DBP deficiency at neonatal. Severe hypotonia was seen

5.3.1.3 D-Bifunctional Protein (DBP) Deficiency

HSD17B4 encodes DBP which catalyzes the second and third steps of peroxisomal fatty acid β-oxidation. In addition to straight-chain fatty acids, bile acids and branched fatty acids are oxidized by this enzyme, therefore, the biochemical findings show accumulation of VLCFA, D/THCA, phytanic, and pristanic acids. Patients with DBP deficiency are in a more severe condition than those with ACOX1 deficiency and are characterized by facial dysmorphism (Fig. 5.6d), hypotonia from the neonatal period (Fig. 5.6e), poor feeding, hepatomegaly, convulsions since the neonatal period, and die within 2 years of life. These clinical findings seem to resemble those of the severe phenotypes of ZS patients (Fig. 5.2a, b). Peroxisomes in DBP deficient patients also present a larger shape than usual, similar to that of peroxisomes in ACOX1 deficient patients (Funato et al. 2006). In DBP deficiency, as well, broad phenotypes, including the Perrault syndrome, which is characterized by ovarian malformation, hearing loss, and cerebellar ataxia (Pierce et al. 2010), and sensorineural hearing loss, progressive cerebellar ataxia and subclinical retinitis pigmentosa (McMillan et al. 2012) have been reported.

5.3.1.4 Sterol Carrier Protein X (SCPx) Deficiency

The SCPx enzyme encoded by *SCP2* exerts thiolase activity in the last step of peroxisomal β-oxidation and oxidizes branched-chain fatty acids. Patients with SCPx deficiency have elevated levels of pristanic and phytanic acids, and D/THCA whereas normal VLCFA levels, and clinical manifestations of leukoencephalopathy with dystonia and motor neuropathy (Ferdinandusse et al. 2006).

5.3.1.5 2-Methylacyl-CoA Racemase (AMACR) Deficiency

AMACR is a peroxisomal enzyme that catalyzes the conversion of 2R-pristanoyl-CoA and 25R-D/THCA to their (S)-stereoisomers. Consequently, the enzymatic defect causes accumulation of plasma pristanic acid and D/THCA in patients with various clinical symptoms, such as adult-onset sensorimotor neuropathy (Ferdinandusse et al. 2000). Furthermore, a homozygous mutation in the *AMACR* was identified in an infant with defect in bile acid synthesis and increased levels of THCA (Setchell et al. 2003).

5.3.1.6 Acyl-CoA-Binding Domain-Containing Protein 5 (ACBD5) Deficiency

ACBD5 is a peroxisomal membrane protein with a cytosolic acyl-CoA binding domain. A variant of *ACBD5* in three siblings characterized by cone-rod dystrophy, developmental delay, spastic paraparesis, and white matter disease was identified by autozygome analysis followed by exome sequencing (Abu-Safieh et al. 2013). Next, Ferdinandusse et al. identified another patient with a homozygous deleterious indel mutation in *ACBD5* presenting progressive leukodystrophy, syndromic cleft palate, ataxia, retinal dystrophy, and accumulation of VLCFA due to impaired peroxisomal β-oxidation (Ferdinandusse et al. 2017).

5.3.2 *Impaired Bile Acids Synthesis*

5.3.2.1 Acyl-CoA Oxidase 2 (ACOX2) Deficiency

ACOX2 is a peroxisomal branched-chain acyl-CoA oxidase participating in bile acid synthesis. A patient with *ACOX2* deficiency identified by WES presented intermittently elevated transaminase levels, liver fibrosis, mild ataxia, and cognitive impairment (Vilarinho et al. 2016). The patient showed increased D/THCA levels in plasma and urine, whereas no increase in branched-chain fatty acids, phytanic acid, and pristanic acid was noticed.

5.3.2.2 Peroxisomal Membrane Protein, 70KD (PMP70) (ABCD3) Deficiency

ABCD3 encodes a PMP70 involved in the transport of branched-chain fatty acids and C27 bile acids into the peroxisomes. A patient with a homozygous 1758-bp deletion in *ABCD3* had accumulation of D/THCA, and increased C26/C22 and C24/C22 ratios owing to low levels of C22:0 whereas normal phytanic and pristanic acids in the plasma levels (see Table 7.1). On the contrary, measurement of peroxisomal beta-oxidation activities in fibroblasts from the patient revealed decreased beta-oxidation of pristanic acid, whereas that of C26:0 was normal. Peroxisomes in PMP70 deficient patients present a larger shape than usual and fewer in number. The patient manifested hepatosplenomegaly with severe liver dysfunction, but normal developmental milestones (Ferdinandusse et al. 2015).

5.3.2.3 Bile Acid-CoA: Amino Acid N-Acyltransferase (BAAT) Deficiency

BAAT transfers bile acid moiety from the acyl-CoA thioester to either glycine or taurine. Hence, bile acids conjugated glycine or taurine are decreased in the body fluids of patients deficient in BAAT. Patients' phenotype shows familial hypercholanemia characterized by elevated bile acid serum concentrations, itching, and fat malabsorption (Carlton et al. 2003).

5.3.3 Impaired α-Oxidation of Fatty Acids

5.3.3.1 Phytanoyl-CoA Hydroxylase (PHYH) Deficiency (Refsum Disease)

Refsum disease is characterized by an increase in phytanic acid due to deficiency of PHYH, localized in the peroxisomes. Phytanic acid is converted to pristanic acid by α-oxidation and then is subjected to β oxidation; therefore, pristanic acid levels in the patients are not increased. Many patients with Refsum disease have been reported in UK and Northern Europe and develop symptoms, such as retinitis pigmentosa, polyneuropathy (atrophy of lower limb muscles, muscle weakness, sensory paralysis), and cerebellar ataxia at the age of 1–50 years. Treatment is based on a diet that severely restricts dairy products rich in phytanic acid, and also meat and fats derived from cows, sheep, goats, etc.

5.3.4 *Impaired Plasmalogen Biosynthesis*

(a) Dihydroxyacetone phosphate acyltransferase (GNPAT) deficiency (RCDP type 2)
(b) Alkyl-dihydroxyacetone phosphate synthase (AGPS) deficiency (RCDP type 3)

The first and second steps of plasmalogen biosynthesis are performed in peroxisomes by GNPAT (PTS1 protein) and AGPS (PTS2 protein), respectively. Clinical findings in both deficiencies revealed an RCDP phenotype, including rhizomelic shortening of upper extremities, typical facial appearance, cataract, dwarfism, and severe mental retardation. Biochemically, both types of patients only show decreased levels of plasmalogens, whereas plasma phytanic acid levels are normal.

(c) Fatty acyl-CoA reductase 1 (FAR1) deficiency (RCDP type 4)

Mutations of *FAR1* involved in plasmalogen biosynthesis in peroxisomes were identified by WES in two families affected by severe intellectual disability, early-onset epilepsy, microcephaly, congenital cataracts, growth retardation and spasticity (Buchert et al. 2014). This disease was later named as RCDP type 4, although showing no typical RCDP phenotype.

5.3.5 *Impaired Hydrogen Peroxide Metabolism*

5.3.5.1 Catalase Deficiency (Acatalasemia, Hypocatalasemia)

Acatalasemia (Takahara disease) a metabolic disorder characterized by a total or near total loss of catalase activity in erythrocytes was first discovered by Takahara in patients with progressive oral gangrene (Takahara and Miyamoto 1948). Patients with hypocatalasemia have heterozygous mutations of the *catalase* gene and manifest half-normal levels of catalase activity and no obvious clinical symptoms, however, studies on Hungarian patients with hypocatalasemia showed increased occurrence of type 2 diabetes (see Sect. 5.5.2).

5.3.6 *Impaired Glyoxylate Metabolism*

5.3.6.1 Hyperoxaluria Type 1 (Alanine: Glyoxylate Aminotransferase Deficiency)

Primary hyperoxaluria type 1 (PH1) is a glyoxylate metabolism disorder caused by a deficiency of alanine: glyoxylate aminotransferase (AGT) present in liver peroxisomes. Glyoxylic acid is a precursor of oxalic acid, and due to the deficiency in AGT, an enzyme converting glyoxylate to glycine, a large amount of oxalic acid is produced, and insoluble calcium oxalate is deposited on the whole body organs, including the kidney. After successive renal colic and hematuria, typical symptoms of urinary calculus, the disease progressed to nephrocalcinosis and renal failure, and most of the cases resulted in end-stage renal failure. In mild cases, administration of vitamin B6, a coenzyme of AGT, can be effective. Early liver transplantation for replacement of the AGT enzyme is considered an effective curative treatment. Kidney transplantation may be necessary in cases of renal failure, when recovery is not expected.

5.3.6.2 Glycolate Oxidase 1 (GOX1) Deficiency

The hydroxy-acid oxidase 1 (*HAO1*) gene encodes glycolate oxidase 1 (GOX1), which catalyzes the oxidation of glycolate to glyoxylate in the peroxisomes of hepatocytes. Frishberg et al. reported a patient with a homozygous splicing site mutation in *HAO1* who manifested a persistent and markedly increased urinary glycolate excretion; normal excretion of oxalate, citrate and glycerate; and no obvious renal symptoms. This observation suggested that substrate reduction might be targeted for the development of novel approaches for the treatment of PH1 (Frishberg et al. 2014).

5.4 Adrenoleukodystrophy (ALD)

Adrenoleukodystrophy (ALD) is the most common PD characterized by demyelination of the cerebral white matter and adrenal dysfunction. ALD is an X-linked inherited disease attributed to mutations in the *ABCD1* gene, and its product, ALDP/ABCD1, a peroxisomal membrane protein. ALDP possesses an ATP-binding cassette region at the C-terminus involved in the import of saturated VLCFA into the peroxisomes, leading to β-oxidation of the saturated VLCFA. Therefore, a dysfunction of ALDP/ABCD1 results in the accumulation of saturated VLCFA in the tissues and plasma. Various clinical phenotypes exist in ALD, such as the childhood cerebral ALD (CCALD), adolescent cerebral ALD (AdolCALD), adult cerebral ALD (ACALD), adrenomyeloneuropathy (AMN), olivo-ponto-cerebellar type of ALD (OPCALD) and Addison only with no genotype-phenotype correlation. Even female carriers of the mutated gene present occasionally mild spinal symptoms with age. The prognosis of cerebral ALD is generally very poor and many patients risk becoming bedridden within a few years if they remain without effective treatment. Hematopoietic stem cell transplantation (HSCT) is currently the only curative approach, which can prevent the progression of brain deterioration; however, HSCT is only effective for patients in the early stages of cerebral ALD (Peters et al. 2004). Therefore, not only is early diagnosis just after disease onset critical, but presymptomatic diagnosis is also essential in order to prevent the progression of cerebral ALD.

5.4.1 *Epidemiology and Phenotypes*

In the United States, one ALD patient in 21,000 newborn boys and one ALD carrier in 14,000 newborn girls have been reported (Bezman et al. 2001). Furthermore, in New York, over 700,000 newborns were screened; 45 babies (22 boys and 23 girls) were identified as having ALD, suggesting that the birth-incidence of ALD could be 1:15,000 (https://adrenoleukodystrophy.info/clinical-diagnosis/newborn-screening). At least 4.1% of individuals with ALD have de novo mutations (Wang et al. 2011); therefore, the mothers of male probands may not be carriers. There are various phenotypes, shown below, which do not correlate with genotypes; however, the incidence of each phenotype varies from country to country, probably due to different genetic backgrounds.

5.4.1.1 Childhood Cerebral ALD (CCALD)

CCALD is the most common phenotype and is characterized by the progressive deterioration of the intellectual, psychic, visual, and gait characteristics at the age of onset between 3 and 10 years. The prognosis of CCALD is generally very poor and many patients become bedridden within a few years.

5.4.1.2 Adolescent Cerebral ALD (AdolCALD)

AdolCALD has symptoms similar to CCALD at the age of onset between 11 and 21 years but tends to progress more slowly.

5.4.1.3 Adrenomyeloneuropathy (AMN)

AMN is a noninflammatory distal axonopathy which develops after puberty with gait disturbance, rectal bladder dysfunction, and impotence. AMN progresses slowly, however, it may develop to cerebral ALD.

5.4.1.4 Adult Cerebral ALD (ACALD)

ACALD develops after adulthood and presents personality changes, intellectual deterioration and psychiatric symptoms, therefore, sometimes is misdiagnosed as psychosis or dementia. Clinical progression varies and can lead to bedridden patients after a few years.

5.4.1.5 Olivo-Ponto-Cerebellar Type of ALD (OPCALD)

Cerebellar ataxia is the main symptom of patients with OPCALD, with most of them being diagnosed in Japan. Some patients with OPCALD may develop cerebral ALD.

5.4.1.6 Addison Only

Addison disease develops between the age of 2 years and adulthood with symptoms of adrenal insufficiency, such as unexplained vomiting, weight loss, and pigmentation. The youngest patient showing abnormal adrenal function was reported to be 7 months old. There was no significant difference in the VLCFA values of the plasma among patients with adrenal insufficiency (Huffnagel et al. 2019). It should be noted that most patients with an "Addison only" phenotype may progress to AMN and/or cerebral ALD and most male patients with ALD have adrenal insufficiency regardless of their phenotype.

5.4.1.7 Symptomatic Female

Some female carriers have symptoms like those of AMN patients including gait disorder, sensory disturbance, and fecal incontinence. Symptoms rarely appear before the age of 20, and the incidence increases with age. In details, 18% of women under 40 and 88% of women over 60 years of age present light neurological signs (Engelen et al. 2014). In female carriers, the adrenal dysfunction is rare and cerebral ALD is even rarer.

5.4.2 *Diagnostic Methods*

(a) Very long chain fatty acids (VLCFA)
Male patients show increased saturated VLCFA plasma levels such as C26:0, C25:0, C24:0. There is no correlation between the rate of accumulation of VLCFA and clinical severity. Fifteen percent to 20% of female carriers have normal VLCFA levels, although increased VLCFA levels are observed in the majority (Kemp et al. 2001). Therefore, *ABCD1* mutation analysis should be performed in all suspected female carriers even with normal VLCFA levels.
(b) Brain MRI
In cerebral ALD, CT and T2 weighted MRI imaging show a low density and a high signal region, respectively, coinciding with the site of demyelination in the cerebral white matter (Fig. 5.7a). The distribution of demyelination is common in the white matter of the occipital lobe, around the lateral ventricle in the white matter of the parietal lobe, and in the splenium of the corpus callosum; however,

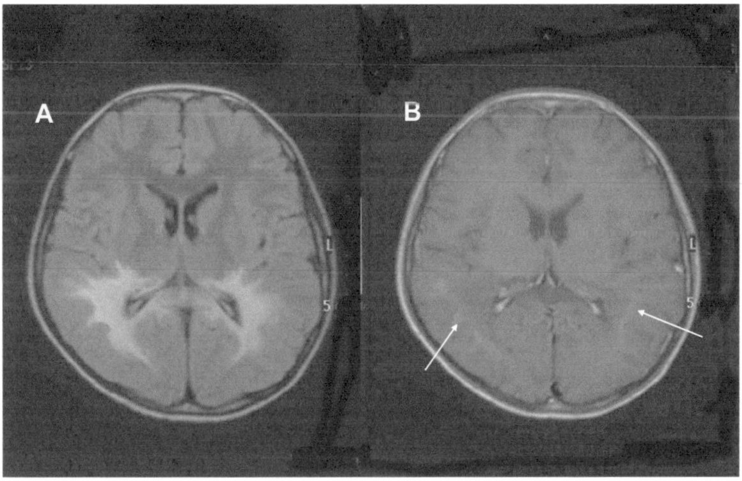

Fig. 5.7 Brain MRI of adolescent cerebral ALD at the age of 13 years. (**a**) FLAIR image of brain MRI. High signal regions in white matter of occipital lobe with subcortical region. (**b**) Gadolinium enhancement. Contrast effect is recognized at the sites with active phase of neuroinflammation

in some cases, it initiates from the white matter of the frontal lobe. At sites with active neuroinflammation, the contrast effect is recognized by gadolinium (Gd) enhancement (Fig. 5.7b). In AMN and OPCALD, abnormal T2 weighted MRI findings are mainly observed in the pyramidal tract, cerebellum, and spinocerebellar tract.

(c) Adrenal function test

The lifetime risk of adrenal insufficiency in male patients with ALD is nearly 80% (Huffnagel et al. 2019). Even without adrenal insufficiency symptoms, elevated plasma ACTH levels or low response to rapid ACTH loading test are observed. Adrenal insufficiency affects the prognosis of ALD patients, and a recent study on the natural history of adrenal insufficiency in ALD recommends adrenal testing every 4–6 months for patients aged ≤10 years, annual testing for those aged 11–40 years, and testing on demand for those aged >40 years (Huffnagel et al. 2019).

(d) *ABCD1* mutation analysis

The mutations in the *ABCD1* are diverse, as more than 750 different mutations have been identified (https://adrenoleukodystrophy.info/mutations-and-variants-in-abcd1). There is no genotype-phenotype correlation, even in female carriers. As 15–20% of female carriers display normal VLCFA levels (Kemp et al. 2001), *ABCD1* mutation analysis is recommended for female carrier detection.

(e) Pathological findings

In autopsy of patients with cerebral ALD, demyelination of the white matter, gliosis, and infiltration of the inflammatory cells around the blood vessel are recognized in the cerebral lesion.

(f) Neurophysiological and psychological findings (see Chap. 12)

5.4.3 *Differential Diagnosis*

5.4.3.1 Diseases to Be Differentiated in ALD of Boys

Attention deficit hyperactivity disorder (ADHD), learning disability, psychosomatic disorder, strabismus, blurred vision, hearing loss, Addison's disease, brain tumor, subacute sclerosing panencephalitis (SSPE), and other leukodystrophies.

5.4.3.2 Diseases to Be Differentiated in ALD of Adults

Familial spastic paraplegia, multiple sclerosis, psychosis, dementia, spinocerebellar degeneration, Addison's disease, brain tumor, malignant lymphoma, and other leukodystrophies.

5.4.4 *Treatment*

(a) Steroid replacement therapy
Adrenal insufficiency significantly affects prognosis; therefore, it is necessary to evaluate the adrenal function of all male patients, including asymptomatic and post HSCT patients. Corticosteroid replacement therapy should be initiated when necessary, however it should not affect the cerebral and spinal cord lesions.

(b) Lorenzo's oil
Lorenzo's oil, a blend of 4:1 mixture of glycerol trioleate and glycerol trierucate reduces VLCFA in plasma, whereas it does not affect the natural course of the disease after the onset of cerebral symptoms. It has also been tried in the treatment of presymptomatic or AMN patients; however, its efficiency was not defined.

(c) HSCT
HSCT is the only curative approach, that when performed early can prevent the progression of brain involvement in CCALD and AdolCALD. Raymond et al. recently reported that prognosis of early HSCT was clearly improved when survival was assessed without the major functional disabilities considered as a relevant treatment goal, rather than solely assessing overall survival as an indicator of treatment success (Raymond et al. 2018). Because of that, it is essential to suspect ALD as soon as possible and to obtain a prompt diagnosis. Kato et al. reported that allogeneic HSCT with reduced intensity conditioning for ALD patients was safely performed without major transplant-related complications even in symptomatic patients (Kato et al. 2018). Furthermore, the effectiveness of transplantation has been reported even in ACALD, through a retrospective analysis of the feasibility, toxicity, and long-term neurological outcome of 14 adult males treated with allogeneic HSCT in four European centers (Kühl et al. 2017).

There was a report that 3 out of 5 cases developed myelopathy in a long-term follow-up study of patients transplanted at an early stage. It is suggested that although the inhibitory effect on the progression of inflammation in the cerebral type is recognized in HSCT, the effect of inhibiting the onset of AMN may not be recognized (van Geel et al. 2015).

(d) AMN and symptomatic female
For myelopathy in AMN and symptomatic females, there has been no effective therapy available yet, therefore, physical therapy and antispasmodic drugs are the main treatments. Studies using *Abcd1* knockout mice have revealed that oxidative stress may be involved in axonal degeneration of AMN, hence, the examinations on antioxidant drugs are ongoing (López-Erauskin et al. 2011).

(e) HSC gene therapy
ALD patients at the early stages of the cerebral-type of disease were administered Lenti-D gene therapy, where autologous $CD34^+$ cells transduced with Lenti-D lentiviral vector were injected in patients as a phase II-III safety and efficacy study (Eichler et al. 2017). Based on results, the FDA in the US has granted the Breakthrough Therapy designation to Lenti-D™ for treating patients with the cerebral type of ALD on May 23, 2018.

(f) Further therapeutic strategies for ALD (see Chap. 8)

5.4.5 *Presymptomatic Diagnosis and Newborn Screening*

5.4.5.1 Presymptomatic Diagnosis

Patients after cerebral ALD onset have limitations in HSCT effect; hence, in order to improve prognosis, it is important to identify patients before the onset of disease, by familial analysis of the probands. This is also important for improving the prognosis of adrenal insufficiency. Furthermore, as the onset of symptoms cannot be predicted, it is necessary to present a system of long-term follow-up (Engelen et al. 2012) (in detail, see Sect. 7.4.2 and Fig. 7.3).

5.4.5.2 Newborn Screening (NBS)

In New York, neonatal screening for ALD was initiated on December 30, 2013. During the first 3 years, over 700,000 newborns were screened in New York and 45 babies with ALD, including 22 boys and 23 girls, were identified (https://adrenoleukodystrophy.info/clinical-diagnosis/newborn-screening). Later, testing was conducted in many states in the United States. In an effort to arrive to a steady effect on overcoming ALD, it is essential to establish a precise diagnostic system even for female ALD patients and other PD, a genetic counseling system and a long-term follow up system for patients found by NBS.

5.4.6 Pathophysiology

5.4.6.1 Elucidated Facts and Unresolved Issues

Dysfunction of ALDP/ABCD1 due to mutated *ABCD1* in ALD patients, which is a peroxisomal membrane transporter causes impaired β-oxidation of saturated VLCFA resulting in the accumulation of VLCFA in the tissues and plasma. Therefore, diagnosis of ALD can be confirmed by elevated saturated VLCFA in plasma and detection of *ABCD1* mutations, however there is no correlation between genotypes and various phenotypes. Cerebral ALD is an inflammatory demyelinating disease, whereas AMN is a non-inflammatory distal axonopathy, and some AMN patients can develop cerebral ALD. We cannot predict phenotypes and prognosis in presymptomatic patients, therefore, it is now difficult to perform HSCT before cerebral ALD onset. *Abcd1* knockout mice exhibited only minor neurologic symptoms without inflammatory demyelination (Pujol et al. 2002), whereas there was a recent report that chimpanzee naturally developed cerebral ALD (Curiel et al. 2017). Furthermore, the function of ALDP/ABCD1 as a transporter in the peroxisomal membrane and the pathophysiology caused by accumulated VLCFA are not completely understood.

5.4.6.2 Task to Be Solved

The most important task should be the development of a phenotype prediction diagnosis method for medical intervention before the onset of the disease. For that purpose, the search for modifier factors causing cerebral ALD onset is essential and can lead to the development of a cerebral-type onset in a mice model. This model can help to elucidate the mechanism of onset of inflammatory demyelination as well as the therapeutic mechanism of HSCT, leading to new treatments and optimal transplantation methods for patients with cerebral ALD. It is also important to clarify the underlying biochemical and molecular pathology of ALDP/ABCD1, including synthesis and β-oxidation of VLCFA, for the discovery of new approaches for successful preventive treatment.

5.4.7 Current and Future Prospective

Currently, early diagnosis is the most important factor in conquering ALD; hence, it is important to spread information regarding the first symptoms of ALD widely and provide a prompt diagnostic system to detect VLCFA levels and *ABCD1* mutations. We have developed a prompt ALD diagnostic system that provides results on VLCFA values and *ABCD1* mutations within a few days (see Sect. 7.4.1). In addition, presymptomatic diagnosis and neonatal screening combined with genetic counseling and a long-term follow-up system may have to be adopted as a national strategy. Furthermore, in an effort to improve the prognosis of patients diagnosed before disease onset, it is important to develop a phenotype prediction method, as well as further therapeutic approaches.

5.5 Role of Peroxisome in Cancer and Age-Related Diseases

Research on the pathology of patients with PD, including the prototype of ZS, has greatly contributed to the elucidation of the physiological functions of peroxisomes in humans. Furthermore, the spread of WES in recent years has led to the discovery of further variants of known genetic diseases and also newly identified PD (see Sect. 7.3); hence, the concept of PD has been expanding.

On the other hand, in age-related diseases, such as diabetes, cancer and neurodegenerative disorders, the association with peroxisomal function has been suggested long ago, through disease models and genetic, epidemiological, and biochemical research. For example, there have been reports on the induction of liver cancer in rodents by peroxisome proliferators, increased occurrence of type 2 diabetes in Hungarian patients with hypocatalasemia, reduced plasmalogen levels in postmortem brain tissues of patients with Alzheimer's disease *etc*. Furthermore, recent studies suggest that peroxisomal function may be altered with aging and could contribute to these age-related diseases (Cipolla and Lodhi 2017).

It is well known that mitochondrial dysfunction may be involved in the onset and progression of age-related diseases *via* reactive oxygen species (ROS). Peroxisomes produce ROS during the process of fatty acid oxidation, and also contain catalase, an enzyme that reduces ROS (see Sect. 4.5). Peroxisomal functions are performed in cooperation with the function of other organelles, including mitochondria (see Sect. 4.6), thus it can be difficult to evaluate the exact role of peroxisomes independently.

In a report of age-related changes in peroxisomes of human cells, aging compromised PTS1 protein import, affecting the critical anti-oxidant enzyme catalase, which led to an increased load of ROS, further reduction of peroxisomal protein import, and exacerbation of aging effects (Legakis et al. 2002). Furthermore, the analysis of peroxisome dynamics in mammalian cells also suggested heterogeneity in peroxisomal import ability with age (Huybrechts et al. 2009). In this section, we pay particular attention to the relationship between age-related diseases and PD.

5.5.1 *Role of Peroxisomes in Neurodegenerative Diseases*

PD patients themselves exhibit neurologic symptoms, such as white matter degeneration, cerebellar ataxia, and developmental regression, as described above (see Sects. 5.2–5.4). Peroxisomes are involved in the biosynthesis of plasmalogens that are rich in myelin sheaths (Wanders and Poll-The 2017); therefore, patients with ZSD and RCDP manifest decreased levels of plasmalogens in plasma and tissues (see Table 7.1), which may be related to myelination deficits in these patients (Bams-Mengerink et al. 2006). Moreover, a mouse model of RCDP type 2 (*Gnpat*$^{-/-}$), which completely lacked plasmalogens, showed defects in myelination in the cerebellum (Teigler et al. 2009). On the other hand, although no impaired plasmalogen synthesis was seen in ALD (see Sect. 5.4), not only increased VLCFA levels, but also reduced plasmalogen levels and increased ROS levels were observed in the white matter of the brains of cerebral ALD patients (Khan et al. 2008).

In Alzheimer's disease (AD), Han et al. reported a dramatic decrease in plasmalogen contents in white matter at a very early stage, which indicated that plasmalogen defects may play an important role in AD pathogenesis and suggested that altered plasmalogen contents may contribute to neurodegeneration, synapse loss, and synaptic dysfunction in AD (Han et al. 2001). Recently, it was reported that oral administration of scallop-derived purified plasmalogens may improve cognitive functions of mild AD (Fujino et al. 2017).

In Parkinson's disease, reduced levels of plasmalogen were reported in lipid rafts isolated from the cortical gray matter of patients (Fabelo et al. 2011). Furthermore, Zellweger model mice (*Pex2*$^{-/-}$, *Pex5*$^{-/-}$ and *Pex13*$^{-/-}$) exhibited increased α-synuclein phosphorylation, oligomerization, and inclusion body formation (Yakunin et al. 2010). These findings are seen in the pathology of patients with Parkinson's disease. Later, Wang et al., using a *pex3* yeast mutant, reported that a defect in peroxisomal biogenesis prevents the binding of alpha-synuclein to lipid droplets in lipid-loaded yeast (Wang et al. 2013). These *PEX* are disease-causing gene for PBD (see Fig. 5.1).

5.5.2 *Role of Peroxisomes in Diabetes*

In a study on Hungarian patients with acatalasemia, the frequency of occurrence of type 2 diabetes is high in patients with hypocatalasemia who are heterozygous for a mutation in the catalase gene, which detoxifies cells from hydrogen peroxide (Nagy et al. 2015). Increased levels of ROS are a key factor involved in the pathogenesis of type 2 diabetes, and oxidative stress is thought to promote pancreatic β-cell dysfunction and contribute to type 2 diabetes. On the contrary, recent studies have shown that β-cells have the capacity to detoxify hydrogen peroxide through a thioredoxin reductase-dependent mechanism and are not as sensitive to oxidative damage as was previously thought (Stancill et al. 2019).

5.5.3 *Role of Peroxisomes in Cancer*

There are many reports on peroxisomal function in cancer. Some of them show decreased peroxisomal function in many tumors, whereas others show the requirement of peroxisomal function for efficient tumor growth. These contradictory findings might be a result of tumor heterogeneity (Islinger et al. 2018). Recent studies revealed that the overexpression of a tumor suppressor, phospholipase A/acyltransferase (PLA/AT)-3, inhibited the binding of PEX19 to peroxisomal membrane proteins, resulting in the specific disappearance of peroxisomes and decrease in levels of plasmalogen. PLA/AT-3 inhibited the binding of PEX19 to various peroxisomal membrane proteins, such as PEX3 and PEX11β (see Fig. 5.1), which suggested that PLA/AT-3 may be involved in a novel regulatory mechanism of peroxisomal biogenesis (Uyama et al. 2015). Moreover, Asare et al. found that the imbalance in epidermal differentiation resulting from PEX11β deficiency and peroxisome mislocalization in mitosis was caused by the inability of basal stem cells to orient their spindle perpendicularly to the underlying basement membrane (Asare et al. 2017). Further studies on the metabolic function, proliferation, and division of peroxisomes in cells might elucidate their role in cancer development and proliferation by clarifying the dynamics of organelles in cell growth and differentiation.

Acknowledgments The author would like to thank Dr. Yasuyuki Suzuki for his guidance and collaboration at Gifu University, Dr. Ronald JA Wanders for his helpful comment, and Dr. A. Goto, Dr. K. Maeda, Dr. S. Matsui, Dr. K. Hibino, Dr. Y. Tanabe and Dr. R. Ikema for providing data of the patients, and further thank many patients and their parents for providing the resources.

The author memorializes the death of the late Professor Emeritus Takashi Hashimoto with deep grief. He was our great pioneer in the field of peroxisomal research, and has passed away on April 18th, 2019 at the age of 86. We would like to deeply grateful to him for his great contribution to the elucidation of peroxisomal disorders.

Ethics Statement The diagnostic studies of PD are approved by the Ethical Committee of the Graduate School of Medicine, Gifu University.
Conflict of Interest The author declares no conflict of interest.

References

Abu-Safieh L, Alrashed M, Anazi S et al (2013) Autozygome-guided exome sequencing in retinal dystrophy patients reveals pathogenetic mutations and novel candidate disease genes. Genome Res 23:236–247

Asare A, Levorse J, Fuchs E (2017) Coupling organelle inheritance with mitosis to balance growth and differentiation. Science 355:eaah4701

Bacino C, Chao YH, Seto E, Lotze T, Xia F, Jones RO, Moser A, Wangler MF (2015) A homozygous mutation in PEX16 identified by whole-exome sequencing ending a diagnostic odyssey. Mol Genet Metab Rep 5:15–18

Bams-Mengerink AM, Majoie CB, Duran M, Wanders RJ, Van Hove J, Scheurer CD, Barth PG, Poll-The BT (2006) MRI of the brain and cervical spinal cord in rhizomelic chondrodysplasia punctate. Neurology 66:798–803

Barøy T, Koster J, Strømme P et al (2015) A novel type of rhizomelic chondrodysplasia punctata, RCDP5, is caused by loss of the PEX5 long isoform. Hum Mol Genet 24:5845–5854

Bezman L, Moser AB, Raymond GV, Rinaldo P, Watkins PA, Smith KD, Kass NE, Moser HW (2001) Adrenoleukodystrophy: incidence, new mutation rate, and results of extended family screening. Ann Neurol 49:512–517

Bjørgo K, Fjær R, Mørk HH, Ferdinandusse S, Falkenberg KD, Waterham HR, Øye AM, Sikiric A, Amundsen SS, Kulseth MA, Selmer K (2017) Biochemical and genetic characterization of an unusual mild PEX3-related Zellweger spectrum disorder. Mol Genet Metab 121:325–328

Blomqvist M, Ahlberg K, Lindgren J, Ferdinandusse S, Asin-Cayuela J (2017) Identification of a novel mutation in PEX10 in a patient with attenuated Zellweger spectrum disorder: a case report. J Med Case Reports 11:218

Bowen P, Lee CSN, Zellweger H, Lindenberg R (1964) A familial syndrome of multiple congenital defects. Bull Johns Hopkins Hosp 114:402–414

Braverman N, Chen L, Lin P, Obie C, Steel G, Douglas P, Chakraborty PK, Clarke JTR, Boneh A, Moser A, Moser H, Valle D (2002) Mutation analysis of PEX7 in 60 probands with rhizomelic chondrodysplasia punctata and functional correlations of genotype with phenotype. Hum Mutat 20:284–297

Braverman NE, Raymond GV, Rizzo WB, Moser AB, Wilkinson ME, Stone EM, Steinberg SJ, Wangler MF, Rush ET, Hacia JG, Bose M (2016) Peroxisome biogenesis disorders in the Zellweger spectrum: an overview of current diagnosis, clinical manifestations, and treatment guidelines. Mol Genet Metab 117:313–321

van den Brink DM, Brites P, Haasjes J, Wierzbicki AS, Mitchell J, Lambert-Hamill M, de Belleroche J, Jansen GA, Waterham HR, Wanders RJA (2003) Identification of PEX7 as the second gene involved in Refsum disease. Am J Hum Genet 72:471–477

Buchert R, Tawamie H, Smith C, Uebe S, Innes AM, Al Hallak B, Ekici AB, Sticht H, Schwarze B, Lamont RE, Parboosingh JS, Bernier FP, Abou Jamra R (2014) A peroxisomal disorder of severe intellectual disability, epilepsy, and cataracts due to fatty acyl-CoA reductase 1 deficiency. Am J Hum Genet 95:602–610

Carlton VEH, Harris BZ, Puffenberg EG, Batta AK, Knisely AS, Robinson DL, Strauss KA, Shneider BL, Lim WA, Salen G, Morton DH, Bull LN (2003) Complex inheritance of familial hypercholanemia with associated mutations in TJP2 and BAAT. Nat Genet 34:91–96

Cipolla CM, Lodhi IJ (2017) Peroxisomal dysfunction in age-related diseases. Trends Endocrinol Metab 28:297–308

Cuesta A, Pedrola L, Sevilla T, Garcia-Planells J, Chumillas MJ, Mayordomo F, LeGuern E, Marin I, Vilche JJ, Palau F (2002) The gene encoding ganglioside-induced differentiation-associated protein 1 is mutated in axonal Charcot-Marie-Tooth type 4A disease. Nat Genet 30:22–24

Curiel J, Steinberg SJ, Bright S, Snowden A, Moser AB, Eichler F, Dubbs HA, Hacia JG, Ely JJ, Bezner J, Gean A, Vanderver A (2017) X-linked adrenoleukodystrophy in a chimpanzee due to an ABCD1 mutation reported in multiple unrelated humans. Mol Genet Metab 122:130–133

Demaret T, Varma S, Stephenne X, Smets F, Scheers I, Wanders R, Van Maldergem L, Reding R, Sokal E (2018) Living-donor liver transplantation for mild Zellweger spectrum disorder: up to 17 years follow-up. Pediatr Transplant 22:e13112

Ebberink MS, Csanyi B, Chong WK et al (2010) Identification of an unusual variant peroxisome biogenesis disorder caused by mutations in the PEX16 gene. J Med Genet 47:608–615

Ebberink MS, Koster J, Visser G, van Spronsen F, Stolte-Dijkstra I, Smit GP, Fock JM, Kemp S, Wanders RJ, Waterham HR (2012) A novel defect of peroxisome division due to a homozygous non-sense mutation in the PEX11-beta gene. J Med Genet 49:307–313

Eichler F, Duncan C, Musolino PL et al (2017) Hematopoietic stem-cell gene therapy for cerebral adrenoleukodystrophy. N Engl J Med 377:1630–1638

Engelen M, Kemp S, de Visser M, van Geel BM, Wanders RJ, Aubourg P, Poll-The BT (2012) X-linked adrenoleukodystrophy (X-ALD): clinical presentation and guidelines for diagnosis, follow-up and management. Orphanet J Rare Dis 7:51

Engelen M, Barbier M, Dijkstra IM, Schür R, de Bie RM, Verhamme C, Dijkgraaf MG, Aubourg PA, Wanders RJ, van Geel BM, de Visser M, Poll-The BT, Kemp S (2014) X-linked adrenoleukodystrophy in women: a cross-sectional cohort study. Brain 137:693–706

Fabelo N, Martín V, Santpere G, Marín R, Torrent L, Ferrer I, Díaz M (2011) Severe alterations in lipid composition of frontal cortex lipid rafts from Parkinson's disease and incidental Parkinson's disease. Mol Med 17:1107–1118

Falkenberg KD, Braverman NE, Moser AB et al (2017) Allelic expression imbalance promoting a mutant PEX6 allele causes Zellweger spectrum disorder. Am J Hum Genet 101:965–976

Ferdinandusse S, Denis S, Clayton PT, Graham A, Rees JE, Allen JT, McLean BN, Brown AY, Vreken P, Waterham HR, Wanders RJ (2000) Mutations in the gene encoding peroxisomal alpha-methylacyl-CoA racemase cause adult-onset sensory motor neuropathy. Nat Genet 24:188–191

Ferdinandusse S, Kostopoulos P, Denis S, Rusch H, Overmars H, Dillmann U, Reith W, Haas D, Wanders RJ, Duran M, Marziniak M (2006) Mutations in the gene encoding peroxisomal sterol carrier protein X (SCPx) cause leukencephalopathy with dystonia and motor neuropathy. Am J Hum Genet 78:1046–1052

Ferdinandusse S, Barker S, Lachlan K, Duran M, Waterham HR, Wanders RJ, Hammans S (2010) Adult peroxisomal acyl-coenzyme A oxidase deficiency with cerebellar and brainstem atrophy. J Neurol Neurosurg Psychiatry 81:310–312

Ferdinandusse S, Jimenez-Sanchez G, Koster J et al (2015) A novel bile acid biosynthesis defect due to a deficiency of peroxisomal ABCD3. Hum Mol Genet 24:361–370

Ferdinandusse S, Falkenberg KD, Koster J, Mooyer PA, Jones R, van Roermund CWT, Pizzino A, Schrader M, Wanders RJA, Vanderver A, Waterham HR (2017) ACBD5 deficiency causes a defect in peroxisomal very long-chain fatty acid metabolism. J Med Genet 54:330–337

Frishberg Y, Zeharia A, Lyakhovetsky R, Bargal R, Belostotsky R (2014) Mutations in HAO1 encoding glycolate oxidase cause isolated glycolic aciduria. J Med Genet 51:526–529

Fujino T, Yamada T, Asada T, Tsuboi Y, Wakana C, Mawatari S, Kono S (2017) Efficacy and blood plasmalogen changes by oral Administration Of Plasmalogen In Patients With Mild Alzheimer's disease and mild cognitive impairment: a multicenter, randomized, double-blind, placebo-controlled trial. EBioMedicine 17:199–205

Funato M, Shimozawa N, Nagase T, Takemoto Y, Suzuki Y, Imamura Y, Matsumoto T, Tsukamoto T, Kojidani T, Osumi T, Fukao T, Kondo N (2006) Aberrant peroxisome morphology in peroxisomal beta-oxidation enzyme deficiencies. Brain Dev 28:287–292

van Geel BM, Poll-The BT, Verrips A, Boelens JJ, Kemp S, Engelen M (2015) Hematopoietic cell transplantation does not prevent myelopathy in X-linked adrenoleukodystrophy: a retrospective study. J Inherit Metab Dis 38:359–361

Gerber S, Charif M, Chevrollier A et al (2017) Mutations in DNM1L, as in OPA1, result in dominant optic atrophy despite opposite effects on mitochondrial fusion and fission. Brain 140:2586–2596

Goldfischer S, Moore CL, Johnson AB, Spiro AJ, Valsamis MP, Wisniewski HK, Ritch RH, Norton WT, Rapin I, Gartner LM (1973) Peroxisomal and mitochondrial defects in the cerebro-hepatorenal syndrome. Science 182:62–64

Han X, Holtzman DM, McKeel DW Jr (2001) Plasmalogen deficiency in early Alzheimer's disease subjects and in animal models: molecular characterization using electrospray ionization mass spectrometry. J Neurochem 77:1168–1180

Huber N, Guimaraes S, Schrader M, Suter U, Niemann A (2013) Charcot-Marie-Tooth disease-associated mutants of GDAP1 dissociate its roles in peroxisomal and mitochondrial fission. EMBO Rep 14:545–552

Huffnagel IC, Laheji FK, Aziz-Bose R, Tritos NA, Marino R, Linthorst GE, Kemp S, Engelen M, Eichler F (2019) The natural history of adrenal insufficiency in X-linked adrenoleukodystrophy: an international collaboration. J Clin Endocrinol Metab 104:118–126
Huybrechts SJ, Van Veldhoven PP, Brees C, Mannaerts GP, Los GV, Fransen M (2009) Peroxisome dynamics in cultured mammalian cells. Traffic 10:1722–1733
Islinger M, Voelkl A, Fahimi HD, Schrader M (2018) The peroxisome: an update on mysteries 2.0. Histochem Cell Biol 150:443–471
Kallijärvi J, Avela K, Lipsanen-Nyman M, Ulmanen I, Lehesjoki AE (2002) The TRIM37 gene encodes a peroxisomal RING-B-box-coiled-coil protein: classification of Mulibrey Nanism as a new peroxisomal disorder. Am J Hum Genet 70:1215–1228
Kato K, Maemura R, Wakamatsu M et al (2018) Allogeneic stem cell transplantation with reduced intensity conditioning for patients with adrenoleukodystrophy. Mol Genet Metab Rep 18:1–6
Kemp S, Pujol A, Waterham HR, van Geel BM, Boehm CD, Raymond GV, Cutting GR, Wanders RJ, Moser HW (2001) ABCD1 mutations and the X-linked adrenoleukodystrophy mutation database: role in diagnosis and clinical correlations. Hum Mutat 18:499–515
Khan M, Singh J, Singh I (2008) Plasmalogen deficiency in cerebral adrenoleukodystrophy and its modulation by lovastatin. J Neurochem 106:1766–1779
Kühl JS, Suarez F, Gillett GT, Hemmati PG, Snowden JA, Stadler M, Vuong GL, Aubourg P, Köhler W, Arnold R (2017) Long-term outcomes of allogeneic haematopoietic stem cell transplantation for adult cerebral X-linked adrenoleukodystrophy. Brain 140:953–966
Kumar KR, Wali GM, Kamate M, Wali G, Minoche AE, Puttick C, Pinese M, Gayevskiy V, Dinger ME, Roscioli T, Sue CM, Cowley MJ (2016) Defining the genetic basis of early onset hereditary spastic paraplegia using whole genome sequencing. Neurogenetics 17:265–270
Lazarow PB, De Duve C (1976) A fatty acyl-CoA oxidizing system in rat liver peroxisomes; enhancement by clofibrate, a hypolipidemic drug. Proc Natl Acad Sci U S A 73:2043–2046
Legakis JE, Koepke JI, Jedeszko C, Barlaskar F, Terlecky LJ, Edwards HJ, Walton PA, Terlecky SR (2002) Peroxisome senescence in human fibroblasts. Mol Biol Cell 13:4243–4255
López-Erauskin J, Fourcade S, Galino J, Ruiz M, Schlüter A, Naudi A, Jove M, Portero-Otin M, Pamplona R, Ferrer I, Pujol A (2011) Antioxidants halt axonal degeneration in a mouse model of X-adrenoleukodystrophy. Ann Neurol 70:84–92
Matsui S, Funahashi M, Honda A, Shimozawa N (2013) Newly identified milder phenotype of peroxisome biogenesis disorder caused by mutated PEX3 gene. Brain Dev 35:842–848
Matsunami M, Shimozawa N, Fukuda A, Kumagai T, Kubota M, Chong PF, Kasahara M (2016) Living-donor liver transplantation from a heterozygous parent for infantile Refsum disease. Pediatrics 137:e20153102
McMillan HJ, Worthylake T, Schwartzentruber J et al (2012) Specific combination of compound heterozygous mutations in 17beta-hydroxysteroid dehydrogenase type 4 (HSD17B4) defines a new subtype of D-bifunctional protein deficiency. Orphanet J Rare Dis 7:1–9
Mignarri A, Vinciguerra C, Giorgio A, Ferdinandusse S, Waterham H, Wanders R, Bertini E, Dotti MT, Federico A (2012) Zellweger spectrum disorder with mild phenotype caused by PEX2 gene mutations. JIMD Rep 6:43–46
Nagy T, Paszti E, Kaplar M, Bhattoa HP, Goth L (2015) Further acatalasemia mutations in human patients from Hungary with diabetes and microcytic anemia. Mutat Res 772:10–14
Ohba C, Osaka H, Iai M (2013) Diagnostic utility of whole exome sequencing in patients showing cerebellar and/or vermis atrophy in childhood. Neurogenetics 14:225–232
Peters C, Charnas LR, Tan Y et al (2004) Cerebral X-linked adrenoleukodystrophy: the international hematopoietic cell transplantation experience from 1982 to 1999. Blood 104:881–888
Pierce SB, Walsh T, Chisholm KM, Lee MK, Thornton AM, Fiumara A, Opitz JM, Levy-Lahad E, Klevit RE, King MC (2010) Mutations in the DBP-deficiency protein HSD17B4 cause ovarian dysgenesis, hearing loss, and ataxia of Perrault Syndrome. Am J Hum Genet 87:282–288
Pujol A, Hindelang C, Callizot N, Bartsch U, Schachner M, Mandel JL (2002) Late onset neurological phenotype of the X-ALD gene inactivation in mice: a mouse model for adrenomyeloneuropathy. Hum Mol Genet 11:499–505

Raas-Rothschild A, Wanders RJ, Mooijer PA, Gootjes J, Waterham HR, Gutman A, Suzuki Y, Shimozawa N, Kondo N, Eshel G, Espeel M, Roels F, Korman SH (2002) A PEX6-defective peroxisomal biogenesis disorder with severe phenotype in an infant, versus mild phenotype resembling Usher syndrome in the affected parents. Am J Hum Genet 70:1062–1068

Ratbi I, Falkenberg KD, Sommen M et al (2015) Heimler syndrome is caused by hypomorphic mutations in the peroxisome-biogenesis genes PEX1 and PEX6. Am J Hum Genet 97:535–545

Raymond GV, Aubourg P, Paker A et al (2018) Survival and functional outcomes in boys with cerebral adrenoleukodystrophy with and without hematopoietic stem cell transplantation. Biol Blood Marrow Transplant S1083-8791:30603–30607

Régal L, Ebberink MS, Goemans N, Wanders RJ, De Meirleir L, Jaeken J, Schrooten M, Van Coster R, Waterham HR (2010) Mutations in PEX10 are a cause of autosomal recessive ataxia. Ann Neurol 68:259–263

Renaud M, Guissart C, Mallaret M, Ferdinandusse S, Cheillan D, Drouot N, Muller J, Claustres M, Tranchant C, Anheim M, Koenig M (2016) Expanding the spectrum of PEX10-related peroxisomal biogenesis disorders: slowly progressive recessive ataxia. J Neurol 263:1552–1558

Sá MJ, Rocha JC, Almeida MF, Carmona C, Martins E, Miranda V, Coutinho M, Ferreira R, Pacheco S, Laranjeira F, Ribeiro I, Fortuna AM, Lacerda L (2016) Infantile Refsum Disease: influence of dietary treatment on plasma phytanic acid levels. JIMD Rep 26:53–60

Schabhüttl M, Wieland T, Senderek J, Baets J, Timmerman V, De Jonghe P, Reilly MM, Stieglbauer K, Laich E, Windhager R, Erwa W, Trajanoski S, Strom TM, Auer-Grumbach M (2014) Whole-exome sequencing in patients with inherited neuropathies: outcome and challenges. J Neurol 261:970–982

Setchell KD, Heubi JE, Bove KE, O'Connell NC, Brewsaug T, Steinberg SJ, Moser A, Squires RH Jr (2003) Liver disease caused by failure to racemize trihydroxycholestanoic acid: gene mutation and effect of bile acid therapy. Gastroenterology 124:217–232

Sevin C, Ferdinandusse S, Waterham H, Wanders RJ, Aubourg P (2011) Autosomal recessive cerebellar ataxia caused by mutations in the PEX2 gene. Orphanet J Rare Dis 6:8

Shimozawa N (2011) Molecular and clinical findings and diagnostic flowchart of peroxisomal diseases. Brain Dev 33:770–776

Shimozawa N, Tsukamoto T, Suzuki Y, Orii T, Shirayoshi Y, Mori T, Fujiki Y (1992) A human gene responsible for Zellweger syndrome that affects peroxisome assembly. Science 255:1132–1134

Shimozawa N, Tsukamoto T, Nagase T, Takemoto Y, Koyama N, Suzuki Y, Komori M, Osumi T, Jeannette G, Wanders RJ, Kondo N (2004) Identification of a new complementation group of the peroxisome biogenesis disorders and PEX14 as the mutated gene. Hum Mutat 23:552–558

Smith CE, Poulter JA, Levin AV, Capasso JE, Price S, Ben-Yosef T, Sharony R, Newman WG, Shore RC, Brookes SJ, Mighell AJ, Inglehearn CF (2016) Spectrum of PEX1 and PEX6 variants in Heimler syndrome. Eur J Hum Genet 24:1565–1571

Stancill JS, Broniowska KA, Oleson BJ, Naatz A, Corbett JA (2019) Pancreatic β-cells detoxify H2O2 through the peroxiredoxin/thioredoxin antioxidant system. J Biol Chem 294:4843

Takahara S, Miyamoto H (1948) Three cases of progressive oral gangrene due to lack of catalase in the blood. Nippon Jibiinkoka Gakkai Kaiho 51:163

Teigler A, Komljenovic D, Draguhn A, Gorgas K, Just WW (2009) Defects in myelination, paranode organization and Purkinje cell innervation in the ether lipid-deficient mouse cerebellum. Hum Mol Genet 18:1897–1908

Tran C, Hewson S, Steinberg SJ, Mercimek-Mahmutoglu S (2014) Late-onset Zellweger spectrum disorder caused by PEX6 mutations mimicking X-linked adrenoleukodystrophy. Pediatr Neurol 51:262–265

Uyama T, Kawai K, Kono N, Watanabe M, Tsuboi K, Inoue T, Araki N, Arai H, Ueda N (2015) Interaction of phospholipase A/acyltransferase-3 with Pex19p: a possible involvement in the down-regulation of peroxisomes. J Biol Chem 290:17520–17534

Ventura MJ, Wheaton D, Xu M, Birch D, Bowne SJ, Sullivan LS, Daiger SP, Whitney AE, Jones RO, Moser AB, Chen R, Wangler MF (2016) Diagnosis of a mild peroxisomal phenotype with next-generation sequencing. Mol Genet Metab Rep 9:75–78

Vilarinho S, Sari S, Mazzacuva F, Bilgüvar K, Esendagli-Yilmaz G, Jain D, Akyol G, Dalgiç B, Günel M, Clayton PT, Lifton RP (2016) ACOX2 deficiency: a disorder of bile acid synthesis with transaminase elevation, liver fibrosis, ataxia, and cognitive impairment. Proc Natl Acad Sci U S A 113:11289–11293

Wanders RJ, Poll-The BT (2017) Role of peroxisomes in human lipid metabolism and its importance for neurological development. Neurosci Lett 637:11–17

Wang Y, Busin R, Reeves C et al (2011) X-linked adrenoleukodystrophy: ABCD1 de novo mutations and mosaicism. Mol Genet Metab 104:160–166

Wang S, Horn PJ, Liou LC, Muggeridge MI, Zhang Z, Chapman KD, Witt SN (2013) A peroxisome biogenesis deficiency prevents the binding of alpha-synuclein to lipid droplets in lipid-loaded yeast. Biochem Biophys Res Commun 438:452–456

Wang RY, Monuki ES, Powers J, Schwartz PH, Watkins PA, Shi Y, Moser A, Shrier DA, Waterham HR, Nugent DJ, Abdenur JE (2014) Effects of hematopoietic stem cell transplantation on acyl-CoA oxidase deficiency: a sibling comparison study. J Inherit Metab Dis 37:791–799

Wang W, Xia ZJ, Farré JC, Subramani S (2017) TRIM37, a novel E3 ligase for PEX5-mediated peroxisomal matrix protein import. J Cell Biol 216:2843–2858

Waterham HR, Koster J, van Roermund CW, Mooyer PA, Wanders RJ, Leonard JV (2007) A lethal defect of mitochondrial and peroxisomal fission. N Engl J Med 356:1736–1741

Yakunin E, Moser A, Loeb V, Saada A, Faust P, Crane DI, Baes M, Sharon R (2010) Alpha-Synuclein abnormalities in mouse models of peroxisome biogenesis disorders. J Neurosci Res 88:866–876

Yamashita T, Mitsui J, Shimozawa N et al (2017) Ataxic form of autosomal recessive PEX10-related peroxisome biogenesis disorders with a novel compound heterozygous gene mutation and characteristic clinical phenotype. J Neurol Sci 375:424–429

Yoon G, Malam Z, Paton T, Marshall CR, Hyatt E, Ivakine Z, Scherer SW, Lee K-S, Hawkins C, Cohn RD (2016) Lethal disorder of mitochondrial fission caused by mutations in DNM1. J Pediatr 171:313–316

Chapter 6
Model Organisms for Understanding Peroxisomal Disorders

Shigeo Takashima and Nobuyuki Shimozawa

Abstract Peroxisomal disorders are congenital human diseases caused by the dysfunction of peroxisomes, which are small vesicular organelles distributed in the cytoplasm. In patients suffering from these disorders, multiple defects manifest in a variety of tissues and organs, such as the brain, spinal cord, peripheral nerves, eyes, kidneys, liver, spleen, and bone. A number of biological metabolites are synthesized and degraded in the peroxisomes, such that the metabolites fluctuate severely in patients with peroxisomal disorders. A link between peroxisomal metabolites and symptoms of peroxisomal disorders has long been suspected; however, we have only limited knowledge about the pathology of this disease in humans due to the rarity of peroxisomal disorders. To overcome this problem, model organisms of peroxisomal disorders were established and studied in detail. These models successfully recapitulate the major human symptoms and have become powerful tools to understand the biological basis of the disease pathology and the development of therapeutic strategies against it.

Keywords Peroxisomal disorders · Peroxisome · Very long chain fatty acid · Phytanic acid · Bile acid intermediates · Plasmalogens · Mouse · Fruit fly · Zebrafish · Model organism

6.1 Introduction

A wide variety of model organisms of multiple animal species, such as the mouse, zebrafish, and fruit fly, are available to evaluate peroxisomal disorders (Table 1). To overcome the limited pathological information available for human patients, model

S. Takashima (✉)
Division of Genomics Research, Life Science Research Center, Gifu University, Gifu city, Gifu, Japan
e-mail: staka@gifu-u.ac.jp

N. Shimozawa
Division of Genomics Research, Life Science Research Center, Gifu University, Gifu, Japan

Department of Pediatrics, Gifu University School of Medicine, Gifu, Japan
e-mail: nshim@gifu-u.ac.jp

T. Imanaka, N. Shimozawa (eds.), *Peroxisomes: Biogenesis, Function, and Role in Human Disease*, https://doi.org/10.1007/978-981-15-1169-1_6

Table 1 List of selected mutant animal models for studying peroxisomal disorders

Organism	Disease type	Affected gene	Type of mutation	Phenotypes	Biochemical hallmark	References
Models of peroxisome biogenesis disorder (PBD)						
Mouse	Zellweger spectrum disorder (ZSD)	*Pex1*	G844D substitution	Preadult mortality. Progressive psychomotor retardation	VLCFA and bile acid intermediates increase. Plasmalogen decrease. Decreased branched fatty acid oxidation in fibroblasts	Hiebler et al. (2014)
Mouse	ZSD	*Pex2*	Generalized KO (in different back grounds)	Growth retardation. Hypotonia. Sucking disability. Neonatal mortality. Neural migration defect. Cerebellar anomaly. Abnormal Purkinje cells. Fatty liver. Neuronal lipidosis. Hepatic steatosis	VLCFA increase. Plasmalogen decrease	Faust and Hatten (1997), Faust et al. (2001), Faust (2003), Keane et al. (2007)
Mouse	ZSD	*Pex5*	Generalized KO (and conditional rescue)	Growth retardation. Sucking disability. Neonatal mortality. Neural migration defect. Hepatic steatosis. Abnormal mitochondria in liver	VLCFA increase, Plasmalogen decrease	Baes et al. (1997), Janssen et al. (2003)
Mouse	ZSD	*Pex5*	Conditional KO in CNS	Growth retardation. Early death. Delayed cortical neuronal migration. Delayed cerebellar development. Abnormal Purkinje cells. Neuronal lipidosis. Gliosis. Myelin anomaly. Cataract	VLCFA increase in brain. Plasmalogen and DHA decrease in brain	Krysko et al. (2007), Hulshagen et al. (2008)

(continued)

Table 1 (continued)

Organism	Disease type	Affected gene	Type of mutation	Phenotypes	Biochemical hallmark	References
Mouse	ZSD	*Pex5*	Conditional KO in liver	Growth retardation. Neonatal mortality. Abnormal hepatic mitochondria. Neural migration defect (severer than CNS KO). Abnormal cerebellum. Abnormal Purkinje cells	Unchanged plasmalogen level in liver	Krysko et al. (2007)
Mouse	ZSD	*Pex10*	Generalized KO	Progressive locomotive defect in embryo. Sucking disability. Neonatal mortality. Growth retardation. Schwann cell defects in peripheral nerves	VLCFA increase. Plasmalogen decrease. Bile acid intermediates increase	Hanson et al. (2014)
Mouse	–	*Pex11α*	Generalized KO	Increased body weight. High triglyceride levels. Hepatic steatosis	Normal peroxisomal metabolites	Li et al. (2002b), Weng et al. (2013)
Mouse	ZSD	*Pex11β*	Generalized KO	Growth retardation. Neonatal mortality. Sucking disability. Hypotonia. Neural migration defect	Slight increase of VLCFAs and slight decrease of plasmalogen	Li et al. (2002a)
Mouse	ZSD	*Pex13*	Generalized KO	Growth retardation. Neonatal Mortality. Sucking disability. Hypotonia. Neural migration defect. Hepatic steatosis. Abnormal mitochondria in liver	VLCFA increase. Plasmalogen decrease. Decreased branched fatty acid oxidation in fibroblasts	Maxwell et al. (2003)
Mouse	RCDP type1	*Pex7*	Generalized KO	Growth retardation. Cataracts. Incomplete and delayed ossification. Neonatal mortality. Neural migration defect in cerebrum. Male and female sterility. Testicular atrophy	Plasmalogen decrease. VLCFA increase in neonate	Brites et al. (2003), Brites et al. (2009), da Silva et al. (2014)

(continued)

Table 1 (continued)

Organism	Disease type	Affected gene	Type of mutation	Phenotypes	Biochemical hallmark	References
Mouse	RCDP type1	*Pex7*	Hypomorph	Slight growth retardation. Delayed ossification. Cataracts	Plasmalogen decrease. Normal VLCFAs	Braverman et al. (2010)
Fruit fly	ZSD	*Pex1*	Generalized KO	Larval death. Locomotive defect. Developmental defects in CNS and PNS	Unexamined	Mast et al. (2011)
Fruit fly	ZSD	*Pex2*	Generalized KO	Delayed growth. Male sterility. Decreased female fertility. Defective spermatogenesis	Unexamined	Chen et al. (2010)
Fruit fly	ZSD	*Pex3*	Generalized KO	Larval death. Growth retardation	Longer acyl chain length in triacylglycerol	Faust et al. (2014), Nakayama et al. (2011)
Fruit fly	ZSD	*Pex10*	Generalized KO	Delayed growth. Male sterility. Decreased female fertility. Defective spermatogenesis	VLCFA increase	Chen et al. (2010)
Fruit fly	ZSD	*Pex12*	Generalized KO?	Defective spermatogenesis	Unexamined	Chen et al. (2010)
Fruit fly	ZSD	*Pex16*	Generalized KO	Male sterile. Abnormal spermatogenesis. Locomotive defects. Affected optic lobe development	VLCFA increase	Nakayama et al. (2011)
Fruit fly	ZSD	*Pex19*	Generalized KO	Larval-pupal death in zygotic mutants. Embryonic death in maternal-zygotic mutants. Locomotive defects. Increased cell death in CNS	VLCFA increase. Decrease of shorter fatty acids. Increased free fatty acids	Bülow et al. (2018), Sellin et al. (2018)
Models of single enzyme deficiencies						
Mouse	ACOX1 deficiency	*Acox1*	Generalized KO	Growth retardation. Male and female sterility. Hepatic steatosis. Viable	VLCFA increase	Fan et al. (1996), Fan et al. (1998), Sheridan et al. (2011)

(continued)

Table 1 (continued)

Organism	Disease type	Affected gene	Type of mutation	Phenotypes	Biochemical hallmark	References
Mouse	D BP deficiency	*Hsd17b4*	Generalized KO	Growth retardation. Increased neonatal death. Cataracts. Male sterile. Cerebellar atrophy with demyelination. Purkinje cell death	VLCFA increase. Bile acid intermediates (DHCA, THCA) increase. Branched fatty acid increase	Baes et al. (2000), Huyghe et al. (2006a, b), De Munter et al. (2016)
Mouse	SCPx deficiency	*Scp2*/ *SCPx*	Generalized KO	Normal and fertile	Phytanic acid and pristanic acid increase on phytol diet	Seedorf et al. (1998), Atshaves et al. (2007)
Mouse	RCDP type2	*Gnpat*	Generalized KO	Growth retardation. Cataracts. Increased mortality (especially in males). Testicular atrophy. Cerebellar defects with ataxia, abnormal myelination, Purkinje cell misinnervation	Plasmalogen decrease. DHA decrease	Rodemer et al. (2003), Komljenovic et al. (2009), Teigler et al. (2009), da Silva et al. (2014)
Mouse	RCDP type3	*Agps*	Hypomorph	Cataracts. Male sterility. Testicular atrophy. Shortening of humerus bones (rhizomelia)	Ether lipid decrease	Liegel et al. (2011), Liegel et al. (2014)
Mouse	X-ALD	*Abcd1*	Generalized KO	Locomotive defects (AMN-like)	VLCFA increase	Kobayashi et al. (1997), Lu et al. (1997), Forss-Petter et al. (1997), Pujol et al. (2002)
Zebrafish	X-ALD	*Abcd1*	Generalized KO	Defect in adrenal gland development. Locomotive defects. Reduced oligodendrocyte precursor cells in CNS. Hypomyelination. Increased cell death in brain	VLCFA increase	Strachan et al. (2017)

(continued)

Table 1 (continued)

Organism	Disease type	Affected gene	Type of mutation	Phenotypes	Biochemical hallmark	References
Mouse	AMACR deficiency	*AMACR*	Generalized KO	No symptomatic phenotype. Development of liver failure on a phytol diet	Phytanic acid and pristanic acid increase. Decrease of bile acids. Increase of bile acid intermediates (DHCA, THCA)	Savolainen et al. (2004)

organisms are practically important to comprehensively understand the pathogenesis of the disease. In this section, we focus on studies using model organisms for studying peroxisomal disorders based on the similarity of their phenotypes and symptoms with those of human patients. The benefits and limitations of these model organism studies are also discussed.

6.2 Model Organisms for Peroxisome Biogenesis Disorder (PBD)

Peroxisome biogenesis disorder (PBD) manifests as multiple symptoms in various organs, including the brain, eyes, liver, kidneys, and bones. Zellweger spectrum disorder (ZSD) is a subgroup of PBD, where the peroxisomal function is entirely affected. ZSD is classified into Zellweger syndrome (ZS), neonatal adrenoleukodystrophy, and infantile Refsum disease, based on severe to mild disease severity. Rhizomelic chondrodysplasia punctata (RCDP), type 1 and type 5, comprise another subgroup where the defect is limited to selected metabolic pathways (see Sect. 5.1.2). The defect in the peroxisome biogenesis causing PBD resulted from a mutation of the *Pex* genes. *Pex* genes have been well-conserved throughout vertebrate evolution and hence share high sequence similarity at the nucleic acid and amino acid levels among species. Such high similarity implies the conserved function of proteins encoded by *Pex* genes across the animal kingdom; therefore, model organisms carrying inactivation of *Pex* genes would be useful to study human PBD. In order to understand the pathogenesis of PBD in detail, several animal models have been established, including those in mice, zebrafish, and fruit flies, by genetically altering the *Pex* genes. The following sections describe the phenotypic and biochemical details, as well as the proposed mechanisms of pathogenesis of PBD model animals.

6.2.1 PBD Mouse Models

In the models established thus far, eight *Pex* genes have been inactivated in mice, namely *Pex1, Pex2, Pex5, Pex7, Pex10, Pex11α, Pex11β,* and *Pex13* genes (Table 1). In the first attempt, knockout (KO) mice with the gene inactivation in the whole body were produced; however, they had higher lethal rates after birth, which hindered detailed analysis of the disease. Alternatively, hypomorphic or conditional KO mice were generated to overcome this problem. The mice successfully recapitulated some of the important human symptoms and their pathology was studied.

6.2.2 Pex KO Mice Lacking Peroxisomes in Whole Body: the Model for Zellweger Syndrome (ZS) (Pex2, Pex5, Pex13 KO Mouse)

The most severe form of PBD is ZS, where functional peroxisomes do not form, thereby impairing metabolism in peroxisomes. Patients with peroxisomal disorders exhibit symptoms in multiple organs as mentioned above (also see Sect. 5.1.1). The mouse models for ZS were first reported in 1997 independently by Baes and colleagues, and Faust and Hatten (Baes et al. 1997; Faust and Hatten 1997), where *Pex5* and *Pex2 genes*, respectively, were targeted for gene KO. In 2003, Maxwell et al. reported a *Pex13* KO mouse (Maxwell et al. 2003). These three mouse models showed impairment of peroxisome biogenesis and an absence of functional peroxisomes in the entire body during most of the embryonic development. They demonstrated a fluctuation in the biochemical substances as in human patients, such as accumulation of very long chain fatty acids (VLCFAs) and a decrease in the plasmalogen level. The KO mice mimicked several symptoms of ZS in human patients, such as growth retardation and sucking disabilities; they were found to be severely hypotonic as well. The brains of these animals exhibited neuronal migration defects in the cerebral cortex, consistent with a display of a diffuse laminar structure of the brain. Lipid droplets in the hepatocytes of the *Pex5* and *Pex13* KO mice were reminiscent of the fatty liver observed in human patients. Cerebellar anomalies in neuronal migration or in Purkinje cell differentiation could not be assessed because of early death in the mutant animals, as the process of cerebellar neuronal migration and Purkinje cell differentiation occurs in the later neonatal stages. General growth retardation with hypotonia and cerebral neuronal migration defects were the common features manifested among all these KO animals.

Further analysis was carried out using *Pex2* mutant mice by altering their genetic background (Faust et al. 2001; Faust 2003; Keane et al. 2007). Most of the *Pex2* mutant mice with the original C57BL/6 × 129 Sv background died on the day of birth, while the mice with129SvEv or Swiss Webster × 129SvEv background could survive a few weeks. Interestingly, additional phenotypes appeared in long-lived mutant animals with the latter genetic backgrounds. In the cerebellum, abnormal organ morphology with impaired dendritic patterns of the Purkinje cells was seen in mutants (Faust et al. 2001; Faust 2003). Severe hepatic steatosis also appeared in this mouse model (Keane et al. 2007). The positive influence of bile acid administration on the liver phenotype was examined using this model animal. The prolonged life span allowed the appearance of more phenotypes similar to those found in humans; however, the detailed genetic differences influencing the phenotypic differences between the mouse strains remain unknown.

6.2.3 Conditional KO and Conditional Rescue in Mouse Models

In order to understand the effect of peroxisomes on organ-specific dysfunction for neuronal phenotypes, *Pex5* was manipulated in an organ-specific manner (Krysko et al. 2007; Hulshagen et al. 2008) and the brain phenotype was analyzed. To disrupt peroxisome biogenesis specifically in the brain or liver, brain (with *Nestin* promoter) or liver (*Alfp promoter*)-expressing Cre drivers were used. Similarly, organ-specific promoters (*Nestin* for brain and *AFP* for liver) were fused with wild-type *Pex5* and overexpressed in the whole-body *Pex5* KO background to rescue peroxisomal biogenesis in an organ-specific manner (Janssen et al. 2003). Interestingly, elimination of the liver peroxisomes was responsible for neural migration defects in the cerebral cortex and Purkinje cell differentiation in the cerebellum. On the other hand, brain-specific elimination of peroxisomes had a limited effect. These models suggest that the peroxisomal metabolites circulating in the body, rather than those produced locally, are responsible for the neuronal phenotype in the brain. Therefore, subtraction or supplementation approaches are more feasible for human patients through venous injection or oral intake, rather than brain-targeted intervention.

6.2.4 Other KO Mice (Pex1, Pex10, Pex11α, Pex11β)

Several pathological states of other *Pex* genes were also studied. Progressive failure of the peripheral nerves in PBD was studied in the *Pex10* KO mouse (Hanson et al. 2014). Schwann cell defects and loss of axon integrity leading to failure of the peripheral nervous system were found in the *Pex 10* KO mouse model, which were overlooked in preceding studies that used other *Pex* KO mouse models. The pathological mechanism proposed by *Pex 10* KO mouse explains the development of similar symptoms in the milder form of human PBD.

There are three *Pex11* genes (*Pex11α*, *Pex11β*, and *Pex11γ*) in humans and mice, all of which are involved in peroxisomal division. Depletion of any of them results in impaired peroxisomal division, leading to the formation of dysfunctional peroxisomes (see Chap. 2). Mutant mice with a deletion of *Pex11α* or *Pex11β* gene, respectively, were established (Li et al. 2002a, b). Among them, *Pex11β* KO mouse manifested ZSD-like phenotype (Li et al. 2002a). The neonates had high lethal rates and their growth was retarded. They demonstrated sucking inability and hypotonia, as well as neuronal migration defects in the neocortex. The import of peroxisomal matrix proteins was largely normal and the metabolic fluctuations were very weak in *Pex11β* KO mice. A slight elevation of VLCFAs was seen in the liver, and a slight decrease in plasmalogen level was observed in the brain, suggesting a possible causal relationship of these metabolites with the symptoms manifested in human patients.

A *Pex1* mutant mouse model was established for a specific purpose. Amino acid substitution, G843D, is the most frequent mutation in human *PEX1*, and is the most frequent cause of ZSD in humans. G843D mutation falls into the category of a milder form of ZSD, in which the patients can survive up to adulthood, but manifest progressive psychomotor retardation. The equivalent amino acid substitution, G844D, was introduced in order to produce an equivalent disease model in mice for this specific mutation (Hiebler et al. 2014). Mice with a G844D mutation were found to be viable at birth. Overall, about 40% of them died before adulthood. Besides growth retardation, they also showed visual defects as reported in human patients. Human PEX1 protein with G843D mutation could be ameliorated by chaperone-like chemicals in fibroblasts (Zhang et al. 2010). The G844D mouse can be used as an animal model to test similar drugs and other agents that could be therapeutic in human patients carrying G843D mutation.

6.2.5 *Models of Rhizomelic Chondrodysplasia Punctata (RCDP) Type 1*

RCDP type 1 and type 5 comprise another subgroup of PBD with limited metabolic defects, in which *PEX7* and *PEX5L* (codes for long isoform of PEX5 protein), respectively, are inactivated. In patients with RCDP type1 and type5, the transport of PTS2 proteins with peroxisome targeting signal 2 (PTS2), into peroxisomal matrix is disturbed (Fig. 1), because PEX7 and PEX5L are necessary for the peroxisomal import of PTS2 proteins (see Sect. 5.1.2). *Pex7* KO mouse model of RCDP type 1 has been established and analyzed (Brites et al. 2003; Braverman et al. 2010). Mice with *Pex7* KO in whole body (Brites et al. 2003) showed phenotypes similar to the symptoms observed in human patients such as growth retardation, congenital cataracts, and incomplete or delayed endochondral ossification (intramembrane ossification is unaffected). Moreover, the mice showed increased neonatal mortality and mild neural migration defects in the cerebral cortex. Rhizomelia and punctate joint calcification (chondrodysplasia punctata) did not develop in the *Pex7* KO mice, in contrast to observations in human patients. The biochemical fluctuations seen in human patients were also reported in the mutant mice. Plasmalogen level was considerably reduced due to the failure in import of alkyl-dihydroxyacetone phosphate synthase (Agps; a PTS2 protein and a key enzyme in plasmalogen synthesis) in peroxisomal matrix

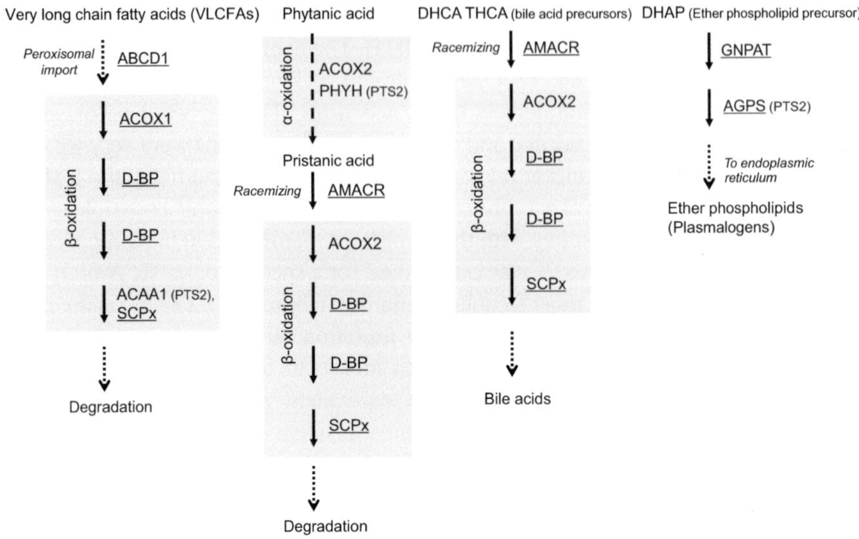

Fig. 1 Peroxisomal metabolic processes in the degradation of VLCFAs and branched-chain fatty acids (phytanic acid and pristanic acid), as well as the synthesis of bile acids and ether phospholipids. Enzymes related to each step are shown, and the enzymes with corresponding mutant models described in the text are underlined

(Brites et al. 2003). Phytanic acid level increased on the dietary intake of phytol (a precursor of phytanic acid) because of the loss of phytanoyl-CoA hydroxylase (Phyh) (another PTS2 protein targeted to peroxisomal matrix) in the peroxisomes, which is required for the α-oxidation of phytanic acid. The loss of Phyh in the peroxisomal matrix results from defective Pex7-Pex5L chaperoning. The amount of VLCFAs was mildly increased in neonates, considered to be caused by the loss of thiolase with PTS2 (*Acaa1* gene product) in the peroxisomal matrix. However, VLCFA accumulation was normalized by adulthood, presumably by the compensative action of Scpx, another protein with thiolase activity possessing PTS1. In accordance with this, levels of VLCFAs are unaffected in the plasma of human patients but are increased in blood cells (Schutgens et al. 1993), which might have lower SCPx expression (Brites et al. 2003). Most of the mutant mice exhibited neonatal mortality, while approximately 30% of them survived to adulthood. The adult mutant mice, however, were infertile, and males developed testicular atrophy, highlighting the possible importance of plasmalogen for gonadal development and maintenance. Due to our limited histological knowledge of human systems, such as the central nervous system, the cause of psychomotor retardation in patients is not fully understood. The phenotype observed in the brain of *Pex7* KO mice provides an important insight into the cause of psychomotor retardation in human patients. The *Pex7* KO mouse, as described above, recapitulated the symptoms seen in human patients and revealed previously unknown effects of this disease, such as testicular atrophy.

Hypomorphic *Pex7* mutant mice were also established (Braverman et al. 2010). The mice had a normal lifespan, could survive to adulthood, and were fertile. They still, however, developed remarkable symptoms in the bones and eyes and were also slightly growth retarded compared to their non-mutant littermates; they showed delayed endochondral ossification in the limbs and underdevelopment of the skeleton and developed cataracts with strong lens fiber anomaly. As in the *Pex7* KO mice shown above, the mutant mice showed no signs of rhizomelia and chondrodysplasia punctata. Biochemically, the *Pex7* hypomorphic mutant mice showed decreased plasmalogen level and normal levels of VLCFAs; this is in contrast to the *Pex7* KO mice, which showed transiently high level of VLCFAs as discussed above (Brites et al. 2003). Investigation of the hypomorphic *Pex7* mutant allele revealed that the bones and eyes are highly sensitive to the reduction in plasmalogen level, suggesting that plasmalogens are required for proper development of these organs.

6.2.6 PBD Zebrafish Models

So far there is no *Pex*-gene knockout zebrafish model available. However, the functions of the *Pex* gene products have been studied using Morpholino oligo-mediated gene knockdown (KD). Morpholino is a nucleotide analogue, and its antisense oligomer can bind with mRNA and pre-mRNA to inhibit translation or proper splicing, respectively, thus disturbing the function of the target gene. While the technique itself worked, *Pex3* KD and *Pex13* KD did not eliminate peroxisomes successfully (Krysko et al. 2010). This is possibly due to the long life of the already existing Pex3 and Pex13 proteins after the inhibition of their translation, or because of the slow turnover of existing peroxisomes in zebrafish. As zebrafish embryonic development takes only a few days, it would be difficult to eliminate peroxisomes during this short period by gene KD. Moreover, the use of Morpholino sometimes induces non-specific pleiotropic effects; hence, the use of a large amount of Morpholino is not preferred. Alternatively, the direct gene editing methods using the CRISPR/Cas9 and TALEN systems can be used to establish a PBD disease model in zebrafish.

The difficulty of gene KD as well as gene KO in zebrafish embryos also arises from the evidence that relatively high amounts of maternally deposited mRNAs and proteins exist in zebrafish oocytes compared to mouse oocytes. This feature is accompanied with some advantages and disadvantages. Even if the zygotic gene is totally disrupted, maternal counterparts can compensate for the loss of zygotic gene products so that the effect of zygotic mutation is weakened, and the resulting phenotypes might be too subtle to be examined during the embryonic and early larval period. However, such maternal contribution can circumvent the lethality of the zygotic gene loss and support the survival of mutants beyond the embryonic and early larval period. Consequently, the effect of the gene loss can be analyzed in later developmental stages even in adults; for instance, the long-lived mutant animal is helpful in studying *Pex* gene losses in a mouse model (see above). Taken together, the zebrafish would be an ideal model organism to study the pathology of peroxisomal diseases.

6.2.7 PBD Fruit Fly Models

Disease-related *Pex* genes are common between humans and fruit flies (*Drosophila melanogaster*) (with the exception of *Pex26*, which is lost in the fruit fly), and their functional similarity was confirmed using a fly cell line (Mast et al. 2011). The fruit fly community has a large stock of genomic mutants and RNAi (RNA interference) lines, thus making this animal model highly advantageous for genetic studies. To model PBD, mutant phenotypes were studied for *pex1*, *pex2*, *pex3*, *pex10, pex12, pex16, and pex19* genes using genomic mutants and RNAi lines.

The phenotype of *pex2*, *pex10*, and *pex12* mutant flies predominantly demonstrated defective spermatogenesis (Chen et al. 2010). The mutant flies were all viable up to the adult stage and showed slightly delayed growth and infertility, despite the total elimination of *pex2* and *pex10* gene products (the molecular detail of *pex12* mutant was not mentioned). VLCFAs accumulated along with aging, confirming that peroxisomes are also required for VLCFA breakdown in this species, as in mice and humans. The defective spermatogenesis phenotype was enhanced as the amount of VLCFAs increases, proposing VLCFAs to be pathogenic agents affecting sperm development; this was similarly observed in some mouse models of PBD (see above sections) and other peroxisomal diseases (see below). The neuronal phenotypes, including locomotive defects, were not reported in these mutant flies, while other *Pex* mutants exhibited them (see below).

In contrast to the above non-lethal *pex* mutations, the *pex1* mutation exhibited high lethality during the early larval period after hatching (Mast et al. 2011). *Pex1* mutants showed growth retardation and hence were much smaller than their wild-type siblings. Locomotion in the mutant larvae was also strongly affected despite the morphologically normal muscle development. Furthermore, the architecture of the CNS and peripheral nervous system (PNS) was strongly disturbed during the embryonic development, which could explain the hypotonic larvae. Conversely, muscle dysfunction was also suggested as the cause of the larval hypotonia (Faust et al. 2014; see below).

pex3-null mutants completely lose their peroxisomes in larval Malpighian tubules (fruit fly equivalent of kidneys), oenocytes (liver equivalent), and, possibly, other organs, although there are no detailed reports available (Nakayama et al. 2011; Faust et al. 2014). The *pex3* mutant larvae were small and died in the larval period, similarly to the *pex1* mutants. Only a few individuals escaped this lethal period and emerged as adult flies, but they exhibited severe growth retardation and a short lifespan (Faust et al. 2014). Accumulation of VLCFAs was not directly analyzed, but the extension of acyl chains in triacylglycerol was observed, again confirming that fruit fly peroxisomes are involved in VLCFA breakdown. However, because other peroxisomal metabolites have not been examined, the detailed role of peroxisomes remains unclear. Interestingly, the muscle-specific *pex3* KD using RNAi, resulted in reduced muscle function, suggesting that the muscular hypotonia of the *pex3* mutants, as well as human PBD patients, could, at least in part, be of muscular origin and not of neurological origin. However, further investigations are required to answer this pathologically important question.

pex16 null mutant flies were viable up to the adult stage, expressing low levels of remnant peroxisome like structures (Nakayama et al. 2011). VLCFAs accumulated in these mutants, revealing the requirement of peroxisomes for VLCFA breakdown. Male mutant flies showed infertility with abnormal spermatogenesis, as in the case of *pex2*, *pex10*, and *pex12* mutants (see above), while females could produce eggs. Both male and female mutants exhibited locomotive defects and abnormal development of optic lobes (the brain structure lie behind the compound eyes). However, progression of the locomotive defect was not observed, suggesting that the defect was likely to be originated during brain (optic lobe) development. The neurological and locomotive phenotypes could be rescued by forced expression of the wild-type *pex16* gene in the brain neurons or even in the fat body, indicating that the circulating metabolites are important for the normal development of the brain. This finding is similar to the results obtained from the liver specific *Pex5* KO or *Pex5* rescue experiments using mouse models (see above).

pex19 mutant flies also showed high mortality (Bülow et al. 2018). Most zygotic mutants died before reaching the adult stage, while maternal-zygotic mutants, which lack the possible maternal deposition of *Pex19* mRNA and proteins in the eggs, died at the embryonic stage. This suggests that peroxisomes remain for some time after the elimination of zygotic *pex19*. Some *pex19* mutant survivors showed low locomotive activity, accompanied by increased cell death (apoptosis) in the CNS. This phenotype could be rescued with non-CNS overexpression of the wild-type *pex19*, again highlighting the role of circulating metabolites on the CNS phenotypes (Bülow et al. 2018). As with other *pex* mutant flies, *pex19* mutants accumulated VLCFAs. Moreover, they showed a reduced amount of medium chain fatty acids (MCFAs) such as C12:0 and C14:0 and increased free fatty acids (FFAs). Interestingly, the lethality of the *pex19* mutation (and also of some other *pex* mutations) could be reduced by the dietary administration of MCFA-containing oils, such as coconut oil (Sellin et al. 2018). This rescue was considered due to a normalization of mitochondrial lesions through the reduction of mitotoxic FFAs, which were abnormally elevated upon peroxisomal dysfunction and subsequent MCFA reduction. Because abnormal mitochondria were observed in some *Pex* mutant mouse models, the amount of MCFAs and FFAs should be addressed in the mouse model.

6.3 Model Organisms for Single Enzyme Deficiencies

As many metabolic pathways are affected in PBD, multiple metabolites have fluctuated in patients. Disruption of a specific pathway could aid in the identification of metabolite(s) that induce a distinct human symptom, such as neurological atrophy. Accordingly, several model organisms, which were deficient in an enzyme involved in the peroxisomal metabolic pathway, were studied (Table 1). In these model organisms, the metabolism of limited peroxisomal metabolites is affected.

6.3.1 ACOX1 Deficiency Mouse Model

ACOX1 is the first enzyme in the VLCFA β-oxidation process (Fig. 1). In patients with ACOX1 deficiency, β-oxidation of VLCFAs is affected, which results in the accumulation of VLCFAs as the primary biochemical hallmark. In patients, symptoms include neonatal muscular hypotonia, hepatomegaly, leukodystrophy in cerebrum and cerebellum, and general growth retardation (see Sect. 5.2.1). The phenotypes of targeted *Acox1* knock-out mice were milder than the symptoms observed in human patients (Fan et al. 1996, 1998; Sheridan et al. 2011). The mutant mice exhibited growth retardation and infertility but were fully viable. Severe hepatic steatosis was observed in the mutant mice, which subsequently led to hepatocellular death and focal proliferation (regeneration). Steatosis could result in hepatic adenomas and carcinomas. Enhanced level of VLCFA in *Acox1* mutant mice was suspected to induce the associated phenotypes. In human patients, liver lipidosis also develops and in some cases cell death and focal proliferation are observed. However, hepatic adenomas and carcinomas are uncommon in human patients. There were no other comparable phenotypes observed between human patients and the mutant mice. Psychomotor defects were not reported in the mutant mice and the defects in their brain were undescribed. The reason for the phenotypic divergence between human patients and *Acox1* KO mice is unknown. However, it can be attributed to the physiological differences between humans and mice.

6.3.2 D-BP Deficiency Mouse Model

D-bifunctional protein (D-BP, encoded by *HSD17B4* gene) is involved in the second and third steps of β-oxidation of VLCFAs and pristanic acid (obtained by the α-oxidation of phytanic acid), as well as in the bile acid synthesis (Fig. 1). Patients with D-BP deficiency accumulate VLCFAs, branched fatty acids (on dietary intake), and bile acid intermediates, such as 3α,7α-dihydroxycholestanoic acid (DHCA) and 3α,7α,12α-trihydroxycholestanoic acid (THCA). Patients exhibit symptoms such as hypotonia from birth, ocular and hearing impairments, liver abnormality, abnormal brain development with neural migration defect, and hypomyelination and demyelination (see Sect. 5.2.1). The symptoms of D-BP deficient patients are more severe compared to patients with ACOX1 deficiency, reflecting the involvement of D-BP in multiple metabolic pathways. D-BP KO mice accumulated VLCFAs, bile acid intermediates, and branched fatty acids as in human patients (Baes et al. 2000; Huyghe et al. 2006a, b). They showed marked growth retardation and were highly weak in the neonatal stage. However, a considerable number of mutant mice (around 70%) survived to adulthood. The females were fertile and most of the males were infertile (Huyghe et al. 2006b). Although no neural migration defects were observed during brain development, aged mutant mice had defects in the locomotory behavior and exhibited signs of neuroinflammation. Studies on conditional KO mice revealed that the loss of D-BP is responsible for cerebellar atrophy with demyelination, neuroinflammation, and Purkinje cell death. Cataract emerged as an eye phenotype (Baes et al. 2000). However, D-BP KO mice failed to exhibit some human symptoms, including cerebral atrophy, which again highlights the physiological differences between mice and humans.

6.3.3 SCPx Deficiency Mouse Model

The function of sterol carrier protein X (SCPx) was studied using KO mice (Seedorf et al. 1998) long before the first human patient was identified (Ferdinandusse et al. 2006). *SCP2* gene encodes two proteins: SCPx and sterol carrier protein 2 (SCP2), as a result of transcription initiation from two independently regulated promoters. SCPx protein is a peroxisome-residing thiolase enzyme and mediates the final step of peroxisomal β-oxidation (Fig. 1). On the other hand, SCP2 is thought to be a cytosolic lipid transfer protein. Because another thiolase, 3-ketoacyl-CoA thiolase encoded by *ACAA1* gene, can compensate for the loss of SCPx in VLCFA β-oxidation, loss of SCPx function induces accumulation of branched fatty acids, i.e. phytanic acid, pristanic acid, and bile acid intermediates (DHCA and THCA) as a biochemical hallmark. The *Scp2/Scpx* KO mice were normal and fertile, but degradation of the branched fatty acid was considerably impaired, which increased the level of phytanic acid and pristanic acid markedly on high fat or high phytol diets. When *Scp2/Scpx* KO mice were fed with high phytol diet, they exhibited ataxia and signs of neuropathy (Seedorf et al. 1998). Scpx-specific null mice were also studied (Atshaves et al. 2007) and found to be phenotypically normal and viable. However, dietary intake of phytol led to the accumulation of phytanic acid and subsequently hepatic lesions. Notably, the degree of this effect was gender related so that male KO mice were more sensitive to this treatment than female counterparts. *SCPx* deficiency in humans is extremely rare, therefore the symptomatic details are not fully clear. Nevertheless, *Scp2/Scpx* deficient mice exhibit phenotypes similar to human symptoms and are an attractive disease model.

6.3.4 GNPAT and AGPS Mutant Mice: RCDP Type 2 and Type3 Models

Dihydroxyacetone phosphate acyltransferase (GNPAT) and AGPS exclusively constitute the first and second steps of ether phospholipid synthesis in peroxisomes (Fig. 1). Depletion of either protein results in the reduction of ether phospholipids, typically plasmalogens. In humans, GNPAT and AGPS deficiencies cause RCDP type 2 and RCDP type 3, respectively. *Gnpat* KO mice exhibit phenotypes similar to the human symptoms, such as growth retardation, cataract, and high mortality in early life. However, rhizomelia and chondrodysplasia punctata were not observed (Rodemer et al. 2003; Teigler et al. 2009). The mutant males showed infertility with testicular atrophy and were associated with early death compared to their female siblings. Biochemical analysis revealed severe depletion of plasmalogens and a reduction of docosahexaenoic acid (DHA; abundant in plasmalogens) in the brain. Locomotory defects with cerebellar ataxia were also observed. In the cerebellum, multiple defects such as foliation, myelination, and Purkinje cell innervation defects were observed. In a spontaneous hypomorphic mutant for *Agps* (*blind sterile 2*, *bs2*), cataract and male sterility with testicular atrophy were observed, along with the decrease in the level of ether phospholipids (Liegel et al. 2011), suggesting that the lens and testes are highly sensitive to the loss of ether phospholipids/plasmalogens. The *bs2* mutation on *Agps* gene also showed increased fetal lethality, however, some survived to adulthood, though their growth rate was slow. Remarkably, they demonstrated a shortening of humerus bones (but not of femur bones) partially recapitulating rhizomelia observed in human patients (Liegel et al. 2014), which was not clearly observed in other RCDP model mice (see above). The partial rhizomelia seen in this mouse model suggests that rhizomelia is the not the consequence of defects in early bone formation but occurs due to the defects in bone growth during later development.

6.3.5 X-ALD Model Mouse

X-linked adrenoleukodystrophy (X-ALD) is caused by a mutation in *ABCD1* gene on the X chromosome. *ABCD1* encodes ATP-binding cassette (ABC) transporter type D1 protein embedded in the peroxisomal membrane, and imports VLCFA-CoAs into the peroxisomal matrix for their β-oxidation (Fig. 1). X-ALD patients (mostly males) exhibit an accumulation of VLCFAs, resulting from defective VLCFA degradation, which is the primary consequence of this disease. Progressive cerebral leukodystrophy develops in some patients; this is categorized as childhood, adolescent, and adult cerebral ALD (CCALD, AdolCALD, and ACALD). Locomotor ataxia with axonopathy is another common symptom that develops in adult patients (adrenomyeloneuropathy, AMN). Adrenal insufficiency (Addison's disease) and cerebellar ataxia are also common in patients. Female carriers often develop AMN-like symptoms in old age. X-ALD is the most frequent peroxisomal disease, with an incidence of 1 in 17,000 births; therefore, understanding its pathology, especially that of the cerebral ALD, is highly desired (see Sect. 5.3). However, leukodystrophy did not manifest in the *Abcd1* KO X-ALD mouse model (Kobayashi et al. 1997; Lu et al. 1997; Forss-Petter et al. 1997). AMN-like phenotypes (axonopathy) appeared in aged KO mice, suggesting that AMN could be the basic form of this disease (Pujol et al. 2002). It is possible that there is a secondary modifier that causes demyelination followed by leukodystrophy in the cerebral form of human X-ALD patients, which might be absent in the X-ALD mouse model.

6.3.6 X-ALD Zebrafish Model

Because there are no sex chromosomes in zebrafish, the X-ALD model was produced through the biallelic disruption of the zebrafish *abcd1* gene (Strachan et al. 2017). X-ALD zebrafish model showed phenotypes, such as developmental defects in the adrenal grand, reduced oligodendrocyte precursors in the CNS (brain and spinal cord), increased cell death in the brain, hypomyelination of the spinal cord, and decreased overall survival rate. Moreover, VLCFAs were elevated in the tissues. The X-ALD zebrafish model also showed decreased swimming activity in the larval and adult stages. The neuronal phenotype of the X-ALD zebrafish model was somewhat stronger compared to that of the X-ALD mouse model, however, unlike the X-ALD mouse model, AMN-like progressive axonopathy was not observed in the X-ALD zebrafish model. Developmental reduction of the oligodendrocyte precursors, some of which consequently form a myelin sheath, was observed as a new aspect of X-ALD using the zebrafish model and needs to be confirmed in other research models as it might be related to the emergence of leukodystrophy and axonopathy in human patients.

6.3.7 AMACR Deficiency Mouse Model

The racemase enzyme encoded by *AMACR* gene is required for the racemization of α-methyl-branched fatty acids such as pristanic acid, DHCA, and THCA, from *R* configuration into *S* configuration in order to make them processable in the subsequent β-oxidation cascade (Fig. 1). In human *AMACR* deficiency, patients accumulate pristanic acid and phytanic acid, as well as DHCA and THCA with *R* configuration. The symptoms in patients include liver cholestasis and late onset sensorimotor polyneuropathy. *Amacr* KO mice were established, and the accumulation of pristanic acid, phytanic acid, DHCA and THCA was confirmed (Savolainen et al. 2004). The mutant mice, however, did not show any symptoms unless a phytol diet was supplied. Liver failure, but not neuropathic phenotype, developed on a phytol diet in mutant mice, so that the elimination of phytol is suggested for patients with *AMACR* deficiency.

6.4 Conclusions

Studies of peroxisomal disorders using model organisms have provided many clues to understand the pathology of this devastating disease group. Most of the studies have been conducted using a mouse model because of the close relationship between mice and humans in terms of the basic body plan and physiology; however, other model organisms, such as the zebrafish and the fruit fly are emerging as alternative model systems. Here, we have summarized the major models of peroxisomal disorders. There are several additional models that have been examined but have not been listed here. Mutant mice for some of the integral and membrane peroxisomal proteins, such as Catalase, Abcd3 (also known as Pxmp1/PMP70), Peroxisomal membrane protein 2 (Pxmp2/PMP22), were established (Shaffer and Preston 1990; Ferdinandusse et al. 2015; Rokka et al. 2009). In the nematode, *Caenorhabditis elegans*, some mutants for peroxisomal genes were studied, which might be informative in comparing the consequences of defective peroxisomal functions between lower and higher eukaryotes (Thieringer et al. 2003). In the model organisms, not every human symptom is reproduced. However, it is possible to study specific symptoms that are common between humans and a model organism to reveal their fundamental biological backgrounds. Each organism has its experimental advantages, such as the high genetic tractability and quick generation time of the fruit fly, experimental manipulability and availability for chemical assays of the zebrafish, and the mammalian physiology of the mouse; therefore, the nature of peroxisomal disorders can be examined from multiple angles using this variety of animal models. By combining the experimental outcomes from each research model, we could comprehensively reveal the yet unknown pathology of peroxisomal disorders and, hopefully, develop beneficial treatment strategies in the near future.

References

Atshaves BP, McIntosh AL, Landrock D, Payne HR, Mackie JT, Maeda N, Ball J, Schroeder F, Kier AB (2007) Effect of SCP-x gene ablation on branched-chain fatty acid metabolism. Am J Physiol Gastrointest Liver Physiol 292:G939–G951

Baes M, Gressens P, Baumgart E, Carmeliet P, Casteels M, Fransen M, Evrard P, Fahimi D, Declercq PE, Collen D et al (1997) A mouse model for Zellweger syndrome. Nat Genet 17:49–57

Baes M, Huyghe S, Carmeliet P, Declercq PE, Collen D, Mannaerts GP, Van Veldhoven PP (2000) Inactivation of the peroxisomal multifunctional protein-2 in mice impedes the degradation of not only 2-methyl-branched fatty acids and bile acid intermediates but also of very long chain fatty acids. J Biol Chem 275:16329–16336

Braverman N, Zhang R, Chen L, Nimmo G, Scheper S, Tran T, Chaudhury R, Moser A, Steinberg S (2010) A Pex7 hypomorphic mouse model for plasmalogen deficiency affecting the lens and skeleton. Mol Genet Metab 99:408–416

Brites P, Motley AM, Gressens P, Mooyer PAW, Ploegaert I, Everts V, Evrard P, Carmeliet P, Dewerchin M, Schoonjans L et al (2003) Impaired neuronal migration and endochondral ossification in Pex7 knockout mice: a model for rhizomelic chondrodysplasia punctata. Hum Mol Genet 12:2255–2267

Brites P, Mooyer PA, El Mrabet L, Waterham HR, Wanders RJ (2009) Plasmalogens participate in very-long-chain fatty acid-induced pathology. Brain 132:482–492

Bülow MH, Wingen C, Senyilmaz D, Gosejacob D, Sociale M, Bauer R, Schulze H, Sandhoff K, Teleman AA, Hoch M et al (2018) Unbalanced lipolysis results in lipotoxicity and mitochondrial damage in peroxisome-deficient. Mol Biol Cell 29:396–407

Chen H, Liu Z, Huang X (2010) Drosophila models of peroxisomal biogenesis disorder: peroxins are required for spermatogenesis and very-long-chain fatty acid metabolism. Hum Mol Genet 19:494–505 da Silva TF, Eira J, Lopes AT, Malheiro AR, Sousa V, Luoma A, Avila RL, Wanders RJ, Just WW, Kirschner DA et al (2014) Peripheral nervous system plasmalogens regulate Schwann cell differentiation and myelination. J Clin Invest 124:2560–2570

De Munter S, Verheijden S, Vanderstuyft E, Malheiro AR, Brites P, Gall D, Schiffmann SN, Baes M (2016) Early-onset Purkinje cell dysfunction underlies cerebellar ataxia in peroxisomal multifunctional protein-2 deficiency. Neurobiol Dis 94:157–168

Fan CY, Pan J, Chu R, Lee D, Kluckman KD, Usuda N, Singh I, Yeldandi AV, Rao MS, Maeda N et al (1996) Hepatocellular and hepatic peroxisomal alterations in mice with a disrupted peroxisomal fatty acyl-coenzyme A oxidase gene. J Biol Chem 271:24698–24710

Fan CY, Pan J, Usuda N, Yeldandi AV, Rao MS, Reddy JK (1998) Steatohepatitis, spontaneous peroxisome proliferation and liver tumors in mice lacking peroxisomal fatty acyl-CoA oxidase. Implications for peroxisome proliferator-activated receptor alpha natural ligand metabolism. J Biol Chem 273:15639–15645

Faust PL (2003) Abnormal cerebellar histogenesis in PEX2 Zellweger mice reflects multiple neuronal defects induced by peroxisome deficiency. J Comp Neurol 461:394–413

Faust PL, Hatten ME (1997) Targeted deletion of the PEX2 peroxisome assembly gene in mice provides a model for Zellweger syndrome, a human neuronal migration disorder. J Cell Biol 139:1293–1305

Faust PL, Su HM, Moser A, Moser HW (2001) The peroxisome deficient PEX2 Zellweger mouse: pathologic and biochemical correlates of lipid dysfunction. J Mol Neurosci 16:221–289

Faust JE, Manisundaram A, Ivanova PT, Milne SB, Summerville JB, Brown HA, Wangler M, Stern M, McNew JA (2014) Peroxisomes are required for lipid metabolism and muscle function in Drosophila melanogaster. PLoS One 9:e100213

Ferdinandusse S, Kostopoulos P, Denis S, Rusch H, Overmars H, Dillmann U, Reith W, Haas D, Wanders RJA, Duran M et al (2006) Mutations in the gene encoding Peroxisomal Sterol Carrier Protein X (SCPx) Cause Leukencephalopathy with Dystonia and motor neuropathy. Am J Hum Genet 78:1046–1052

Ferdinandusse S, Jimenez-Sanchez G, Koster J, Denis S, Van Roermund CW, Silva-Zolezzi I, Moser AB, Visser WF, Gulluoglu M, Durmaz O et al (2015) A novel bile acid biosynthesis defect due to a deficiency of peroxisomal ABCD3. Hum Mol Genet 24:361–370

Forss-Petter S, Werner H, Berger J, Lassmann H, Molzer B, Schwab MH, Bernheimer H, Zimmermann F, Nave KA (1997) Targeted inactivation of the X-linked adrenoleukodystrophy gene in mice. J Neurosci Res 50:829–543

Hanson MG, Fregoso VL, Vrana JD, Tucker CL, Niswander LA (2014) Peripheral nervous system defects in a mouse model for peroxisomal biogenesis disorders. Dev Biol 395:84–95

Hiebler S, Masuda T, Hacia JG, Moser AB, Faust PL, Liu A, Chowdhury N, Huang N, Lauer A, Bennett J et al (2014) The Pex1-G844D mouse: a model for mild human Zellweger spectrum disorder. Mol Genet Metab 111:522–532

Hulshagen L, Krysko O, Bottelbergs A, Huyghe S, Klein R, Van Veldhoven PP, De Deyn PP, D'Hooge R, Hartmann D, Baes M (2008) Absence of functional peroxisomes from mouse CNS causes dysmyelination and axon degeneration. J Neurosci 28:4015–4027

Huyghe S, Schmalbruch H, De Gendt K, Verhoeven G, Guillou F, Van Veldhoven PP, Baes M (2006a) Peroxisomal multifunctional protein 2 is essential for lipid homeostasis in sertoli cells and male fertility in mice. Endocrinology 147:2228–2236

Huyghe S, Schmalbruch H, Hulshagen L, Van Veldhoven P, Baes M, Hartmann D (2006b) Peroxisomal multifunctional protein-2 deficiency causes motor deficits and glial lesions in the adult central nervous system. Am J Pathol 168:1321–1334

Janssen A, Gressens P, Grabenbauer M, Baumgart E, Schad A, Vanhorebeek I, Brouwers A, Declercq PE, Fahimi D, Evrard P et al (2003) Neuronal migration depends on intact peroxisomal function in brain and in extraneuronal tissues. J Neurosci 23:9732–9741

Keane MH, Overmars H, Wikander TM, Ferdinandusse S, Duran M, Wanders RJ, Faust PL (2007) Bile acid treatment alters hepatic disease and bile acid transport in peroxisome-deficient PEX2 Zellweger mice. Hepatology 45:982–997

Kobayashi T, Shinnoh N, Kondo A, Yamada T (1997) Adrenoleukodystrophy protein-deficient mice represent abnormality of very long chain fatty acid metabolism. Biochem Biophys Res Commun 232:631–636

Komljenovic D, Sandhoff R, Teigler A, Heid H, Just WW, Gorgas K (2009) Disruption of blood-testis barrier dynamics in ether-lipid-deficient mice. Cell Tissue Res 337:281–299

Krysko O, Hulshagen L, Janssen A, Schütz G, Klein R, De Bruycker M, Espeel M, Gressens P, Baes M (2007) Neocortical and cerebellar developmental abnormalities in conditions of selective elimination of peroxisomes from brain or from liver. J Neurosci Res 85:58–72

Krysko O, Stevens M, Langenberg T, Fransen M, Espeel M, Baes M (2010) Peroxisomes in zebrafish: distribution pattern and knockdown studies. Histochem Cell Biol 134:39–51

Li X, Baumgart E, Dong GX, Morrell JC, Jimenez-Sanchez G, Valle D, Smith KD, Gould SJ (2002a) PEX11alpha is required for peroxisome proliferation in response to 4-phenylbutyrate but is dispensable for peroxisome proliferator-activated receptor alpha-mediated peroxisome proliferation. Mol Cell Biol 22:8226–8240

Li X, Baumgart E, Morrell JC, Jimenez-Sanchez G, Valle D, Gould SJ (2002b) PEX11 beta deficiency is lethal and impairs neuronal migration but does not abrogate peroxisome function. Mol Cell Biol 22:4358–4365

Liegel RP, Ronchetti A, Sidjanin DJ (2014) Alkylglycerone phosphate synthase (AGPS) deficient mice: models for rhizomelic chondrodysplasia punctate type 3 (RCDP3) malformation syndrome. Mol Genet Metab Rep 1:299–311

Liegel R, Chang B, Dubielzig R, Sidjanin DJ (2011) Blind sterile 2 (bs2), a hypomorphic mutation in Agps, results in cataracts and male sterility in mice. Mol Genet Metab 103:51–59

Lu JF, Lawler AM, Watkins PA, Powers JM, Moser AB, Moser HW, Smith KD (1997) A mouse model for X-linked adrenoleukodystrophy. Proc Natl Acad Sci U S A 94:9366–9371

Mast FD, Li J, Virk MK, Hughes SC, Simmonds AJ, Rachubinski RA (2011) A Drosophila model for the Zellweger spectrum of peroxisome biogenesis disorders. Dis Model Mech 4:659–672

Maxwell M, Bjorkman J, Nguyen T, Sharp P, Finnie J, Paterson C, Tonks I, Paton BC, Kay GF, Crane DI (2003) Pex13 Inactivation in the mouse disrupts peroxisome biogenesis and leads to a Zellweger syndrome phenotype. Mol Cell Biol 23:5947–5957

Nakayama M, Sato H, Okuda T, Fujisawa N, Kono N, Arai H, Suzuki E, Umeda M, Ishikawa HO, Matsuno K (2011) Drosophila carrying Pex3 or Pex16 mutations are models of zellweger syndrome that reflect its symptoms associated with the absence of peroxisomes. PLoS One 6:e22984

Pujol A, Hindelang C, Callizot N, Bartsch U, Schachner M, Mandel JL (2002) Late onset neurological phenotype of the X-ALD gene inactivation in mice: a mouse model for adrenomyeloneuropathy. Hum Mol Genet 11:499–505

Rodemer C, Thai TP, Brugger B, Kaercher T, Werner H, Nave KA, Wieland F, Gorgas K, Just WW (2003) Inactivation of ether lipid biosynthesis causes male infertility, defects in eye development and optic nerve hypoplasia in mice. Hum Mol Genet 12:1881–1895

Rokka A, Antonenkov VD, Soininen R, Immonen HL, Pirilä PL, Bergmann U, Sormunen RT, Weckström M, Benz R, Hiltunen JK (2009) Pxmp2 is a channel-forming protein in Mammalian peroxisomal membrane. PLoS One 4:e5090

Savolainen K, Kotti TJ, Schmitz W, Savolainen TI, Sormunen RT, Ilves M, Vainio SJ, Conzelmann E, Hiltunen JK (2004) A mouse model for alpha-methylacyl-CoA racemase deficiency: adjustment of bile acid synthesis and intolerance to dietary methyl-branched lipids. Hum Mol Genet 13:955–965

Schutgens RBH, Bouman IW, Nijenhuis AA, Wanders RJA, Frumau MEJ (1993) Profiles of very-long-chain fatty acids in plasma, fibroblasts, and blood cells in Zellweger syndrome, X-linked adrenoleukodystrophy, and rhizomelic chondrodysplasia punctata. Clin Chem 39:1632–1637

Seedorf U, Raabe M, Ellinghaus P, Kannenberg F, Fobker M, Engel T, Denis S, Wouters F, Wirtz KWA, Wanders RJA et al (1998) Defective peroxisomal catabolism of branched fatty acyl coenzyme A in mice lacking the sterol carrier protein-2/sterol carrier protein-x gene function. Genes Dev 12:1189–1201

Sellin J, Wingen C, Gosejacob D, Senyilmaz D, Hänschke L, Büttner S, Meyer K, Bano D, Nicotera P, Teleman AA et al (2018) Dietary rescue of lipotoxicity-induced mitochondrial damage in Peroxin19 mutants. PLoS Biol 16:e2004893

Shaffer JB, Preston KE (1990) Molecular analysis of an acatalasemic mouse mutant. Biochem Biophys Res Commun 173:1043–1050

Sheridan R, Lampe K, Shanmukhappa SK, Putnam P, Keddache M, Divanovic S, Bezerra J, Hoebe K (2011) Lampe1: an ENU-germline mutation causing spontaneous hepatosteatosis identified through targeted exon-enrichment and next-generation sequencing. PLoS One 6:e21979

Strachan LR, Stevenson TJ, Freshner B, Keefe MD, Miranda Bowles D, Bonkowsky JL (2017) A zebrafish model of X-linked adrenoleukodystrophy recapitulates key disease features and demonstrates a developmental requirement for abcd1 in oligodendrocyte patterning and myelination. Hum Mol Genet 26:3600–3614

Teigler A, Komljenovic D, Draguhn A, Gorgas K, Just WW (2009) Defects in myelination, paranode organization and Purkinje cell innervation in the ether lipid-deficient mouse cerebellum. Hum Mol Genet 18:1897–1908

Thieringer H, Moellers B, Dodt G, Kunau W-H, Driscoll M (2003) Modeling human peroxisome biogenesis disorders in the nematode Caenorhabditis elegans. J Cell Sci 116:1797–1804

Weng H, Ji X, Naito Y, Endo K, Ma X, Takahashi R, Shen C, Hirokawa G, Fukushima Y, Iwai N (2013) Pex11α deficiency impairs peroxisome elongation and division and contributes to nonalcoholic fatty liver in mice. Am J Physiol Endocrinol Metab 304:E187–E196

Zhang R, Chen L, Jiralerspong S, Snowden A, Steinberg S, Braverman N (2010) Recovery of PEX1-Gly843Asp peroxisome dysfunction by small-molecule compounds. Proc Natl Acad Sci U S A 107:5569–5574

Chapter 7
Diagnosis of Peroxisomal Disorders

Nobuyuki Shimozawa

Abstract Conventionally, diagnosis of peroxisomal disorders (PD) has been performed by enzymatic and genetic analysis after measurement of peroxisomal metabolites in blood. With the progress of genome analysis technology accompanying the advent of the next generation sequencer, including whole-exome sequencing (WES), the approach of undiagnosed diseases by comprehensive genomic analysis has become a global trend. In PD also, not only new phenotypes of known pathogenic genes but also newly identified PD have been reported by WES, mainly in the patients with undiagnosed neurological diseases. Therefore, PD diagnostic system combining screening of peroxisomal metabolites and WES should be useful for efficient diagnosis.

On the one hand, diagnosis of adrenoleukodystrophy (ALD) should be done as soon as possible, as early hematopoietic stem cell transplantation is critically important for improving the prognosis of brain-type ALD. We have established a rapid diagnostic system combined with measurement of very long chain fatty acids and detection of *ABCD1* mutations. Furthermore, the pre-symptomatic diagnosis by family analysis of probands as well as newborn screening, combined with long-term follow-up system are also important for overcoming this intractable disease. In this chapter, we describe the efficient and prompt approach for diagnosis of PD and ALD and introduce our diagnostic system in Japan.

Keywords Peroxisomal disorders · VLCFA · Plasmalogen · Phytanic acid · Next generation sequencing · Whole-exome sequencing · Undiagnosed diseases · Chromatography/mass spectrometry · Adrenoleukodystrophy · Genetic testing · Presymptomatic diagnosis · Carrier detection · *ABCD1* · Newborn screening

N. Shimozawa (✉)
Division of Genomics Research, Life Science Research Center, Gifu University, Gifu, Japan

Department of Pediatrics, Gifu University School of Medicine, Gifu, Japan
e-mail: nshim@gifu-u.ac.jp

T. Imanaka, N. Shimozawa (eds.), *Peroxisomes: Biogenesis, Function, and Role in Human Disease*, https://doi.org/10.1007/978-981-15-1169-1_7

7.1 Introduction

We have been studying the peroxisomal disorders (PD) in a main diagnostic center of Japan for more than 30 years, and until now 77 Japanese patients with peroxisome biogenesis disorders (PBD), 5 with acyl-CoA oxidase 1 (ACOX1) deficiency, 11 with D-bifunctional protein deficiency (DBP), 2 with rhizomelic chondrodysplasia punctata (RCDP) type 2, 1 with RCDP type 3 and 208 Japanese families with adrenoleukodystrophy (ALD) were diagnosed (Takashima et al. 2019) (Table 7.1). In the meantime, we have identified the first discovered responsible gene for PBD, *PEX2* (Shimozawa et al. 1992), and thereafter, 12 genetic complementary groups in Zellweger spectrum disorders (ZSD) and several disease-causing genes of PD (Shimozawa et al. 2004).

In the conventional diagnostic system, after screening for peroxisomal metabolites such as saturated very long-chain fatty acids (VLCFA) and plasmalogen in serum or plasma of patients by gas chromatography/mass spectrometry (GCMS) analysis (Takemoto et al. 2003), the disease-causable genes were detected by biochemical, morphological and molecular analysis. However, in recent years, PD with broad phenotypes and newly identified genotypes have been evaluated due to advanced analytical techniques of peroxisomal metabolism and the appearance of next-generation sequencing (NGS). Therefore, it is necessary to develop an efficient diagnostic system combining the conventional methods (Shimozawa 2011) with liquid chromatography-mass spectrometry (LCMS) analysis and NGS. Furthermore, in symptomatic cerebral ALD it is essential to diagnose as prompt as possible for the improvement of the affected patients' prognosis. Here, we introduce the current status of PD diagnosis and our system for making a broad and prompt diagnosis in Japan.

7.2 A conventional Diagnostic System for Typical Cases of Peroxisomal Disorders

For the diagnosis of PD, we evaluate the peroxisomal metabolites including VLCFA, phytanic acid, docosahexaenoic acid (DHA), and plasmalogen in serum or plasma of patients suspected for PD by clinical examination, using GCMS analysis (Takemoto et al. 2003). Further diagnosis is proceeded with reference to clinical features and abnormal findings of peroxisomal metabolites (Table 7.1). When impaired β-oxidation of fatty acids including ALD, ACOX1, DBP, sterol carrier protein X (SCPx), 2-Methylacyl-CoA racemase (AMACR), and acyl-CoA-binding domain-containing protein 5 (ACBD5) deficiencies, and RCDP and Refsum disease are typically suspected by the clinical findings and peroxisomal metabolites abnormalities, we proceed to genetic testing of the tentative disease-causing gene by Sanger sequencing. In cases of suspected ZSD and similar phenotypic findings as ZSD such as ACOX1 and DBP deficiencies, we choose further biochemical and

Table 7.1 Peroxisomal metabolites abnormalities in diagnosis

	VLCFA	Phytanic acid[a]	Pristanic acid[a]	Bile acids D/THCA	Plasmalogens	peroxisomes in fibroblasts (shape)	Japanese patients[b] (life span)
A. Peroxisome biogenesis disorders (PBD)							
Zellweger spectrum disorders							
Zellweger syndrome	↑	N ~↑	N ~↑	↑	↓	–	60 (1–21 m)
Neonatal adrenoleukodystrophy	↑	N ~↑	N ~↑	↑	↓	–/mosaic/↓	4 (20–33 m)
Infantile Refsum disease	N ~↑	N ~↑	N ~↑	N ~↑	N ~↓	mosaic/↓	3 (>6y, >33y, 36y)
Broad phenotypes	N ~↑	N ~↑	N ~↑	N ~↑	N	+/↓	6 (surviving)
RCDP type 1, 5	N	↑	N	N	↓	+	type 1: 4 (0 m-)
PEX7 causing Refsum disease	N	↑	N	N	N	+	
Mulibrey nanism						+	
PEX11β deficiency	N	N	N	N	N	↓ (elongated)	
DNM1L deficiency	↑	N	N			↓ (varied in size)	
GDAP1 deficiency						elongated?	
B. Single enzyme deficiencies							
Adrenoleukodystrophy	↑	N	N	N	N	+	208 families
Acyl-CoA oxidase 1 deficiency	↑	N	N	N	N	+ (larger & fewer)	5 (3y-)
D-bifunctional protein deficiency	↑	N ~↑	N ~↑	↑	N	+ (larger & fewer)	11 (0y-)
Sterol carrier protein X deficiency	N	↑	↑	↑	N	+	
2-Methylacyl-CoA racemase deficiency	N	↑	↑	↑	N	+	
Acyl-CoA-binding domain-containing protein 5 deficiency	↑	N	N	N	N	+	
Acyl-CoA oxidase 2 deficiency	N	N	N	↑	N	+	
Peroxisomal membrane protein, 70KD (PMP70) (ABCD3) deficiency	[c]	N	N	↑	↓	+ (larger & fewer)	

(continued)

Table 7.1 (continued)

	VLCFA	Phytanic acid[a]	Pristanic acid[a]	Bile acids D/THCA	Plasmalogens	peroxisomes in fibroblasts (shape)	Japanese patients[b] (life span)
Bile acid-CoA: amino acid N-acyltransferase deficiency	N	N	N	[d]	N	+	
Refsum disease	N	↑	N	N	N	+	
RCDP type 2-4	N	N	N	N	↓	+	type 2: 1 (>2y) type 3: 2 (8y,16y)

RCDP rhizomelic chondrodysplasia punctata, *N* normal, *D/THCA* di-/trihydroxycholestanoic acid
[a]Depend on dietary intake
[b]Diagnosis at Gifu University during 1985–mid-2019
[c]Decreased C22:0
[d]Decreased glycine and taurine bile acid conjugates

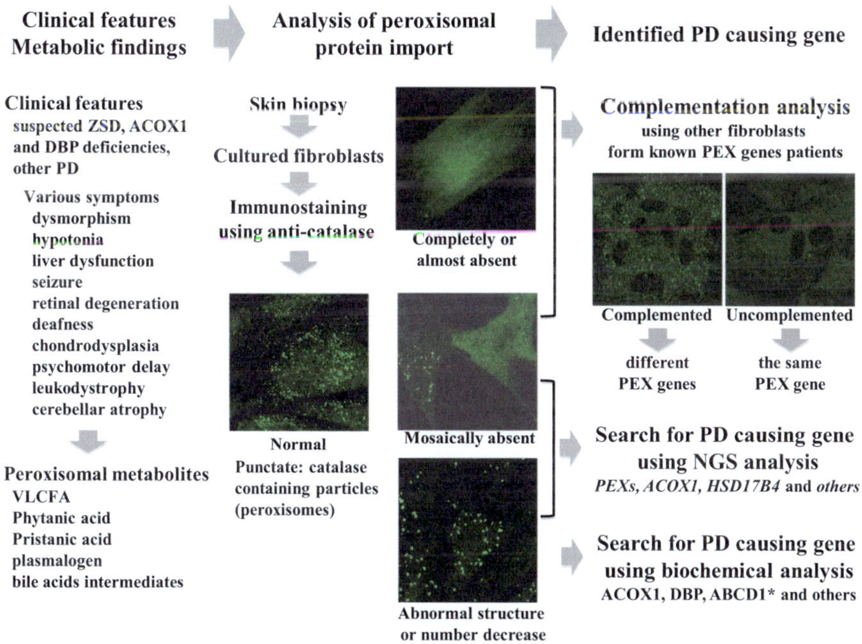

Fig. 7.1 Conventional diagnostic system of peroxisomal disorders in Japan. ABCD1*. Immunoblot and immunostaining analysis using anti ABCD1 (ALDP) antibody can detect patients with contiguous *ABCD1 DXS1357E* deletion syndrome (CADDS) (Corzo et al. 2002). Peroxisomes of fibroblasts from the patients appear normal whereas no detectable ABCD1 protein

immunocytochemical approaches using fibroblasts obtained by skin biopsy from the patients (Shimozawa 2011). When the immunostaining results using anti-catalase antibody which is an excellent diagnostic marker of peroxisomal localization in the fibroblasts show completely impaired peroxisomal biogenesis, we perform complementation analysis in the fibroblasts of the patients with 12 known complementation groups of ZS by cell fusion. When the peroxisomes in the patients with abnormal peroxisomal metabolites showed mosaically absent or aberrant structure and number, we proceed to search for the variants in known disease-caused candidate genes of *PEXs*, *ACOX1*, *HSD17B4* and so on, using NGS analysis, and following confirm with the Sanger sequencing (Fig. 7.1). Even if the fibroblast biochemical analysis was avoided, the genetic diagnosis might be obtained directly by the NGS analysis. However, we would like to recommend analysis of the fibroblasts because biochemical data may provide clarifying pathophysiology and further therapeutic solutions to the patients. Since 1985, we have diagnosed many Japanese patients with PD at the Gifu University (Table 7.1).

7.3 An Advanced Diagnostic System for Broad Cases of Peroxisomal Disorders

With the progress of genomic analysis technology, new genetic diseases as well as phenotypes have been diagnosed among undiagnosed patients even in the PD field; therefore, an advanced diagnostic system with latest technology should be expected.

7.3.1 New Phenotypes of Known Pathogenic Genes Identified by NGS

In *PEX* deficiencies manifesting ZSD, patients with atypical and broad phenotype including neurodegenerative diseases mainly, have been reported by WES as follows: a patient with a mutated *PEX1* presented sensorineural hearing loss, pigmentary retinopathy and normal intellect (Ventura et al. 2016), adult siblings with *PEX3* presented intellectual disability (Bjørgo et al. 2017), slowly progressive cerebellar atrophy due to *PEX10* (Renaud et al. 2016; Yamashita et al. 2017) and sensorineural hearing loss, sensorimotor polyneuropathy, cognitive delay, impaired gross, and fine motor skills, and tremor, and muscle weakness by *PEX10* (Blomqvist et al. 2017), a family with inherited peripheral neuropathies suspected *PEX12* defect (Schabhüttl et al. 2014), and neurodegenerative diseases such as cerebellar atrophy, white matter degeneration, and early-onset hereditary spastic paraplegia by *PEX16* (Ohba et al. 2013; Bacino et al. 2015; Kumar et al. 2016). Furthermore, it has been reported that defects in *PEX1* and *PEX6* occur in Heimler syndrome 1 and 2, respectively, characterized by sensorineural hearing loss, retinitis pigmentosa, amelogenesis imperfecta in secondary teeth, and nail abnormalities (Ratbi et al. 2015; Smith et al. 2016).

In β-oxidation enzyme deficiencies, Perrault syndrome with ovarian malformation, hearing loss, and cerebellar ataxia (Pierce et al. 2010), and cases with hearing loss, cerebellar ataxia, and retinal pigment degeneration (McMillan et al. 2012) have been diagnosed as DBP deficiencies by WES.

7.3.2 Newly Identified Peroxisomal Diseases by NGS

In impaired ether-phospholipid biosynthesis in peroxisome, fatty acyl-CoA reductase 1 (FAR1) deficiency was found by WES as a newly PD with severe mental disability, epilepsy, and cataracts. FAR1, a critical enzyme in plasmalogen biosynthesis reduces fatty acyl-CoAs to the respective fatty alcohols (Buchert et al. 2014). This disease was later named RCDP type 4. RCDP type 5 caused by *PEX5-long isoform* mutation classified as PBD, was also identified by WES and biochemical verification (Barøy et al. 2015). In impaired β-oxidation of fatty acids, ACBD5 deficiency was identified by autozygome analysis followed by exome sequencing

(Abu-Safieh et al. 2013). In impaired bile acids synthesis, acyl-CoA oxidase 2 (ACOX2) deficiency was also discovered by WES in an 8-y-old boy with intermittently elevated transaminase levels, liver fibrosis, mild ataxia, and cognitive impairment (Vilarinho et al. 2016).

7.3.3 *Efficient Diagnostic System Combined Peroxisomal Metabolite Screening and NGS in Japan*

The conventional diagnostic system could be difficult for the diagnosis of patients with mild and atypical phenotypes of PD or mild and subtle metabolic abnormalities of VLCFA, phytanic acid, and plasmalogens. Even though, we have diagnosed so far, several atypical cases of *PEXs* and *HSD17B4* gene defects by detailed biochemical analysis and repeated Sanger sequencing of several candidate genes. Furthermore, our screening system of peroxisomal metabolites has been advanced using LCMS, for the detection of unusual saturated and unsaturated VLCFA, ether phospholipid derivatives, and pristanic acid (Takashima et al. 2017) (Table 7.1). Currently, we are also establishing the screening of di-/trihydroxycholestanoic acid (D/THCA), a bile acid intermediate metabolite of peroxisome for the diagnosis of patients with impaired bile acids synthesis in ACOX2, PMP70, and bile acid-CoA: amino acid N-acyltransferase (BAAT) deficiencies.

Furthermore, in the WES, more than ten thousand variants were detected and each variant filtering according to the disease-cause is a complicated and time-consuming task, therefore, we have been establishing an efficient diagnostic system combining screening of peroxisomal metabolites and WES in collaboration with the Hamamatsu University School of Medicine. Our advanced diagnostic system created can efficiently search for disease-causal mutations in PD related genes by data filtering, focused on PD related genes after WES in the cases exhibiting mild or subtle peroxisomal metabolic abnormalities of VLCFA, phytanic and pristanic acids, plasmalogens, and atypical clinical symptoms (Takashima et al. 2019) (Fig. 7.2). If we cannot get candidate disease-causable gene at this stage, we will investigate further WES analysis using the samples of parents. We are also establishing a more efficient diagnostic system by resequencing the panel targeting PD candidate genes using NGS.

In disease-causing gene mutation search by comprehensive genetic analysis, it is important to verify the relationship between variants and symptoms, while in this system, as the patient has peroxisomal metabolic abnormalities, it is considered that the variants may be disease-causing. On the other hand, there have been several diseases classified as novel peroxisomal diseases due to newly elucidated gene function of previously reported pathogenic genes (*TRIM37, DNM1L* and *GDAP1*) (see Sect. 5.2). Therefore, when the pathological significance of the candidate variant or gene is unknown, it is essential to be validated by functional analysis of the variant or gene, using cultured cells or generation of a model organism by a mouse or zebrafish.

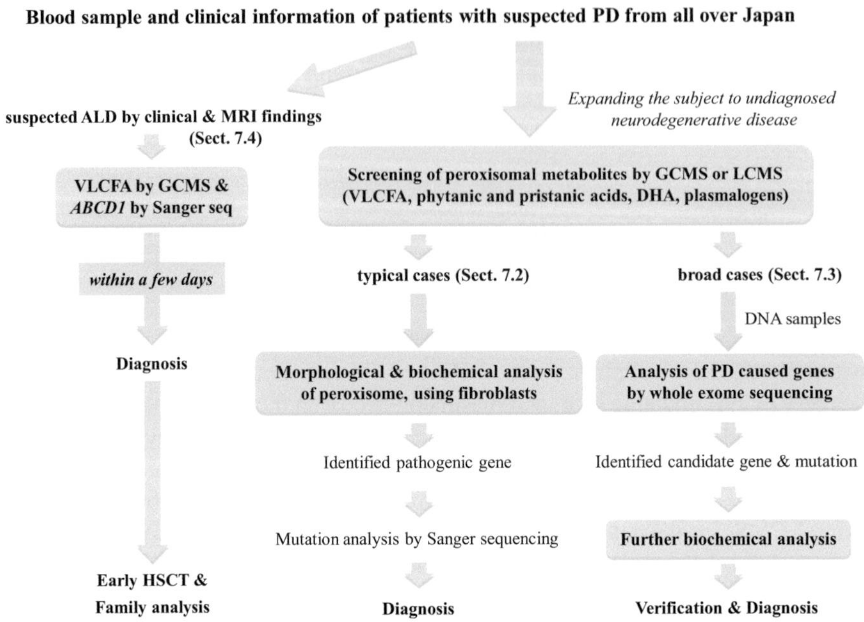

Fig. 7.2 Advanced diagnostic system of peroxisomal disorders in Japan

Expansion of concept, elucidation of the pathology and development of new therapeutic methods in PD will continue to advance by both progress of life science technology and clarifying peroxisomal functions.

7.4 Diagnostic System for ALD

7.4.1 Prompt Diagnostic System in Gifu University

To improve the prognosis of patients with ALD, it is essential to detect ALD immediately after the onset of the disease, therefore, a prompt diagnosis system is very important. We have established a prompt diagnostic system for ALD patients using combining measurement of VLCFA level in the plasma and mutation analysis of *ABCD1* (Shimozawa et al. 2011). Diagnosis can be confirmed ***within a few days***, particularly in cases of cerebral ALD at onset, and hematopoietic stem cell transplantation (HSCT) can be proceeded as soon as possible (Fig. 7.2). We have diagnosed 250 male patients and 187 female carriers with ALD and identified 133 *ABCD1* mutations in 201 Japanese families with ALD until December 2018. To confirm the pathogenic mutations of ALD the mutation database of ALD (https://adrenoleukodystrophy.info/mutations-and-variants-in-abcd1) was very helpful.

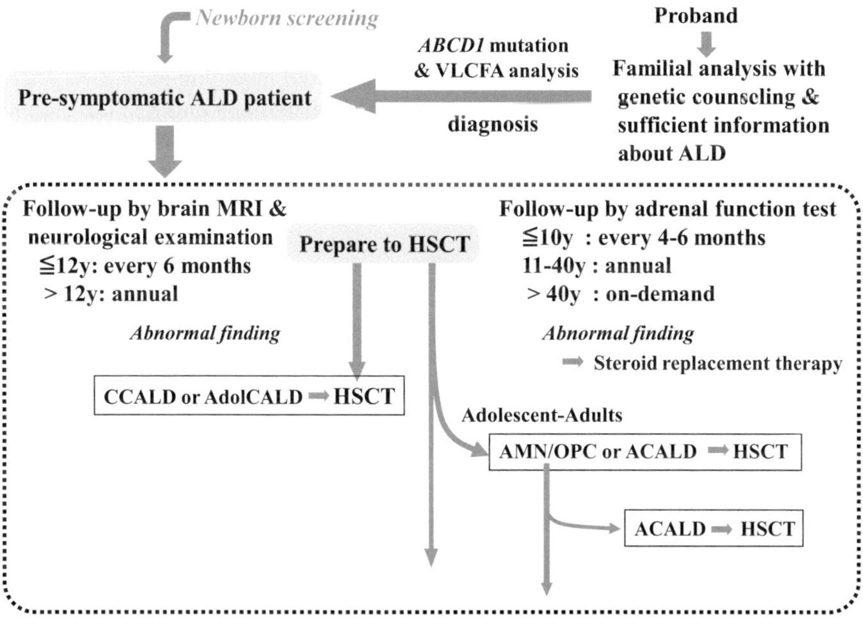

Fig. 7.3 Follow-up flowchart for presymptomatic male ALD patients

7.4.2 Presymptomatic Patients' Diagnosis

The diagnosis of male presymptomatic ALD patients can be useful not only for the early preparation for HSCT but also for the proper administration of steroid hormones to improve the disease outcome. We provide sufficient disease information to the male at risk ALD patients and their families combined with genetic counseling. If consensus is given from the patients at risk and/or their parents, we proceed to the quantification of VLCFA levels and confirm the mutations of the probands. It should be noted that *de novo* mutations occur in approximately 4.1% of probands (Wang et al. 2011). The earliest symptomatic onset age of the cerebral type and adrenal insufficiency is at the age of 2 years, however, as the earliest detection of abnormal adrenal function was 7 months of age (Huffnagel et al. 2019), earlier diagnosis is recommended.

As we cannot predict the phenotypes and prognosis before onset, it is essential to have a consistent and ongoing follow up for the diagnosed presymptomatic male ALD patients (Fig. 7.3). After diagnosis, clinical examination and brain MRI every 6 months is mandatory for the detection of abnormal signs including strabismus and subtle neuropsychological symptoms, and high signal region by FLAIR image until 12 years of age, then brain MRI and neurological examination should be annually performed (Engelen et al. 2012). When any suspected cerebral ALD abnormalities are found at onset, a fast HSCT should be conducted. Adrenal function test is recommended every 4–6 months for patients age ≤10 years, annually for those of 11–40 years, and on-demand testing for those age >40 years without symptoms

(Huffnagel et al. 2019). In the cases of carrier females at risk, genetic counseling is recommended after adulthood or at the time of pregnancy planning.

Newborn screening (NBS) for ALD has been started mainly in the United State (see 5.4.5.2) and should be considered also in Japan, including issues such as genetic counseling, accurate diagnosis and long-term follow up. For the further spread of NBS for ALD, it is important to promote research for developing methods to predict the prognosis of patients before onset.

Ethics Statement The diagnostic studies of PD are approved by the Ethical Committee of the Graduate School of Medicine, Gifu University.
Conflict of Interest The author declares no conflict of interest.

References

Abu-Safieh L, Alrashed M, Anazi S et al (2013) Autozygome-guided exome sequencing in retinal dystrophy patients reveals pathogenetic mutations and novel candidate disease genes. Genome Res 23:236–247

Bacino C, Chao YH, Seto E, Lotze T, Xia F, Jones RO, Moser A, Wangler MF (2015) A homozygous mutation in PEX16 identified by whole-exome sequencing ending a diagnostic odyssey. Mol Genet Metab Rep 5:15–18

Barøy T, Koster J, Strømme P et al (2015) A novel type of rhizomelic chondrodysplasia punctata, RCDP5, is caused by loss of the PEX5 long isoform. Hum Mol Genet 24:5845–5854

Bjørgo K, Fjær R, Mørk HH, Ferdinandusse S, Falkenberg KD, Waterham HR, Øye AM, Sikiric A, Amundsen SS, Kulseth MA, Selmer K (2017) Biochemical and genetic characterization of an unusual mild PEX3-related Zellweger spectrum disorder. Mol Genet Metab 121:325–328

Blomqvist M, Ahlberg K, Lindgren J, Ferdinandusse S, Asin-Cayuela J (2017) Identification of a novel mutation in PEX10 in a patient with attenuated Zellweger spectrum disorder: a case report. J Med Case Rep 11:218

Buchert R, Tawamie H, Smith C, Uebe S, Innes AM, Al Hallak B, Ekici AB, Sticht H, Schwarze B, Lamont RE, Parboosingh JS, Bernier FP, Abou Jamra R (2014) A peroxisomal disorder of severe intellectual disability, epilepsy, and cataracts due to fatty acyl-CoA reductase 1 deficiency. Am J Hum Genet 95:602–610

Corzo D, Gibson W, Johnson K et al (2002) Contiguous deletion of the X-linked adrenoleukodystrophy gene (ABCD1) and DXS1357E: a novel neonatal phenotype similar to peroxisomal biogenesis disorders. Am J Hum Genet 70:1520–1531

Engelen M, Kemp S, de Visser M, van Geel BM, Wanders RJ, Aubourg P, Poll-The BT (2012) X-linked adrenoleukodystrophy (X-ALD): clinical presentation and guidelines for diagnosis, follow-up and management. Orphanet J Rare Dis 7:51

Huffnagel IC, Laheji FK, Aziz-Bose R, Tritos NA, Marino R, Linthorst GE, Kemp S, Engelen M, Eichler F (2019) The natural history of adrenal insufficiency in x-linked adrenoleukodystrophy: an international collaboration. J Clin Endocrinol Metab 104:118–126

Kumar KR, Wali GM, Kamate M, Wali G, Minoche AE, Puttick C, Pinese M, Gayevskiy V, Dinger ME, Roscioli T, Sue CM, Cowley MJ (2016) Defining the genetic basis of early onset hereditary spastic paraplegia using whole genome sequencing. Neurogenetics 17:265–270

McMillan HJ, Worthylake T, Schwartzentruber J et al (2012) Specific combination of compound heterozygous mutations in 17beta-hydroxysteroid dehydrogenase type 4 (HSD17B4) defines a new subtype of D-bifunctional protein deficiency. Orphanet J Rare Dis 7:1–9

Ohba C, Osaka H, Iai M (2013) Diagnostic utility of whole exome sequencing in patients showing cerebellar and/or vermis atrophy in childhood. Neurogenetics 14:225–232

Pierce SB, Walsh T, Chisholm KM, Lee MK, Thornton AM, Fiumara A, Opitz JM, Levy-Lahad E, Klevit RE, King MC (2010) Mutations in the DBP-deficiency protein HSD17B4 cause ovarian dysgenesis, hearing loss, and ataxia of Perrault Syndrome. Am J Hum Genet 87:282–288

Ratbi I, Falkenberg KD, Sommen M et al (2015) Heimler syndrome is caused by hypomorphic mutations in the peroxisome-biogenesis genes PEX1 and PEX6. Am J Hum Genet 97:535–545

Renaud M, Guissart C, Mallaret M, Ferdinandusse S, Cheillan D, Drouot N, Muller J, Claustres M, Tranchant C, Anheim M, Koenig M (2016) Expanding the spectrum of PEX10-related peroxisomal biogenesis disorders: slowly progressive recessive ataxia. J Neurol 263:1552–1558

Schabhüttl M, Wieland T, Senderek J, Baets J, Timmerman V, De Jonghe P, Reilly MM, Stieglbauer K, Laich E, Windhager R, Erwa W, Trajanoski S, Strom TM, Auer-Grumbach M (2014) Whole-exome sequencing in patients with inherited neuropathies: outcome and challenges. J Neurol 261:970–982

Smith CE, Poulter JA, Levin AV, Capasso JE, Price S, Ben-Yosef T, Sharony R, Newman WG, Shore RC, Brookes SJ, Mighell AJ, Inglehearn CF (2016) Spectrum of PEX1 and PEX6 variants in Heimler syndrome. Eur J Hum Genet 24:1565–1571

Shimozawa N, Tsukamoto T, Suzuki Y, Orii T, Shirayoshi Y, Mori T, Fujiki Y (1992) A human gene responsible for Zellweger syndrome that affects peroxisome assembly. Science 255:1132–1134

Shimozawa N, Tsukamoto T, Nagase T, Takemoto Y, Koyama N, Suzuki Y, Komori M, Osumi T, Jeannette G, Wanders RJ, Kondo N (2004) Identification of a new complementation group of the peroxisome biogenesis disorders and PEX14 as the mutated gene. Hum Mutat 23:552–558

Shimozawa N (2011) Molecular and clinical findings and diagnostic flowchart of peroxisomal diseases. Brain and Development 33:770–776

Shimozawa N, Honda A, Kajiwara N, Kozawa S, Nagase T, Takemoto Y, Suzuki Y (2011) X-linked adrenoleukodystrophy: diagnostic and follow-up system in Japan. J Hum Genet 56:106–109

Takashima S, Toyoshi K, Itoh T, Kajiwara N, Honda A, Ohba A, Takemoto S, Yoshida S, Shimozawa N (2017) Detection of unusual very-long-chain fatty acid and ether lipid derivatives in the fibroblasts and plasma of patients with peroxisomal diseases using liquid chromatography-mass spectrometry. Mol Genet Metab 120:255–268

Takashima S, Saitsu H, Shimozawa N (2019) Expanding the concept of peroxisomal diseases and efficient diagnostic system in Japan. J Hum Genet 64:145–152

Takemoto Y, Suzuki Y, Horibe R, Shimozawa N, Wanders RJ, Kondo N (2003) Gas chromatography/mass spectrometry analysis of very long chain fatty acids, docosahexaenoic acid, phytanic acid and plasmalogen for the screening of peroxisomal disorders. Brain and Development 25:481–487

Ventura MJ, Wheaton D, Xu M, Birch D, Bowne SJ, Sullivan LS, Daiger SP, Whitney AE, Jones RO, Moser AB, Chen R, Wangler MF (2016) Diagnosis of a mild peroxisomal phenotype with next-generation sequencing. Mol Genet Metab Rep 9:75–78

Vilarinho S, Sari S, Mazzacuva F, Bilgüvar K, Esendagli-Yilmaz G, Jain D, Akyol G, Dalgiç B, Günel M, Clayton PT, Lifton RP (2016) ACOX2 deficiency: a disorder of bile acid synthesis with transaminase elevation, liver fibrosis, ataxia, and cognitive impairment. Proc Natl Acad Sci 113:11289–11293

Wang Y, Busin R, Reeves C, Bezman L, Raymond G, Toomer CJ, Watkins PA, Snowden A, Moser A, Naidu S, Bibat G, Hewson S, Tam K, Clarke JT, Charnas L, Stetten G, Karczeski B, Cutting G, Steinberg S (2011) X-linked adrenoleukodystrophy: ABCD1 de novo mutations and mosaicism. Mol Genet Metab 104:160–166

Yamashita T, Mitsui J, Shimozawa N et al (2017) Ataxic form of autosomal recessive PEX10-related peroxisome biogenesis disorders with a novel compound heterozygous gene mutation and characteristic clinical phenotype. J Neurol Sci 375:424–429

Chapter 8
Therapeutic Strategies for X-Linked Adrenoleukodystrophy, a Representative Peroxisomal Disorder

Masashi Morita

Abstract X-linked adrenoleukodystrophy (X-ALD) is the most frequent peroxisomal disorder, and is caused by dysfunction of the peroxisomal ABC protein ABCD1. X-ALD patients with the most severe phenotype display cerebral inflammatory demyelination. In X-ALD, VLCFA accumulation, a characteristic feature of all patients, is thought to be the main culprit underlying the pathogenesis. However, the mechanisms by which the VLCFA accumulated in the brain causes demyelinating neurodegeneration have not yet been elucidated. At present, hematopoietic stem cell transplantation (HSCT) at an early symptomatic state is effective in halting disease progression, thus allowing long-term survival. Therefore, early diagnosis and conduct timely transplantation are particularly important to improve the outcome of HSCT. However, HSCT is always associated with significant mortality risk and the difficulty of finding a matching donor. Recently, genetically modified hematopoietic stem cells for *ex vivo* gene therapy have been tested as an alternative option and are expected to eventually become standard treatment for X-ALD. In parallel, the development of therapeutic drugs that can attenuate the symptoms or maintain the asymptomatic stage for patients diagnosed with X-ALD is in progress. To date, many candidate compounds have been reported. In this chapter, we focus on the current state of HSCT and pharmacological treatments, and describe the necessity for newborn screening and the identification of predictive biological markers in X-ALD.

Keywords ATP-binding cassette protein subfamily D1 (ABCD1) · *Ex vivo* gene therapy · X-Linked adrenoleukodystrophy (X-ALD) · Newborn screening · Hematopoietic stem cell transplantation (HSCT) · Very long chain fatty acid (VLCFA)

M. Morita (✉)
Graduate School of Medicine and Pharmaceutical Sciences, University of Toyama, Toyama, Japan
e-mail: masa@pha.u-toyama.ac.jp

T. Imanaka, N. Shimozawa (eds.), *Peroxisomes: Biogenesis, Function, and Role in Human Disease*, https://doi.org/10.1007/978-981-15-1169-1_8

Abbreviations

ABCD1	ATP-binding cassette protein subfamily D
ACALD	Adult cerebral form
AD	Alzheimer's disease
AdolCALD	Adolescent cerebral form
AMN	Adrenomyeloneuropathy
BBB	Blood-brain barrier
BMDM	Bone marrow-derived macrophages
CBT	Cord blood transplantation
CCALD	Childhood cerebral form
CNS	Central nervous system
ELOVL	Elongation of very long chain fatty acid
HSCT	Hematopoietic stem cell transplantation
MD	Multiple sclerosis
MRI	Magnetic resonance imaging
PBD	Peroxisome biogenesis disorder
PD	Parkinson's disease
ROS	Reactive oxygen species
VLCFA	Very long chain fatty acid
X-ALD	X-Linked adrenoleukodystrophy

8.1 Introduction

The peroxisomal diseases are categorized into two forms, peroxisome biogenesis disorders (PBDs) and single enzyme deficiencies (see Chap. 5). The PBDs include Zellweger syndrome, neonatal adrenoleukodystrophy, infantile Refsum disease and rhizomelic chondrodysplasia punctata type 1. Clinically, the PBDs range from severe, frequently lethal disorders to milder, late-onset progressive neurological disease. Zellweger syndrome is the most severe PBD variant, presenting with typical craniofacial dysmorphism and severe neurological abnormalities at birth. Most patients exhibit hepatomegaly and cardiac defects. They commonly fail to thrive in infancy and die of multiple organ failure before 1 year of age. At present, no effective therapeutic approach to the PBDs has been established.

In contrast to the PBDs, single enzyme deficiencies manifest clinical symptoms due to the lack of a specific enzyme or transporter in the peroxisomal metabolic pathways, such as the oxidation of fatty acids and the synthesis of ether-phospholipids. Sixteen diseases have been found thus far as single enzyme deficiency types (see Chap. 5). The clinical phenotype of single peroxisomal enzyme deficiency depends on the specific function of the affected protein in peroxisomal metabolism. Single enzyme deficiencies in peroxisomes induce biochemical abnormalities that frequently result in severe disease.

Peroxisomal diseases are rare and severe conditions. The development of effective treatments is one of the most important challenges in the field. In the case of X-linked adrenoleukodystrophy (X-ALD), the number of patients is relatively large, and analysis of the pathogenesis and development of diagnostic and therapeutic approaches are currently in progress. X-ALD patients develop normally until school age and a portion of the patients do not exhibit any severe symptoms until adulthood. Therefore, it is of critical importance to initiate therapeutic treatment before disease progression in X-ALD. At present, several therapies including allogeneic HSCT and pharmacological treatments are undergoing clinical investigation (https://www.clinicaltrials.gov/).

8.2 X-Linked Adrenoleukodystrophy (X-ALD)

According to the literature (Engelen et al. 2012a; Morita 2007), in 1897 Heubner first described a young boy with rapidly progressive neurologic deterioration consistent with X-ALD who had "diffuse sclerosis" on autopsy. In 1963, ALD was speculated to be an X-linked inherited disease (Fanconi et al. 1963), and in 1970 the name adrenoleukodystrophy was proposed and ultimately the condition came to be known as X-ALD (Blaw 1970). X-ALD was hypothesized to be a lipid storage disease based on the finding of lipid inclusion in several different types of cells, including brain macrophages. In 1976, Igarashi et al. established the biochemical basis for X-ALD by showing VLCFA accumulation in the brain and adrenal glands of patients (Igarashi et al. 1976). In 1993, Aubourg and colleagues identified the *ALD* gene by positional cloning as the gene in which mutations cause X-ALD (Mosser et al. 1993). The *ALD* gene encodes ABCD1.

The estimated incidence of X-ALD is approximately 1:20,000 males, indicating that X-ALD is the most frequent peroxisomal disorder and with a rather high incidence among the inborn error of metabolism (Kemp et al. 2016; Moser et al. 2007). Patients display progressive demyelination in the central nervous system (CNS), adrenal insufficiency, and testicular dysfunction as pathological characteristics. There are several X-ALD phenotypes that are categorized based on the onset and severity of the disease: the childhood cerebral form (CCALD), adolescent cerebral form (AdolCALD), adult cerebral form (ACALD), adrenomyeloneuropathy (AMN), and Addison's disease alone (Table 8.1). CCALD patients, comprising 35–40% of the patients, manifest cerebral inflammatory demyelination, usually in childhood (Kemp et al. 2012), and most of these patients enter into a vegetative state and die within 3 years of symptom onset. In AdolCALD and ACALD, the first symptoms of progressive inflammatory demyelination occur in adolescence or in adulthood. ACALD patients have a risk of developing inflammatory demyelination over time and can manifest severe disability or a vegetative state within a few years of symptom onset. In contrast, AMN patients, comprising approximately 45% of the patients, progress slowly, with symptoms limited to the spinal cord and peripheral nerves (Kohler 2010). It has been reported that approximately 18% of the AMN

Table 8.1 Clinical phenotypes in X-ALD

Phenotype	Abbreviation	Age of onset	Symptoms
Childhood cerebral ALD (~38%)	CCALD	5–10	Progressive inflammatory demyelination in brain
			Progressive behavioral, cognitive and neurological impairments with total disability within 6 months to 2 years, followed by death
Adolecnet cerebral ALD (4–7%)	AdolALD	10–20	Slower progression compared with CCALD, same symptoms
Adult cerebral ALD (2–5%)	ADALD	Over 21	Slower progression compared with CCALD, same symptoms
			Adults who develop cerebral ALD often already have signs of AMN and adrenal insufficiency
Adrenomyeloneuropathy (~45%)	AMN	30 ± 10	Non-inflammatory demyelination in spinal cord and peripheral nerves
			Progressive stiffness and weakness of his legs, impaired vibration sense in the lower limbs, sphincter disturbances, and sexual dysfunction
Addison's disease			No neurological symptoms
			Most ALD patients (80%) already had impaired adrenal function
Female carrier		Over 30	Abnormalities in the spinal cord and nerves in the legs, just like AMN
			The majority of females younger than 30 years are free of neurological symptoms
			Over 80% of female carriers with ALD will develop signs of neurological dysfunction by the age of 60 years

patients transition to the severe cerebral form over a period of 10 years (van Geel et al. 2001). Ultimately, 65% of all X-ALD patients transition to the cerebral form during childhood or adulthood (Moser 1995). No correlation between phenotype and genotype has been reported (Takemoto et al. 2002). Therefore, trigger factors such as environmental, genetic and/or epigenetic factors have been postulated (Heinzer et al. 2003). Head trauma is known to be a trigger of the onset of progressive inflammatory demyelination (Raymond et al. 2010; Budhram and Pandey 2017). Most female patients are asymptomatic carriers, because of the nature of an X-linked disease, and less than 1% present with adrenal insufficiency or cerebral disease (Engelen et al. 2014; Huffnagel et al. 2019).

Dysfunction of the peroxisomal ABCD1 protein caused by mutation of the *ABCD1* gene results in reduced VLCFA β-oxidation in peroxisomes (van Roermund et al. 2008; Wiesinger et al. 2013) (Fig. 8.1). The accumulation of VLCFAs in tissues is a characteristic feature of X-ALD and presumably caused by the coordination

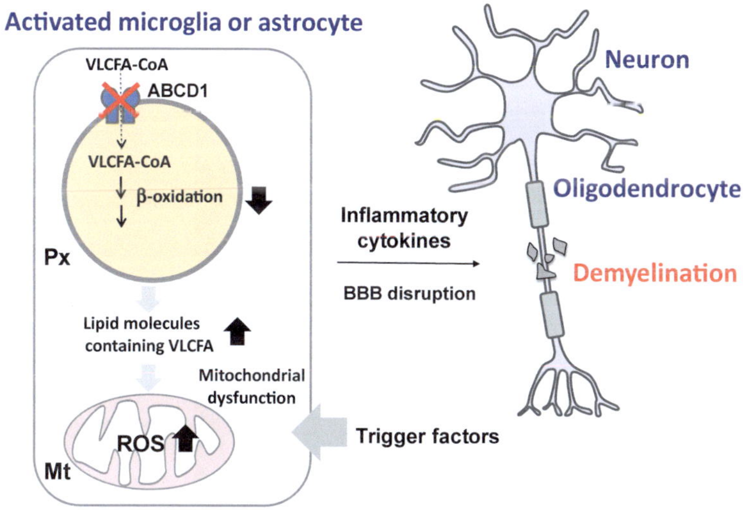

Fig. 8.1 Putative pathogenesis of X-ALD. Dysfunction of the peroxisomal ABCD1 protein caused by mutation of *ABCD1* gene results in the reduction of VLCFA β-oxidation in peroxisomes. The reduction in peroxisomal VLCFA β-oxidation activity together with the stimulation of microsomal VLCFA synthesis leads to the accumulation of lipid molecules containing VLCFA. The precise mechanism by which the accumulated VLCFAs cause demyelinating neurodegeneration has not yet been made clear, but it seems likely that oxidative stress by ROS due to mitochondrial dysfunction is involved in the activation of astrocytes or microglial cells. In addition to the metabolic dysfunction, additional factors that can trigger an inflammatory reaction are required for symptom onset. The production of inflammatory cytokines in these cells leads to BBB disruption and thereby induces the aggravation of inflammation, which might result in the onset of demyelination. VLCFA, very long chain fatty acid (C>22); Px, peroxisomes; Mt, mitochondria

of a reduction in peroxisomal VLCFA β-oxidation and stimulation of microsomal VLCFA synthesis. The mechanism by which the accumulated VLCFA in the brain and spinal cord causes demyelinating neurodegeneration and axonopathy have yet to be determined (Wiesinger et al. 2015; Berger et al. 2014). Presumably, the astrocytes and microglial cells activated by the VLCFA accumulation produce proinflammatory cytokines and reactive oxygen species (ROS) that lead to the cerebral inflammatory symptoms in CCALD and the axonopathy in AMN (Powers et al. 2005; Gilg et al. 2000). However, certain asymptomatic ALD patients do not develop the cerebral symptom despite high VLCFA accumulation, indicating that the accumulation of VLCFA is necessary but not sufficient for symptom development.

The main affected tissues are the brain white matter, spinal cord and adrenal cortex (Kemp et al. 2012). The pathogenic mechanisms underlying the demyelination in CCALD, axonal degeneration in AMN and adrenocortical insufficiency in Addison's disease may be different. Almost all male X-ALD patients develop primary adrenocortical insufficiency and, in these affected patients, adrenal hormone replacements are usually effective. However, hormone replacement therapy does not influence the development or progression of neurological symptoms in CCALD

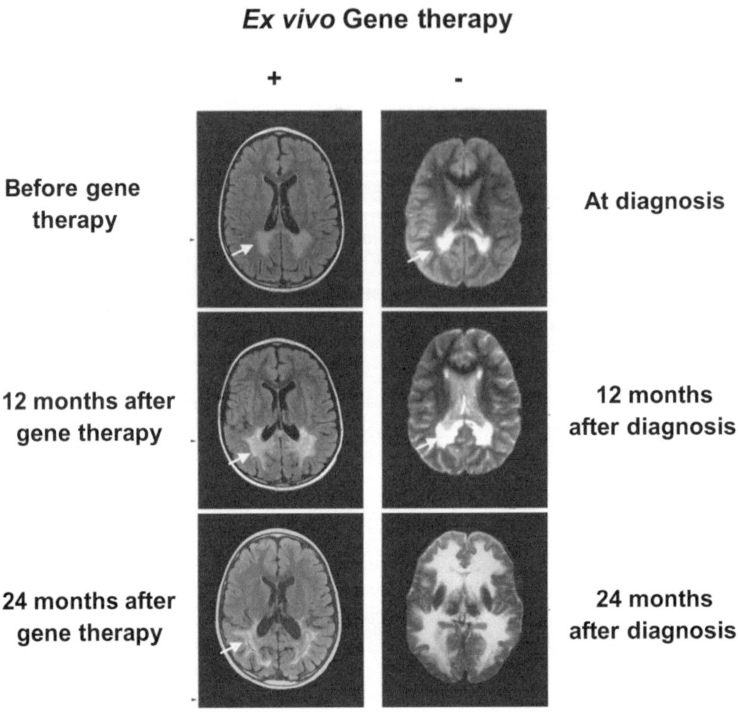

Fig. 8.2 Brain MRIs from CCALD patients before and after *ex vivo* gene therapy. Magnetic resonance imaging (Tagawa et al. 2002) indicates an abnormal hyperintensity signal in CCALD patients. Enhanced gadolinium contrast indicates demyelinating lesions along with inflammation and disruption of the BBB. In an 8-year-old ALD patient without *ex vivo* gene therapy (right), the brain MRI showed continuous progression of cerebral demyelinating lesions up to 24 months after the diagnosis. In contrast, the brain MRI in a patient with *ex vivo* gene therapy (left) showed stabilization or even reversion of the demyelinating lesions. This patient appears to have less extensive demyelinating lesions than the untreated patient (right) at the onset of the treatment with gene therapy, but had lesions with an abnormal hyperintensity signal in the splenium of the corpus callosum, the white matter of the parieto-occipital lobes (white arrows), and the auditory pathways. Although these lesions extended into the white matter of posterior parietal lobes up to 16 months after *ex vivo* gene therapy, they were stable with no further progression at 24 months. (From Cartier et al. (2009). Reproduced with permission from The American Association for the Advancement of Science)

or AMN. Thus, different therapeutic strategies must be considered for the individual case. At present, HSCT is the only therapeutic procedure for CCALD available (Cartier and Aubourg 2010). Recently, autologous HSC-gene therapy has been tested as an alternative option (Cartier et al. 2012) (Fig. 8.2). However, these procedures are only effective in early onset of inflammatory demyelination. Therefore, both early diagnosis by newborn screening and identification of disease-related biological markers that are predictive of the disease phenotype are eagerly anticipated. Furthermore, it is important to develop therapeutic drugs that can attenuate the symptom burden and maintain the asymptomatic stage for as long as possible in patients diagnosed with X-ALD. Drugs that are able to delay or halt the transition to the cerebral form may increase the potential for HSCT manipulation.

It is reported that the chronic axonopathy of AMN does not recover with HSCT because AMN is pathologically distinct from CCALD. In adulthood, all male patients eventually suffer from slowly progressive myelopathy. Furthermore, more than 80% of female carrier patients reportedly develop AMN (Engelen et al. 2014). Therefore, pharmacological treatments are also needed for AMN patients. Many candidate compounds have been reported for X-ALD treatment to date (Morita et al. 2011; Semmler et al. 2008; Pujol 2016). In this section, we focus on the current state of HSCT therapies for cerebral ALD patients and also pharmacological treatments to delay progression.

8.3 Hematopoietic Stem Cell Transplantation (HSCT)

There have been many studies that have demonstrated the therapeutic potential of bone marrow transplantation (BMT) as an effective treatment for neurodegenerative diseases (Mahmood et al. 2007). In X-ALD, HSCT early after symptom onset is reportedly effective in halting symptom progression in CCALD, allowing long-term survival (Aubourg et al. 1990; Kuhl et al. 2018). However, graft versus host disease and prolonged immunodeficiency are frequently associated with HSCT. Recently, utilization of cord blood and genetically modified hematopoietic stem cells for *ex vivo* gene therapy have been considered (Cartier and Aubourg 2010). Here, the current status of HSCT and gene therapy is described and the theory of how HSCT halts pro-inflammatory demyelination is discussed.

8.3.1 Allogeneic HSCT

In 1990, Aubourg et al. first reported that BMT reverses the early neurologic symptom of X-ALD patients (Aubourg et al. 1990). It should be noted that the first BMT in X-ALD patients was performed before identification of the disease-responsible *ABCD1* gene. Subsequently, HSCT has been performed to save X-ALD patients. At present, allogeneic HSCT is the only treatment that effectively halts symptom progression in CCALD. However, HSCT is associated with significant mortality risk because of severe graft rejection and infectious disease. The risk is higher in ACALD than that in CCALD. However, recently the mortality risk of HSCT in CCALD patients has been improved by using reduced intensity conditioning and antioxidant treatment (Miller et al. 2011; Kato et al. 2019).

In addition to the significant mortality risk, finding an HLA-matched donor represents another difficulty in this therapeutic protocol. Occasionally, it becomes too late to perform HSCT due to the failure of donor matching. Recently, van den Broek et al. reported improved survival after cord blood transplantation (CBT) for pre-symptomatic CCALD (van den Broek et al. 2018). Jiang et al. reported a novel approach to CCALD patients in which haplo-identical stem cells are transplanted with an infusion of umbilical cord blood (Jiang et al. 2015). At present, umbilical cord blood grafts are becoming more generally available as a source of donor cells and thus increasingly used for transplantation.

Notably, HSCT is unable to reverse adrenal insufficiency and axonal damage, and hence does not prevent the symptoms of AMN (van Geel et al. 2015). The axonal degeneration observed in AMN is thought to be caused by the oxidative stress due to mitochondrial dysfunction. The abnormal accumulation of VLCFAs in the mitochondrial membrane deranges redox metabolism, which results in oxidative stress. As HSCT does not result in any reduction of the VLCFA level in the CNS, pathological condition in AMN caused by oxidative stress seems to be unimproved. Therefore, HSCT is not recommended in AMN patients without any cerebral involvement, given the risk of the HSCT procedure.

In X-ALD, approximately 65% of the patients diagnosed eventually develop the cerebral form. Therefore, HSCT, that can halt the onset and progression of inflammatory demyelination, is the only treatment for these patients. Further improvement of HSCT is required for improved clinical outcomes (Kuhl et al. 2018).

8.3.2 Ex Vivo *Gene Therapy*

In addition to the increasing availability of CBT, *ex vivo* gene therapy is now becoming a potential therapeutic option for patients with CCALD without any HLA-matched donors (Cartier et al. 2009; Eichler et al. 2017) (Fig. 8.2). In *ex vivo* gene therapy, there are two advantages, as follows. One is that patients do not have to take time to find an HLA-matched donor because the procedure uses the patient's blood stem cells. The other is that there is no risk of graft-versus-host-disease, as the procedure does not require any immunosuppression drugs. In *ex vivo* gene therapy, a lentiviral vector derived from the HIV-1 virus has been used because the vector can be transduced into non-dividing cells such as HSCs. In addition, *ex vivo* transplantation of $CD34^+$ human cells that have the lentivirally transduced human *ABCD1* gene into NOD/SCID mice resulted in long term expression of the ABCD1 protein in perivascular and parenchymal microglia (Benhamida et al. 2003; Asheuer et al. 2004). It is reported that the percentage of ABCD1-positive hematopoietic cells were only 10–15% at 12 and 16 months post-HSCT, but notably, the inflammatory demyelination appeared to have been suppressed.

Clinical investigation with *ex vivo* gene therapy is currently underway in the USA. In a recent ongoing clinical trial, *ex vivo* gene therapy was conducted to CCALD boys with a Loes score less than or equal to 9 and without the availability of an HLA-matched sibling donor. It is reported that the *ex vivo* gene therapy effectively stabilized the disease's progression in 15 of 17 patients more than 2 years after the gene therapy (Eichler et al. 2017). The gene therapy was administered in a clinical trial sponsored by Bluebird Bio. Although follow-up studies are necessary to determine the long-term outcome of this gene therapy, *ex vivo* gene therapy is a promising treatment for asymptomatic CCALD patients. In contrast to asymptomatic CCALD patients, neither *ex vivo* gene therapy nor HSCT is able to rescue symptomatic CCALD patients with advanced disease. Further studies are thus needed to achieve a good clinical outcome for patients with symptomatic CCALD and ACALD.

As another form of gene therapy, Gong et al. reported that a recombinant adeno-associated virus serotype 9 (rAAV9) vector with the *ABCD1* gene was efficiently delivered to the CNS in *Abcd1*-deficient mice by intravenous injections (Gong et al. 2019). In the *Abcd1*-deficient mouse, injection of rAAV9 with the *ABCD1* gene resulted in a 20% decrease of the VLCFA levels in the spinal cord compared to controls. This result suggests that intrathecal AAV vector-mediated gene therapy might be an alternative gene therapy for X-ALD in the future.

8.3.3 How Do Bone Marrow-Derived Macrophages Halt Inflammatory Demyelination?

It is important to determine how HSCT halts cerebral inflammatory demyelination. It seems likely that the donor-derived hematopoietic stem cells engrafted in the brain parenchyma differentiate into macrophage-like cells and suppress neuroinflammation (Cartier et al. 2014; Weber et al. 2014) (Fig. 8.3). In the case of HSCT for lysosomal storage diseases (LSDs), bone marrow-derived macrophages (BMDMs) engrafted in CNS are able to supply deficient lysosomal enzymes to neighboring cells, presumably through exocytosis and endocytosis. The supplemented enzymes reduce the pathogenetic molecules accumulated in the CNS and ameliorate the pathogenesis (Koç et al. 1999). Therefore, the level of the enzyme produced by bone marrow-derived cells is particularly important for the LSDs. In contrast, the ABCD1 protein is a peroxisomal membrane protein and therefore it is unlikely that either the ABCD1 protein or mRNA in BMDMs is transferred into neighboring cells. In our BMT experiments using *abcd1*-deficient mice, the VLCFA plasma level significantly decreased and donor-derived bone marrow cells were engrafted in the brain parenchyma as macrophage-like cells at approximately 45% of the total of the Iba1-positive cells, even at 6 months post BMT (our unpublished data). However, the VLCFA levels in recipient mouse brain was not reduced by BMT, indicating that BMDMs do not contribute to the degradation of the VLCFAs accumulated in the recipient mouse brain. In fact, no decrease in the VLCFA level was observed in the postmortem brain after HSCT. BMDMs might arrest inflammatory demyelination by attenuating metabolic pathways leading to microglial activation.

In certain neuroinflammatory disorders, such as Alzheimer's disease (AD), Multiple sclerosis (MD) and Parkinson's disease (PD), microglia have a central role in the pathogenesis and are thought to be therapeutic targets (Cartier et al. 2014). In

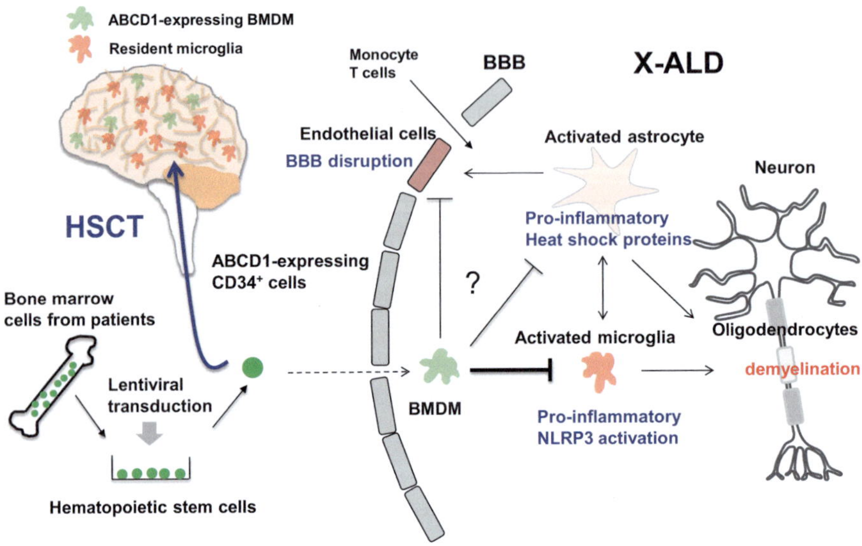

Fig. 8.3 HSCT halts cerebral inflammatory demyelination. In X-ALD, activated microglia and astrocytes play a central role in the onset as well as progression of inflammatory demyelination. Furthermore, the dysfunction of endothelial cells results in the disruption of the BBB, which leads to the entry of peripheral immunocompetent cells (T cells or monocyte/macrophage). In cases of HSCT for CCALD, hematopoietic stem cells (HSCs) are harvested from the bone marrow of the patients, and the CD34$^+$ cells containing HSCs are transduced with a lentiviral vector harboring the *ABCD1* gene. The genetically modified HSCs are then transplanted into patients. The ABCD1-expressed HSCs migrate to the brain parenchyma through the BBB and differentiate into macrophage-like cells. These bone marrow-derived macrophage (BMDMs) may interact with resident microglia, astrocytes or endothelial cells and restore metabolic homeostasis and BBB integrity, which in turn suppresses the neuroinflammation

addition, peroxisomal β-oxidation-deficient mice with knockout of the *Mfp2* gene that encodes multifunctional protein-2 exhibit extensive microgliosis in their gray matter regions of the CNS (Beckers et al. 2017, 2018), while nestin promoter-specific *Mfp2*-deficient mice exhibit minor neuroinflammation. These results suggest that peroxisomal function in microglia may play important roles in the inflammatory state in the CNS. In X-ALD, microglia are recruited in perilesional normal-appearing white matter and microgliosis is observed in inflammatory lesions with little involvement of lymphocyte infiltrates, indicating that microglia play a pivotal role in the onset and progression of proinflammatory demyelination. In HSCT, BMDMs might ameliorate the metabolic dysfunction of microglia or complement the function of microglia (Derecki et al. 2013).

Weinhofer et al. reported that the transition of the pro-inflammatory phase (i.e. the M1-type of microglia) to the anti-inflammatory phase (the M2-type) that halts demyelination is less efficient in *Abcd1*-deficient macrophages (Weinhofer et al. 2018). In macrophages, peroxisomal β-oxidation is reportedly involved in the fine-tuning of M1/M2 polarization (Geric et al. 2018). It thus seems likely that peroxisomal metabolic defects in *Abcd1*-deficient microglia suppress the transition to M2-type microglia. In addition, Jang et al. suggest that NLRP3 activation in microglia may participate in the neuroinflammation observed in X-ALD (Jang et al. 2016). They speculate that an increased level of 25-hydroxycholesterol mediated by up-regulation of cholesterol 25-hydroxylase could act as an endogenous NLRP3 stimulator, which would lead to cerebral inflammation. Thus, it may be possible that BMDMs facilitate the transition of microglia from the pro-inflammatory to anti-inflammatory state or somehow dampen the microglial inflammatory action.

Recently, Musolino et al. reported that an increase in blood-brain barrier (BBB) permeability precedes inflammatory demyelination, suggesting that disruption of the BBB may trigger pro-inflammatory demyelination (Musolino et al. 2015). Microglia and/or astrocytes might contribute to BBB maintenance, and abnormal interaction of *Abcd1*-deficient glial cells with endothelial cells may lead to disruption of the BBB. Upon HSCT, these white matter microvascular alterations normalize within 1 year after treatment in good accordance with the period needed for halting the progression of inflammatory demyelination by HSCT. Orchard et al. suggest that the time to donor myeloid recovery and the extent of donor chimerism correlated significantly with the time required for BBB repair (Orchard et al. 2019). It is thus possible that BMDM may halt the transition to the CCALD form by recovering endothelial cell function and repairing the BBB.

Görtz et al. reported that in the early stages of X-ALD before myelin damage, heat shock proteins (HSPs) were found to be increased, particularly in astrocytes, suggesting that astrocytic stress may play an important role in the activation of microglia and initiation of inflammatory demyelination (Görtz et al. 2018). It is important to determine whether astrocytic stress is attenuated by HSCT.

Taken together, it seems likely that donor-derived BMDMs interact with resident microglia, astrocytes and/or endothelial cells, and thus ameliorate the metabolic abnormality (Yamada et al. 2004), which in turn suppresses neuroinflammation.

8.3.4 Newborn Screening for the Use of HSCT

Newborn screening for early diagnosis is essential to improve the outcome of HSCT for X-ALD (Moser and Fatemi 2018). In X-ALD, it is known that HSCT tends to exacerbate symptoms when performed in advanced states of the disease (Tran et al. 2017; Pierpont et al. 2017). Miller et al. showed that gadolinium enhancement intensity on brain MRI can be scored simply and reproducibly for CCALD and the progressive course of the disease effectively represented based on the Loes MRI scoring system (Miller et al. 2016). Gadolinium enhancement indicates BBB disruption and activation of the inflammatory process. At present, early-stage patients with an MRI severity score of less than 10 are considered to be eligible HSCT (Bladowska et al. 2015). For this reason, early diagnosis to identify asymptomatic males and conduct timely transplantation is particularly important in order to prevent progressive neurodegeneration in this disease. Only newborn screening allows early diagnosis and thus an opportunity to perform HSCT quickly, as soon as the cerebral phenotype is identified.

In 2015, Vogel et al. described a diagnostic protocol, surveillance protocol and treatment guidelines for newborn screening for X-ALD in New York State (Vogel et al. 2015). In 2017, Kemper et al. provided an evidence summary and advisory committee recommendation for newborn screening for X-ALD (Kemper et al. 2017). The data from newborn screening showed the health benefits of earlier HSCT for CCALD. Based on these results, the secretary of the US Department of Health and Human Services recommended the addition of X-ALD to the uniform screening panel for newborn screening programs in 2016. A pilot study of newborn screening in 2018 conducted by North Carolina team demonstrated that the screening assay protocol using a HPLC-MS/MS assay to detect C24:0-LPC and C26:0-LPC successfully identified infants with X-ALD (Taylor and Lee 2019). Among 52,301 specimens, three newborns were confirmed with X-ALD, two as likely carriers or heterozygous females, and two with PBDs. The screening assay protocol was able to identify X-ALD specimens with a high degree of certainty.

Another issue for successful HSCT is that X-ALD lacks pathogenesis-related biological markers, which results in the failure of efforts at early diagnosis. In addition to the accumulation of pathogenic lipids (C24:0-LPC and C26:0-LPC), additional factors that directly or indirectly impact the onset of proinflammatory demyelination remain to be elucidated. At present, the predication of neuropathological staging as a reference for the performance of HSCT entirely relies on the clinical decision reached by neurologists and hematologists. Therefore, in addition to newborn screening, the identification of biomarkers that predict the onset of the cerebral form are greatly anticipated. Natural history studies show that only 40% of boys with X-ALD will develop CCALD in the first two decades of life. Therefore, HSCT is not always beneficial as a preemptive strategy for all boys diagnosed with X-ALD.

Recently, Turk et al. reported that plasma superoxide dismutase (SOD) levels were progressively reduced in patients with AMN or cerebral ALD as compared with healthy controls and heterozygous female carriers, suggesting that the plasma SOD level may be a clinical biomarker for predicting the onset of CCALD (Turk et al. 2017a). Marchetti et al. showed that asymptomatic patients presented with increased plasma levels of a variety of pro-inflammatory cytokines (IL-1β, IL-2, IL-8 and TNF-α) and also anti-inflammatory cytokines (IL-4 and IL-10) (Marchetti et al. 2018), suggesting that an increase of these cytokines level in asymptomatic patients could be a disease-related marker. On the other hand, Lauer et al. provided a model for the *in vivo* assessment of white matter microvascular physiology over time in X-ALD patients and suggested that alteration in white matter microvascular perfusion, which is associated with BBB permeability, may be an early biomarker of disease progression (Lauer et al. 2017). Recently, epigenomic signatures, such as hypermethylation of the genes involved in oligodendrocyte differentiation were reported to be putative biomarkers of disease phenotype (Schlüter et al. 2018).

The finding of predictive biological markers will provide critical information for deciding more easily and appropriately the patients who should undergo HSCT.

8.4 Pharmacological Treatment

Although allogeneic HSCT has become an effective treatment for X-ALD, the number of patients who receive the benefit is limited. Pharmacological therapeutic approaches based on biochemical and pathological characteristics of the disease are thus highly desired. As mentioned above, Igarashi et al. first reported that pathognomonic lipid bodies consisting of cholesterol esterified with VLCFAs were detected in the cerebral white matter of X-ALD patients (Igarashi et al. 1976). In X-ALD, a common biochemical characteristic is the accumulation of lipid molecules containing VLCFAs (such as cholesterol esters, proteolipids, glycerophosphatides and gangliosides) in which the carbon chain length is longer than 22, especially in target organs such as the brain, testes and adrenal glands.

Mutation of *ABCD1* gene leads to a dysfunction of ABCD1 that results in a decrease in VLCFA β-oxidation. The decrease in VLCFA β-oxidation leads to an increase in VLCFA-CoA, which is fatty acid elongation substrate, and the synthesis of phospholipids and cholesterol esters in the ER. As a result, the subsequent abnormal accumulation of VLCFAs directly or indirectly results in oxidative damage and an inflammatory response. Therefore, lowering the VLCFA level as well as suppressing oxidative stress and inflammatory events in the brain and spinal cord are all potentially important strategies in the treatment of X-ALD (Fig. 8.4).

8.4.1 Lowering the VLCFA Level

Despite ABCD1 protein dysfunction, approximately 30% of VLCFA β-oxidation still remains, indicating that an ABCD1-independent pathway is involved in residual VLCFA β-oxidation. ABCD2 might contribute to this residual β-oxidation. On the other hand, it is known that more than 70% of the patient fibroblasts with a missense mutation display either a lack or reduction of the ABCD1 protein because of post-translational degradation. We found that some of the mutant ABCD1 proteins have a capacity to recover function upon a culturing of the mutant cells at a low temperature (Morita et al. 2018). In addition, the increase of microsomal fatty acid elongation activity through the involvement of ELOVL1, 6 and 7 appears to be involved in the VLCFA accumulation (Tsuji et al. 1984; Kemp and Wanders 2010; Baarine et al. 2015a, b; Raas et al. 2019; Morita et al. 2015). Therefore, the stimulation of residual VLCFA β-oxidation, including up-regulation of ABCD2, stabilization of the missense mutant ABCD1 protein, and the suppression of fatty acid elongation activity are all targets for lowering the VLCFA level.

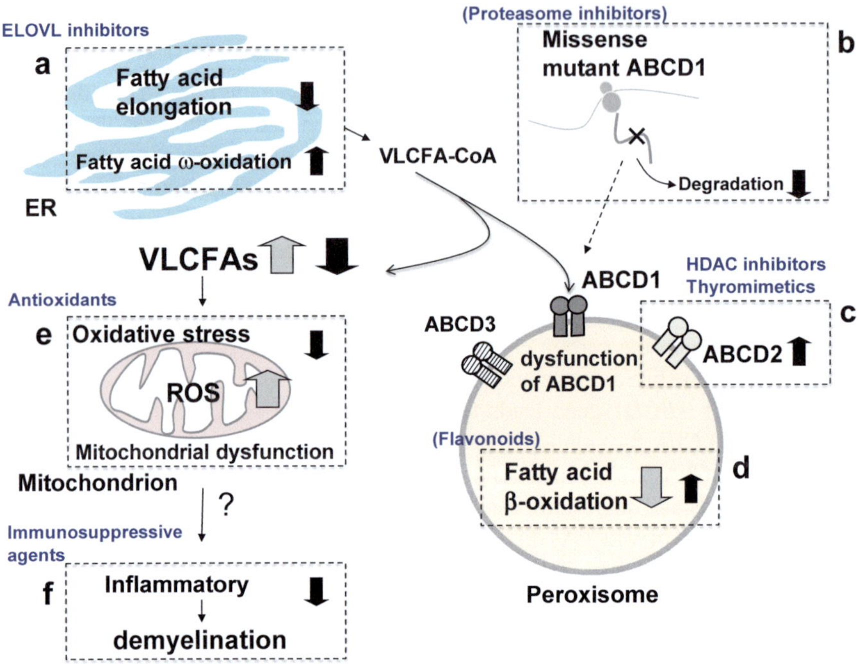

Fig. 8.4 Therapeutic targets and candidate compounds for X-ALD therapy. Mutation of the *ABCD1* gene leads to the dysfunction of ABCD1, which results in decrease in VLCFA β-oxidation in peroxisomes. The failure of substrate transport into peroxisomes leads to an increase in the cellular VLCFA-CoA level, which is further subjected to fatty acid elongation in the ER. As a result, the number of lipid molecules containing VLCFA is increased and the abnormal distribution of the lipid molecules in the mitochondrial membrane results in oxidative stress, which leads to inflammatory demyelination (as indicated by the gray arrow). Many candidate compounds for several different targets have been reported as follows; (**a**) inhibition of microsomal fatty acid elongation (ELOVL inhibitors), (**b**) stabilization and correct subcellular localization of missense mutants of ABCD1 (e.g. proteasome inhibitors), (**c**) induction of ABCD2 expression (HDAC inhibitors and thyromimetics), (**d**) stimulation of residual VLCFA fatty acid β-oxidation (e.g. flavonoids), (**e**) Suppression of oxidative stresses (antioxidants) and (**f**) improvement of inflammatory states (immunosuppressive agents). Any of these might be an effective approach to X-ALD. Chemical compounds that are able to pass through the BBB and act on these targets would be promising drugs for X-ALD therapy. The dashed rectangles show the cellular targets for such candidate compounds (**a** to **f**) and the black arrows the predicated effect of each compound

8.4.1.1 Early Studies

In the 1990s, three compounds for reducing the VLCFA level were reported, Lorenzo's oil (Rizzo et al. 1989), lovastatin (Singh et al. 1998b) and 4-phenylbutyrate (Kemp et al. 1998). Unfortunately, these compounds have no effect in X-ALD. However, these compounds do provide information that is valuable for the development of therapeutic drugs in the disease.

At the end of the 1980s, Augusto Odone found that the C26:0 plasma level of Lorenzo, their 6-year-old son suffering from X-ALD, was reduced to a normal level by the administration of a 4:1 mixture of glyceryl triolate and glyceryl trierucate (Rizzo et al. 1989). The Hollywood movie "Lorenzo's Oil" produced in 1992 is based on the true story of the Odones, two parents seeking a treatment for their son. The mixed oil, subsequently termed Lorenzo's oil (LO), reduced the plasma VLCFA levels within 1 month in most X-ALD patients. Recently, Sassa et al. reported that LO inhibits ELOVL1 by binding to the ELOVL1-substrate complex (Sassa et al. 2014). Unfortunately, in spite of the decrease in the plasma VLCFA level, LO does not prevent the progression of inflammatory demyelination and conversion to the inflammatory form, even when administered before the onset of the demyelination. This is presumably due to its inability to penetrate into the brain through the BBB. However, a possible beneficial effect of LO on the delay of the onset of the CCALD and a slowing of the progression of axonopathy in pure AMN have not yet been excluded (Moser et al. 2005).

In 1998, Lovastatin, a HMG-CoA reductase inhibitor, was reported to reduce the plasma VLCFA level in X-ALD patients (Singh et al. 1998a). However, this effect was largely attributed to the decrease in the level of the LDL-cholesterol containing cholesterol ester with VLCFAs through hepatic up-regulation of the LDL receptor. Therefore, it has been recognized that lovastatin should not be prescribed as a treatment, at least if the goal is to lower VLCFA levels (Engelen et al. 2010). On the other hand, the abnormal accumulation of VLCFAs in X-ALD iPSC derived oligodendrocytes was reduced by treatment with lovastatin, presumably because of the upregulation of *ABCD2* gene expression (Jang et al. 2011), suggesting that lowering cellular cholesterol with lovastatin may reduce the VLCFA level by inducing *ABCD2* expression.

In 1998, Kemp et al. suggested that the histone deacetylase (HDAC) inhibitor 4-phenylbutyrate (4-PBA) stimulates *ABCD2* gene expression, which results in an increase in peroxisomal fatty acid β-oxidation (Kemp et al. 1998). 4-PBA reportedly passes through the BBB and lowers brain VLCFAs in *Abcd1*-deficient mice, but only at very high doses. As 4-PBA is rapidly metabolized in the liver and exhibited tachyphylaxis, it would be difficult to maintain the plasma level of 4-PBA at effective concentrations (McGuinness et al. 2001). In fact, a clinical trial in AMN patients failed to demonstrate efficacy, probably due to the very short half-life of 4-PBA. Although 4-PBA is thus not promising for the treatment of X-ALD, these results teach us other HDAC inhibitors that are structurally more stable might become candidate drugs.

8.4.1.2 Pharmacological Induction of *ABCD2* Gene Expression

A therapeutic strategy that induces *ABCD2* gene expression is a reasonable approach to lowering VLCFAs, because the ABCD2 protein overlaps with the ABCD1 protein in substrate specificity (van Roermund et al. 2011) (Fig. 8.4c). In fact, genetically induced overexpression of the ABCD2 protein is able to restore VLCFA metabolism both *in vitro* and *in vivo* (Netik et al. 1999; Pujol et al. 2004). Weber et al. reported that CD34$^+$-derived monocytes, but not lymphocytes, are metabolically impaired, because ABCD2 is virtually not expressed in monocytes. In HSCT, CD34$^+$-derived macrophages are thought to be involved in the suppression of inflammatory demyelination in the CNS. Thus, compounds that induce *ABCD2* gene expression in the brain and monocytes/macrophages would be promising drugs. To date, several different classes of drugs have been investigated for this purpose, including peroxisome proliferator-activated receptor α agonists, HDAC inhibitors, dehydroepiandrosterone (Gueugnon et al. 2003), thyroid hormone and thyromimetics (Fourcade et al. 2003), liver X receptor antagonists (Weinhofer et al. 2005), retinoid X receptor agonists (Weber et al. 2014), and an AMP-activated protein kinase $\alpha 1$ activator (Singh and Giri 2014). Among them, we focus on the most promising drugs, HDAC inhibitors and thyromimetics.

Following 4-PBA, other HDAC inhibitors including valproic acid and suberoyanilide hydroxamic acid (SAHA), have been reported. Fourcade et al. (2010) reported that valproic acid, a non-specific HDAC inhibitor, decreased the levels of monounsaturated VLCFA (such as C26:1 n-9), but not saturated VLCFA, by inducing *ABCD2* gene expression in X-ALD fibroblasts. Valproic acid is a widely used antiepileptic drug and is able to cross the BBB. They reported that a pilot trial of valproic acid in X-ALD patients resulted in amelioration of the oxidative damage in peripheral blood mononuclear cells. Baarine et al. reported that SAHA, another HDAC inhibitor, decreased the VLCFA level in *Abcd1*-deficient glial cells by inducing *Abcd2* gene expression as well as by inhibiting ELOVL1 activity (Baarine et al. 2015a). SAHA is also able to cross the BBB, and they reported that SAHA corrected mitochondrial defects in glial cells and upregulated *ABCD2* gene expression in human brain hippocampal slice cultures. Taken together, HDAC inhibitors including valproic acid and SAHA have potential as candidate compounds for the development of therapeutic drugs.

Fourcade et al. first reported that thyroid hormone upregulated *ABCD2* gene expression in X-ALD fibroblasts (Fourcade et al. 2003). The *ABCD2* gene promoter has a thyroid hormone responsive element that interacts with the thyroid hormone receptor (TR). Thyromimetic agents are currently among the most promising drugs for X-ALD therapy because they cross the BBB. Sobetirome, a potent TR agonist with selectivity for the TRβ isoform, increases *Abcd2* gene expression and reduces the VLCFA level in the brain of *abcd1*-deficient mice, suggesting that it might be valuable as a therapeutic drug for X-ALD (Hartley et al. 2017). Recently, ethanolamine-derived prodrugs of sobetirome were shown to be better delivered to the brain than unmodified sobetirome, suggesting an improvement in the efficacy of the drug (Ferrara et al. 2017). Clinical trials for sobetirome are presently underway (https://www.clinicaltrials.gov/).

Although ABCD2 can functionally compensate for the absence of ABCD1, it should be noted that the expression profiles of ABCD1 and ABCD2 in tissues, cell types and different developmental stages have strikingly different patterns (Troffer Charlier et al. 1998). In mouse brain, ABCD1 appears to be mainly expressed in glial cells and ABCD2 in neuronal cells. During development, *Abcd1* was shown to be most abundant in the embryonic brain and to gradually decrease during the course of maturation, in contrast to *Abcd2*, which increased in the early postnatal period. Lu et al. reported that the ABCD2 but not ABCD1 protein plays an anti-oxidative stress role in mouse adrenal cells (Lu et al. 2007). It seems likely that ABCD2 plays a role in polyunsaturated fatty acid metabolism and is involved in anti-oxidative activity, at least in the adrenal glands. These results suggest that the physiological functions of ABCD1 and ABCD2 may be different, especially in the brain and adrenal glands. It should be determined if pharmacological induction of *ABCD2* gene expression in the brain is effective in compensating for the *in vivo* physiological function of ABCD1.

Up-regulation of *Abcd2* gene expression by HDAC inhibitors and thyromimetics leads to a decrease in the VLCFA level in *Abcd1*-defcient mouse brain. However, brain VLCFA levels in *Abcd1*-deficeint mice were reduced by only ~20% after oral administration of sobetirome for 12 weeks (~0.4 mg/kg body weight/day). This may be due to the slow turnover of neuronal cells and lipid molecules in the brain. Cholesterol esterase activities toward cholesteryl lignocerate and cerotate in the rat brain are reportedly less than 1% of that toward cholesteryl oleate (Ogino and Suzuki 1981). Therefore, once VLCFA is incorporated into lipid molecules such as cholesterol ester, it is hard to degrade VLCFA via β-oxidation. Taken together, compounds that lower the brain VLCFA level have potential in X-ALD therapy, but prolonged treatment might be needed to achieve a therapeutic outcome.

8.4.1.3 Stimulation of Residual VLCFA β-Oxidation

The VLCFA β-oxidation activity in X-ALD fibroblasts is approximately 30% of that in healthy control subjects. The residual activities seem to be dependent on the ABCD2 protein and/or diffusion of VLCFAs into peroxisomes, followed by the esterification of CoA by peroxisomal acyl-CoA synthetase. In any event, the stimulation of residual VCLFA β-oxidation effectively reduces the VLCFA level (Fig. 8.4d). We previously reported that the flavonoid baicalein trimethyl ether stimulates VLCFA β-oxidation activity in X-ALD fibroblasts (Morita et al. 2005). This flavonoid did not induce *ABCD2* gene expression but did up-regulate the *ACSVL4* gene expression that encodes a peroxisomal acyl-CoA synthetase. We speculate this flavonoid activates peroxisomal VLCFA β-oxidation by accelerating direct fatty acid transport into peroxisomes via acyl-CoA synthetase activity (Morita et al. 2008). These results indicate that stimulation of ABCD1- and ABCD2-independent pathways may be another target for lowering the VLCFA level.

Recently, van Engen et al. identified *CYP4F2* and *CYP4F3* as modifier genes in X-ALD that are involved in the degradation of VLCFAs (van Engen et al. 2016) (Fig. 8.4a). In addition to peroxisomal β-oxidation, VLCFAs are ω-oxidized in microsomes by phylloquinone ω-hydroxylase (CYP4F2) and docosahexaenoic acid ω-hydroxylase (CYP4F3), which generate dicarboxylic-VLCFA. They reported that mutant CYP4F2 decreases the conversion of VLCFA into dicarboxylic-VLCFA by ω-oxidation. Dicarboxylic acid is β-oxidized in peroxisomes via ABCD3 transport.

These results indicate that the stimulation of CYP4F2 may also be an alternative therapeutic target.

8.4.1.4 Stabilization of ABCD1 with a Missense Mutation

More than 340 non-recurrent missense mutations of the *ABCD1* gene have been identified (http://www.x-ald.nl) in X-ALD patients to date. Among these missense mutant ABCD1 proteins, approximately 70% of the ABCD1 proteins are scarcely detected by immunoblotting due to instability (Takahashi et al. 2007) (Fig. 8.4b). However, X-ALD fibroblasts with certain missense mutant ABCD1 proteins appear to have their function restored by incubating them at a low temperature (Zhang et al. 2011). These results indicate that a portion of the missense mutant ABCD1 proteins still have the capacity to recover their function by stabilization. This prompted us to seek therapeutic drugs that are able to stabilize missense mutant ABCD1 proteins (Morita et al. 2018). We have established a fluorescence-based assay method for the screening of chemical libraries and found that bortezomib, a drug used for multiple myeloma patients, induced functional recovery in mutant ABCD1 proteins. At present, bortezomib is not applicable to X-ALD because of its side effects. However, this result provides evidence that pharmaceutical interventions that stabilize a subset of missense mutant ABCD1 proteins may have therapeutic efficacy in X-ALD in the future.

8.4.1.5 Suppression of Excess Fatty Acid Elongation

As in the case with LO treatment, inhibition of fatty acid elongation activity might be effective in reducing VLCFA level (Fig. 8.4a). Engelen et al. found that the PPARα agonist bezafibrate reduces the VLCFA level by inhibiting ELOVL1 in X-ALD fibroblasts (Engelen et al. 2012b). As ELOVL1 is thought to be involved in the accumulation of VLCFAs in X-ALD, this enzyme is a reasonable therapeutic target. To date, however, clinical trials have failed to demonstrate any improved clinical outcome using bezafibrate in X-ALD patients. More potent and specific compounds that are able to inhibit the activity of ELOVL1 might be candidate drugs for X-ALD.

8.5 Suppression of Oxidative Stress

In neurodegenerative diseases, mitochondrial dysfunction and increased ROS production are considered to have toxic effects on cellular activity that lead to neuronal loss and axonal damage. Oxidative damage is detected in the spinal cord of *abcd1*-deficient mice and the post-mortem brain of CCALD patients (Powers et al. 2005; Ranea Robles et al. 2018; Fourcade et al. 2014; Gilg et al. 2000). Nury et al. reported that high levels of 7-ketocholesterol (7KC) due to increased ROS production exerted a deleterious effect on microglial function that might contribute to the development of X-ALD (Nury et al. 2017). Plasma superoxide dismutase (SOD) levels in AMN patients were shown to be decreased during progression from the non-cerebral to cerebral form (CCALD), suggesting that oxidative stress and oxidative damage is likely an early pathogenic factor in X-ALD (Deon et al. 2016). Therefore, administration of antioxidants has been considered a potential therapy (Fig. 8.4e). Especially in AMN, antioxidant treatments would be expected to be effective in suppressing the axonal damage (Pujol 2016; Galino et al. 2011; Lopez-Erauskin et al. 2011), because the axonal degeneration observed in AMN is likely caused by oxidative stress.

It has been reported that *N*-acetyl-L-cysteine (NAC) (Gronemeyer et al. 2013) therapy increases plasma heme oxygenase-1, which has cytoprotective effects and also helps replenish plasma glutathione (GSH) levels in X-ALD patients (Kartha et al. 2015). DNA damage in leukocytes from CCALD patients that might be induced by lipid oxidation is decreased by NAC and the other antioxidant reagents such as Trolox and rosuvastatin (Marchetti et al. 2015). Recently, Turk et al. reported that a polyamidoamine dendrimer conjugated to the anti-oxidant precursor *N*-acetyl cysteine (D-NAC) reduced TNFα and glutamate secretion in macrophages from CCALD patients that were treated with free VLCFAs (C24:0 and C26:0) (Turk et al. 2018). Pujol et al. found that a combination of the antioxidants NAC, α-lipoic acid and vitamin E synergistically halted immunological signs of axonal degeneration in *Abcd1*-deficient mice (Lopez-Erauskin et al. 2011; Pujol 2016). These data suggest that these antioxidants might be useful as adjunctive therapy for X-ALD.

The mechanism by which ABCD1 dysfunction leads to increased oxidative stress has yet to be determined. Presumably, phospholipids containing VLCFAs in the inner mitochondrial membrane might be linked to a disturbance in mitochondrial integrity that in turn leads to increased ROS production. Mitochondrial ROS may oxidize and inhibit the oxidative phosphorylation system, which induces energetic failure and opening of the mitochondrial permeability transition pore (mPTP). Lopez-Erauskin et al. have suggested that mPTP formation may be induced by increased amounts of oxidized cyclophilin D induced by the increased oxidative stress that occurs in X-ALD. Pharmacological blocking of the gate opener cyclophilin D with cyclosporine A has been shown to prevent mitochondrial dysfunction in neurodegenerative disease models. This result indicates that mitochondrial-targeted antioxidants may become therapeutic drugs for X-ALD (Lopez-Erauskin et al. 2012). Mitochondrial-targeted compounds, including pioglitazone, resveratrol and dimethyl fumarate, are reported to recover mitochondrial function in *Abcd1*-deficient mice.

Pioglitazone, an agonist of PPARγ, halts locomotor disability and axonal damage in *Abcd1*-deficient mice (Morato et al. 2013). The oral administration of pioglitazone restored mitochondrial content, reduced oxidative damage, and recovered bioenergetic failure in *Abcd1*-deficient mice. As pioglitazone is widely used as an antidiabetic drug, this drug is a promising therapy for X-ALD. Sirtuin 1, a key regulator of the mitochondrial function, is reportedly impaired in the spinal cord of

Abcd1-deficient mice and in the brain white matter of X-ALD patients (Morató et al. 2015). Resveratrol (3,4′,5-trihydroxystilbene), a polyphenol nonflavonoid compound, reportedly restores sirtuin 1 activity together with normalizing of oxidative stress and mitochondrial function, thereby suppressing axonal degeneration in *Abcd1*-deficient mice, suggesting that resveratrol may be a useful compound in X-ALD. In contrast, Ranea-Robles et al. reported that the response of nuclear factor erythroid 2-like 2 (NRF2), a master regulator of the endogenous antioxidant responses, is impaired in X-ALD fibroblasts (Ranea Robles et al. 2018). They demonstrated that dimethyl fumarate, an NRF2 activator, halted axonal degeneration and locomotor disability in *Abcd1*- or/and *Abcd1*/*Abcd2*-double deficient mice. These results suggest that activation of NRF2 may be a novel target and dimethyl fumarate may be a candidate compound for X-ALD therapy.

In spite of the effectiveness of antioxidants in *Abcd1*-deficient mice, no results are yet available from human trials of antioxidants in AMN patients.

8.6 Suppression of Inflammatory Processes

X-ALD is an inflammatory demyelinating disorder with an infiltration of lymphocytes and monocyte/macrophages that appears similar to MS. In fact, a remarkable infiltration of lymphocytes into the brain reportedly is observed at autopsy in CCALD patients (Ito et al. 2001). However, the pathogenesis of the inflammatory response is different from that in MS. At the outset of cerebral demyelination in X-ALD, microglia and astrocytes appear to be immunologically activated by VLCFA accumulation. The active microglia accumulated around vessels release IL-1β, which abolishes the protective effect of astrocytes on BBB integrity (Musolino et al. 2015). Disruption of BBB integrity lead to interactions between blood-derived immune cells (such as T cells and monocytes/macrophage) and resident glial cells (such as microglia and astrocytes), which in turn lead to the excessive production of proinflammatory cytokines and eventually progressive demyelination in CCALD (Paintlia et al. 2003).

Microglia, macrophages and astrocytes are prominent in active lesions expressing CD1 molecules, the major MHC-unrestricted lipid antigen, which may trigger the T-cell inflammatory response (Ito et al. 2001). In X-ALD, gangliosides or proteolipids in myelin, might be candidate antigens for eliciting auto immunity with VLCFAs (Tagawa et al. 2002). Antigen-presenting cells such as astrocytes and microglia may produce IL-12, which triggers CD4 T-cell in the TH1 response. Activated T cells produce proinflammatory cytokines, including TNFα and interferon-γ (INF-γ). Hence, CCALD can be categorized as a brain autoimmune disease. Thus, immunosuppressive treatments have been attempted in an effort to halt the disease progression (Fig. 8.4f). However, immunosuppressive and/or immunomodulating therapies have not worked in patients with advanced inflammatory demyelination (Naidu et al. 1988; Horvath et al. 2012; Korenke et al. 1997). Blocking CNS infiltration of systemic monocytes using natalizumab and intravenous high-dose immunoglobulins has also failed to show any benefit in CCALD (Turk et al. 2017b; Cappa et al. 1994).

The failure of immunomodulating therapies may be due to the distinct characteristics of disease pathogenesis in CCALD compared with MS. In both diseases, the infiltrated inflammatory cells are mostly macrophages and T-cells. In MS, however, the inflammatory cells are located at the active demyelination edge, whereas they are located behind the active demyelination edge in CCALD. This observation suggests that the inflammatory reaction in CCALD may be secondary to the demyelinating process. In MS, the infiltrated and activated T cells expressing IFN-γ are detected in active demyelinating lesions. In contrast, in CCALD the IFN-γ producing T-cells or INF-γ-stimulated microglia/macrophages are not detected in the active lesions, suggesting that the activated T cells are not directly involved in the inflammatory demyelination. This is why immunosuppression therapy fails to inhibit inflammatory demyelination in CCALD.

In X-ALD, macrophages that have phagocytosed myelin are observed in active lesions (Powers et al. 1992) and display a pro-inflammatory state (the M1 type). Weinhofer et al. reported that the macrophages in the anti-inflammatory state (the M2 type) were significantly lower in CCALD lesions than MS lesions, suggesting that the ability to enter the M2 state is incomplete in *ABCD1*-deficient macrophages, which may result in the inhibition of re-myelination and hence progressive demyelination in CCALD (Weinhofer et al. 2018). Macrophage polarization might be regulated by mitochondrial energic metabolism as well as fatty acid metabolism (Mills and O'Neill 2016; Oishi et al. 2017). Therefore, pharmacological manipulation that can polarize microglia or macrophages so as to enter the anti-inflammatory state may be another crucial therapeutic target for preventing myelin destruction in patients with X-ALD.

8.7 Recent Studies of New Targets

At present, there are no interventions to prevent or slow transition to CCALD except for HSCT and mitigating the symptoms of AMN. However, novel therapeutic studies toward new targets based on continuing studies of the pathogenesis are now underway.

8.7.1 Blood-Brain Barrier

It has been reported that BBB dysfunction is likely to play a role in the pathophysiology of inflammatory demyelination in X-ALD (Melhem et al. 2000; van der Voorn et al. 2011). The BBB consists of endothelial cells, pericytes and astrocytes that are involved in the immune system in the CNS. In CNS diseases, leukocyte infiltration through the BBB is a critical step in neuroinflammation. Recently, Musolino et al.

reported that ABCD1 deficiency in endothelial cells lead to the upregulation of adhesion molecules and a decrease in tight junction proteins, which together result in increased trafficking of leucocytes across the BBB (Musolino et al. 2015). It should be noted that the dysfunction of ABCD1 has a direct effect on endothelial function regardless of VLCFA accumulation. Recently, Lee et al. reported that the block copolymers micelles of poly(ethylene oxide) and poly(propylene oxide) have the capacity to restore the function of the BBB (Lee et al. 2018). They reported that iPSC-induced brain microvascular endothelial cells (iBMECs) from CCALD patients display decreased barrier properties and a greater number of accumulated lipid droplets than did iBMECs from control subjects. The recovery of BBB function by the block copolymers may prevent the onset of inflammatory demyelination and thus be a new therapeutic strategy for CCALD. The levels of matrix metalloproteinases (MMPs) such as MMP10 and TIMP1 were reported to increase according to the severity of neurodegeneration and contribute to the disruption of BBB. It is thus possible that inhibition of MMPs might be another target for reducing BBB disruption (Thibert et al. 2012).

8.7.2 Autophagy

In neurodegenerative diseases such as Alzheimer's, Parkinson's and Huntington's disease, autophagy in the brain is reported to be dysfunctional. Launay et al. reported that autophagy was down-regulated in the brain of X-ALD patients and in spinal cord of *Abcd1*-deficient mice (Launay et al. 2015). This is due to an elevation of mammalian target of rapamycin (mTOR) signaling. They found that temsirolimus, a rapamycin derivative, restored autophagic function and inhibited the degenerative axon process in *Abcd1*- and *Abcd2*-double deficient mice. As temsirolimus is a drug in clinical use for the treatment of renal cell carcinoma, approved by the U.S. Food and Drug Administration (FDA), this drug become a candidate therapeutic drug, especially for AMN.

8.7.3 The Unfolded Protein Response

Launay et al. demonstrated that the bile acid tauroursodeoxycholate (TUDCA) ameliorates axonal degeneration in *Abcd1*- and *Abcd2*-deficient mice by suppressing unfolded protein response (UPR) activation (Launay et al. 2017). UPR activation might be involved in the pathogenesis of X-ALD, presumably due to a redox imbalance. These results suggest that UPR is a novel target in AMN pathogenesis. As TUDCA is able to cross the BBB and has been approved by the FDA for clinical use, it may be a promising drug for AMN.

8.7.4 The Development of Novel Drugs

Despite considerable therapeutic drug development efforts, no effective drugs have been found to date. This is because of the as yet still unexplained aspects of the molecular pathogenesis in this disease, which has made it extremely difficult to identify valuable therapeutic targets. Therefore, the precise mechanisms by which ABCD1 protein deficiency leads to inflammatory demyelination are in urgent need of disclosure.

Animal models that mimic the phenotype of X-ALD patients are required for the development of novel therapeutic drugs. To date, the lack of adequate disease models has proven to be a major obstacle to an elucidation of how ABCD1 dysfunction leads to cerebral demyelination. The *Abcd1*-deficient mouse (Pujol et al. 2002) and *Abcd1-/Abcd2*-double deficient mouse develop axonopathy and locomotor impairment at 20 and 12 months of age (Ferrer et al. 2005), respectively. These mice resemble late-onset axonal degeneration in the spinal cord with which most AMN patients present, providing a model for investigating disease pathogenesis. Certain drugs, such as antioxidants, temsirolimus and TUDCA for AMN might be promising based on the pharmacological test results in these mice. At present, however, there is no adequate model system for CCALD. Even *Abcd1-/Abcd2*-double deficient mice do not display any inflammatory demyelination. The absence of an appropriate model mouse makes the development of therapeutic drugs difficult to achieve. Therefore, the generation of novel animal models with a phenotype similar to CCALD patients is critically required. Recently, Strachan et al. prepared a zebrafish model of X-ALD and demonstrated a developmental requirement for ABCD1 in myelination (Strachan et al. 2017). As zebrafish conserve the genetic pathways for myelin formation, the zebrafish model may afford an opportunity for developing therapeutic drugs for X-ALD in the future.

Recently, directed differentiation of human iPSCs into disease-related cells has become available as a modeling system for the investigation of potential therapeutic interventions. ALD patient-specific iPSC-derived glial cells (Baarine et al. 2015b) or brain microvascular endothelial cells (Lee et al. 2018) have been used for the development of therapeutic drugs. Baarine et al. reported that iPSC-derived astrocytes from CCALD expressed higher levels of proinflammatory cytokines than those from AMN and control astrocytes (Baarine et al. 2015b). As biochemical disease phenotypes can be mimicked by iPSC-derived astrocytes, it constitutes an *in vitro* model for the testing of new therapeutics for X-ALD. Furthermore, the CRISPR-Cas9 and related genome editing systems provide useful model cells for understanding the pathogenesis of X-ALD (Raas et al. 2019).

The process by which novel drugs are developed is expensive and time consuming. Therefore, drug repositioning of existing drugs is one approach to orphan diseases. As the side-effect profiles and related information of the existing drugs is available, the existing drugs are far less time-consuming. Therefore, the usage of existing drugs, like as valproic acid, pioglitazone, temsirolimus and metformin, may provide an opportunity to find novel treatments.

8.8 Concluding Remarks

At present, HSCT is most beneficial for X-ALD patients with a Loes MRI score less than 10 and the availability of a matched bone marrow donor. In the near future, asymptomatic infants will be identifiable through newborn screening and HSCT could be conducted at an asymptomatic stage in all X-ALD males. Furthermore, new drugs that can delay symptom onset or progression may improve the HSCT clinical outcome. Although its high cost has limited its use at present, it is hopeful that *ex vivo* gene therapy will become the standard therapy for X-ALD patients. Furthermore, novel therapeutic drugs that attenuate the impact of the disease in symptomatic X-ALD or AMN patients should be considered.

Acknowledgments The authors thank Professor Tsuneo Imanaka for his comments and suggestions. This work was supported in part by Grants-in-Aid for Scientific Research (C) (16K09961) from the Ministry of Education, Culture, Sports, Science and Technology of Japan. Pacific Edit reviewed the manuscript prior to submission.

References

Asheuer M et al (2004) Human CD34+ cells differentiate into microglia and express recombinant therapeutic protein. Proc Natl Acad Sci U S A 101(10):3557–3562. https://doi.org/10.1073/pnas.0306431101

Aubourg P et al (1990) Reversal of early neurologic and neuroradiologic manifestations of X-linked adrenoleukodystrophy by bone marrow transplantation. N Engl J Med 322(26):1860–1866. https://doi.org/10.1056/NEJM199006283222607

Baarine M et al (2015a) ABCD1 deletion-induced mitochondrial dysfunction is corrected by SAHA: implication for adrenoleukodystrophy. J Neurochem 133(3):380–396. https://doi.org/10.1111/jnc.12992

Baarine M et al (2015b) Functional characterization of IPSC-derived brain cells as a model for X-linked adrenoleukodystrophy. PLoS One 10(11):e0143238. https://doi.org/10.1371/journal.pone.0143238

Beckers L et al (2017) Specific suppression of microgliosis cannot circumvent the severe neuropathology in peroxisomal beta-oxidation-deficient mice. Mol Cell Neurosci 80:123–133. https://doi.org/10.1016/j.mcn.2017.03.004

Beckers L et al (2018) Neuronal dysfunction and behavioral abnormalities are evoked by neural cells and aggravated by inflammatory microglia in peroxisomal beta-oxidation deficiency. Front Cell Neurosci 12:136. https://doi.org/10.3389/fncel.2018.00136

Benhamida S et al (2003) Transduced CD34+ cells from adrenoleukodystrophy patients with HIV-derived vector mediate long-term engraftment of NOD/SCID mice. Mol Ther 7(3):317–324

Berger J et al (2014) Pathophysiology of X-linked adrenoleukodystrophy. Biochimie 98:135–142. https://doi.org/10.1016/j.biochi.2013.11.023

Bladowska J et al (2015) The role of MR imaging in the assessment of clinical outcomes in children with X-linked adrenoleukodystrophy after allogeneic haematopoietic stem cell transplantation. Pol J Radiol 80:181–190. https://doi.org/10.12659/PJR.893285

Blaw M (1970) Melanodermic type leukodystrophy (adrenoleukodystrophy). In: Vinken PJ, Bruyn GW (eds) Neurodystrophies and neurolipidoses. North Holland Publishing Co., Amsterdam, pp. 128–133

Budhram A, Pandey S (2017) Activation of cerebral X-linked adrenoleukodystrophy after head trauma. Can J Neurol Sci 44(5):597–598

Cappa M et al (1994) High dose immunoglobulin IV treatment in adrenoleukodystrophy. J Neurol Neurosurg Psychiatry 57(Suppl):69–70; discussion 71

Cartier N, Aubourg P (2010) Hematopoietic stem cell transplantation and hematopoietic stem cell gene therapy in X-linked adrenoleukodystrophy. Brain Pathol 20(4):857–862. https://doi.org/10.1111/j.1750-3639.2010.00394.x

Cartier N et al (2009) Hematopoietic stem cell gene therapy with a lentiviral vector in X-linked adrenoleukodystrophy. Science 326(5954):818–823. https://doi.org/10.1126/science.1171242

Cartier N et al (2012) Lentiviral hematopoietic cell gene therapy for X-linked adrenoleukodystrophy. Methods Enzymol 507:187–198

Cartier N et al (2014) The role of microglia in human disease: therapeutic tool or target? Acta Neuropathol 128(3):363–380

Deon M et al (2016) Oxidative stress in patients with X-linked adrenoleukodystrophy. Cell Mol Neurobiol 36(4):497–512. https://doi.org/10.1007/s10571-015-0234-2

Derecki NC et al (2013) The role of microglia in brain maintenance: implications for Rett syndrome. Trends Immunol 34(3):144–150. https://doi.org/10.1016/j.it.2012.10.002

Eichler F et al (2017) Hematopoietic stem-cell gene therapy for cerebral adrenoleukodystrophy. N Engl J Med 377(17):1630–1638

Engelen M et al (2010) Lovastatin in X-linked adrenoleukodystrophy. N Engl J Med 362(3):276–277

Engelen M et al (2012a) X-linked adrenoleukodystrophy (X-ALD): clinical presentation and guidelines for diagnosis, follow-up and management. Orphanet J Rare Dis 7:51. https://doi.org/10.1186/1750-1172-7-51

Engelen M et al (2012b) Bezafibrate lowers very long-chain fatty acids in X-linked adrenoleukodystrophy fibroblasts by inhibiting fatty acid elongation. J Inherit Metab Dis 35(6):1137–1145

Engelen M et al (2014) X-linked adrenoleukodystrophy in women: a cross-sectional cohort study. Brain 137(Pt 3):693–706. https://doi.org/10.1093/brain/awt361

Fanconi A et al. (1963) [Addison's disease with cerebral sclerosis in childhood. A hereditary syndrome transmitted through chromosome X?]. Helv Paediatr Acta 18:480–501

Ferrara S et al (2017) Ester-to-amide rearrangement of ethanolamine-derived prodrugs of sobetirome with increased blood-brain barrier penetration. Bioorg Med Chem 25(10):2743–2753

Ferrer I et al (2005) Inactivation of the peroxisomal ABCD2 transporter in the mouse leads to late-onset ataxia involving mitochondria, Golgi and endoplasmic reticulum damage. Hum Mol Genet 14(23):3565–3577. https://doi.org/10.1093/hmg/ddi384

Fourcade S et al (2003) Thyroid hormone induction of the adrenoleukodystrophy-related gene (ABCD2). Mol Pharmacol 63(6):1296–1303. https://doi.org/10.1124/mol.63.6.1296

Fourcade S et al (2010) Valproic acid induces antioxidant effects in X-linked adrenoleukodystrophy. Hum Mol Genet 19(10):2005–2014. https://doi.org/10.1093/hmg/ddq082

Fourcade S et al (2014) Mitochondrial dysfunction and oxidative damage cooperatively fuel axonal degeneration in X-linked adrenoleukodystrophy. Biochimie 98:143–149. https://doi.org/10.1016/j.biochi.2013.09.012

Galino J et al (2011) Oxidative damage compromises energy metabolism in the axonal degeneration mouse model of X-adrenoleukodystrophy. Antioxid Redox Signal 15(8):2095–2107. https://doi.org/10.1089/ars.2010.3877

Geric I et al (2018) Lipid homeostasis and inflammatory activation are disturbed in classically activated macrophages with peroxisomal beta-oxidation deficiency. Immunology 153(3):342–356. https://doi.org/10.1111/imm.12844

Gilg AG et al (2000) Inducible nitric oxide synthase in the central nervous system of patients with X-adrenoleukodystrophy. J Neuropathol Exp Neurol 59(12):1063–1069

Gong Y et al (2019) Intrathecal adeno-associated virus vector-mediated gene delivery for adrenomyeloneuropathy. Hum Gene Ther 30(5):544–555

Görtz AL et al (2018) Heat shock protein expression in cerebral X-linked adrenoleukodystrophy reveals astrocyte stress prior to myelin loss. Neuropathol Appl Neurobiol 44(4):363–376

Gronemeyer T et al (2013) The proteome of human liver peroxisomes: identification of five new peroxisomal constituents by a label-free quantitative proteomics survey. PLoS One 8(2):e57395. https://doi.org/10.1371/journal.pone.0057395

Gueugnon F et al (2003) Dehydroepiandrosterone induction of the Abcd2 and Abcd3 genes encoding peroxisomal ABC transporters: implications for X-linked adrenoleukodystrophy. Adv Exp Med Biol 544:245

Hartley M et al (2017) A thyroid hormone-based strategy for correcting the biochemical abnormality in X-linked adrenoleukodystrophy. Endocrinology 158(5):1328–1338

Heinzer AK et al (2003) Mouse models and genetic modifiers in X-linked adrenoleukodystrophy. Adv Exp Med Biol 544:75–93

Horvath G et al (2012) Failure of repeated cyclophosphamide pulse therapy in childhood cerebral X-linked adrenoleukodystrophy. Neuropediatrics 43(1):48–52

Huffnagel IC et al (2019) Disease progression in women with X-linked adrenoleukodystrophy is slow. Orphanet J Rare Dis 14(1):30. https://doi.org/10.1186/s13023-019-1008-6

Igarashi M et al (1976) Fatty acid abnormality in adrenoleukodystrophy. J Neurochem 26(4):851–860

Ito M et al (2001) Potential environmental and host participants in the early white matter lesion of adreno-leukodystrophy: morphologic evidence for CD8 cytotoxic T cells, cytolysis of oligodendrocytes, and CD1-mediated lipid antigen presentation. J Neuropathol Exp Neurol 60(10):1004–1019

Jang J et al (2011) Induced pluripotent stem cell models from X-linked adrenoleukodystrophy patients. Ann Neurol 70(3):402–409

Jang J et al (2016) 25-Hydroxycholesterol contributes to cerebral inflammation of X-linked adrenoleukodystrophy through activation of the NLRP3 inflammasome. Nat Commun 7:13129–13129

Jiang H et al (2015) Combination of a haploidentical stem cell transplant with umbilical cord blood for cerebral X-linked adrenoleukodystrophy. Pediatr Neurol 53(2):163–165.e161

Kartha R et al (2015) Mechanisms of antioxidant induction with high-dose N-acetylcysteine in childhood cerebral adrenoleukodystrophy. CNS Drugs 29(12):1041–1047

Kato K et al (2019) Allogeneic stem cell transplantation with reduced intensity conditioning for patients with adrenoleukodystrophy. Mol Genet Metab Rep 18:1–6. https://doi.org/10.1016/j.ymgmr.2018.11.001

Kemp S, Wanders R (2010) Biochemical aspects of X-linked adrenoleukodystrophy. Brain Pathol 20(4):831–837. https://doi.org/10.1111/j.1750-3639.2010.00391.x

Kemp S et al (1998) Gene redundancy and pharmacological gene therapy: implications for X-linked adrenoleukodystrophy. Nat Med 4(11):1261–1268. https://doi.org/10.1038/3242

Kemp S et al (2012) X-linked adrenoleukodystrophy: clinical, metabolic, genetic and pathophysiological aspects. Biochim Biophys Acta 1822(9):1465–1474. https://doi.org/10.1016/j.bbadis.2012.03.012

Kemp S et al (2016) Adrenoleukodystrophy - neuroendocrine pathogenesis and redefinition of natural history. Nat Rev Endocrinol 12(10):606–615

Kemper A et al (2017) Newborn screening for X-linked adrenoleukodystrophy: evidence summary and advisory committee recommendation. Genet Med 19(1):121–126

Koç ON et al (1999) Bone marrow-derived mesenchymal stem cells remain host-derived despite successful hematopoietic engraftment after allogeneic transplantation in patients with lysosomal and peroxisomal storage diseases. Exp Hematol 27(11):1675–1681

Kohler W (2010) Leukodystrophies with late disease onset: an update. Curr Opin Neurol 23(3):234–241. https://doi.org/10.1097/WCO.0b013e328338313a

Korenke GC et al (1997) Progression of X-linked adrenoleukodystrophy under interferon-beta therapy. J Inherit Metab Dis 20(1):59–66

Kuhl JS et al (2018) Potential risks to stable long-term outcome of allogeneic hematopoietic stem cell transplantation for children with cerebral X-linked adrenoleukodystrophy. JAMA Netw Open 1(3):e180769. https://doi.org/10.1001/jamanetworkopen.2018.0769

Lauer A et al (2017) ABCD1 dysfunction alters white matter microvascular perfusion. Brain 140(12):3139–3152
Launay N et al (2015) Autophagy induction halts axonal degeneration in a mouse model of X-adrenoleukodystrophy. Acta Neuropathol 129(3):399–415
Launay N et al (2017) Tauroursodeoxycholic bile acid arrests axonal degeneration by inhibiting the unfolded protein response in X-linked adrenoleukodystrophy. Acta Neuropathol 133(2):283–301
Lee CAA et al (2018) Modeling and rescue of defective blood-brain barrier function of induced brain microvascular endothelial cells from childhood cerebral adrenoleukodystrophy patients. Fluids Barriers CNS 15(1):9
Lopez-Erauskin J et al (2011) Antioxidants halt axonal degeneration in a mouse model of X-adrenoleukodystrophy. Ann Neurol 70(1):84–92. https://doi.org/10.1002/ana.22363
Lopez-Erauskin J et al (2012) Oxidative stress modulates mitochondrial failure and cyclophilin D function in X-linked adrenoleukodystrophy. Brain 135(Pt 12):3584–3598. https://doi.org/10.1093/brain/aws292
Lu J-F et al (2007) The role of peroxisomal ABC transporters in the mouse adrenal gland: the loss of Abcd2 (ALDR), Not Abcd1 (ALD), causes oxidative damage. Lab Invest 87(3):261–272
Mahmood A et al (2007) Survival analysis of haematopoietic cell transplantation for childhood cerebral X-linked adrenoleukodystrophy: a comparison study. Lancet Neurol 6(8):687–692. https://doi.org/10.1016/S1474-4422(07)70177-1
Marchetti D et al (2015) Protective effect of antioxidants on DNA damage in leukocytes from X-linked adrenoleukodystrophy patients. Int J Dev Neurosci 43:8–15
Marchetti D et al (2018) Oxidative imbalance, nitrative stress, and inflammation in C6 glial cells exposed to hexacosanoic acid: protective effect of N-acetyl-L-cysteine, trolox, and rosuvastatin. Cell Mol Neurobiol 38(8):1505–1516
McGuinness MC et al (2001) Evaluation of pharmacological induction of fatty acid beta-oxidation in X-linked adrenoleukodystrophy. Mol Genet Metab 74(1–2):256–263
Melhem ER et al (2000) X-linked adrenoleukodystrophy: the role of contrast-enhanced MR imaging in predicting disease progression. AJNR Am J Neuroradiol 21(5):839–844
Miller WP et al (2011) Outcomes after allogeneic hematopoietic cell transplantation for childhood cerebral adrenoleukodystrophy: the largest single-institution cohort report. Blood 118(7):1971–1978. https://doi.org/10.1182/blood-2011-01-329235
Miller WP et al (2016) Intensity of MRI gadolinium enhancement in cerebral adrenoleukodystrophy: a biomarker for inflammation and predictor of outcome following transplantation in higher risk patients. AJNR Am J Neuroradiol 37(2):367–372
Mills EL, O'Neill LA (2016) Reprogramming mitochondrial metabolism in macrophages as an anti-inflammatory signal. Eur J Immunol 46(1):13–21. https://doi.org/10.1002/eji.201445427
Morato L et al (2013) Pioglitazone halts axonal degeneration in a mouse model of X-linked adrenoleukodystrophy. Brain 136(Pt 8):2432–2443. https://doi.org/10.1093/brain/awt143
Morató L et al (2015) Activation of sirtuin 1 as therapy for the peroxisomal disease adrenoleukodystrophy. Cell Death Differ 22(11):1742–1753
Morita M (2007) [Adrenoleukodystrophy: molecular pathogenesis and development of therapeutic agents]. Yakugaku Zasshi 127(7):1059–1064
Morita M et al (2005) Baicalein 5,6,7-trimethyl ether, a flavonoid derivative, stimulates fatty acid beta-oxidation in skin fibroblasts of X-linked adrenoleukodystrophy. FEBS Lett 579(2):409–414. https://doi.org/10.1016/j.febslet.2004.11.102
Morita M et al (2008) Baicalein 5,6,7-trimethyl ether activates peroxisomal but not mitochondrial fatty acid beta-oxidation. J Inherit Metab Dis 31(3):442–449. https://doi.org/10.1007/s10545-008-0857-2
Morita M et al (2011) ABC subfamily D proteins and very long chain fatty acid metabolism as novel targets in adrenoleukodystrophy. Curr Drug Targets 12(5):694–706

Morita M et al (2015) Brain microsomal fatty acid elongation is increased in abcd1-deficient mouse during active myelination phase. Metab Brain Dis 30(6):1359–1367. https://doi.org/10.1007/s11011-015-9701-1

Morita M et al (2018) Stability of the ABCD1 protein with a missense mutation: a novel approach to finding therapeutic compounds for X-linked adrenoleukodystrophy. JIMD Rep. https://doi.org/10.1007/8904_2018_118

Moser HW (1995) Adrenoleukodystrophy. Curr Opin Neurol 8(3):221–226

Moser A, Fatemi A (2018) Newborn screening and emerging therapies for X-linked adrenoleukodystrophy. JAMA Neurol 75(10):1175–1176

Moser HW et al (2005) Follow-up of 89 asymptomatic patients with adrenoleukodystrophy treated with Lorenzo's oil. Arch Neurol 62(7):1073–1080. https://doi.org/10.1001/archneur.62.7.1073

Moser HW et al (2007) X-linked adrenoleukodystrophy. Nat Clin Pract Neurol 3(3):140–151. https://doi.org/10.1038/ncpneuro0421

Mosser J et al (1993) Putative X-linked adrenoleukodystrophy gene shares unexpected homology with ABC transporters. Nature 361(6414):726–730. https://doi.org/10.1038/361726a0

Musolino P et al (2015) Brain endothelial dysfunction in cerebral adrenoleukodystrophy. Brain 138(11):3206–3220

Naidu S et al (1988) Childhood adrenoleukodystrophy. Failure of intensive immunosuppression to arrest neurologic progression. Arch Neurol 45(8):846–848

Netik A et al (1999) Adrenoleukodystrophy-related protein can compensate functionally for adrenoleukodystrophy protein deficiency (X-ALD): implications for therapy. Hum Mol Genet 8(5):907–913

Nury T et al (2017) 7-Ketocholesterol is increased in the plasma of X-ALD patients and induces peroxisomal modifications in microglial cells: potential roles of 7-ketocholesterol in the pathophysiology of X-ALD. J Steroid Biochem Mol Biol 169:123–136

Ogino T, Suzuki K (1981) Specificities of human and rat brain enzymes of cholesterol ester metabolism toward very long chain fatty acids: implication for biochemical pathogenesis of adrenoleukodystrophy. J Neurochem 36(2):776–779

Oishi Y et al (2017) SREBP1 contributes to resolution of pro-inflammatory TLR4 signaling by reprogramming fatty acid metabolism. Cell Metab 25(2):412–427. https://doi.org/10.1016/j.cmet.2016.11.009

Orchard PJ et al (2019) Successful donor engraftment and repair of the blood brain barrier in cerebral adrenoleukodystrophy. Blood. https://doi.org/10.1182/blood-2018-11-887240

Paintlia AS et al (2003) Correlation of very long chain fatty acid accumulation and inflammatory disease progression in childhood X-ALD: implications for potential therapies. Neurobiol Dis 14(3):425–439

Pierpont E et al (2017) Neurocognitive trajectory of boys who received a hematopoietic stem cell transplant at an early stage of childhood cerebral adrenoleukodystrophy. JAMA Neurol 74(6):710–717

Powers JM et al (1992) The inflammatory myelinopathy of adreno-leukodystrophy: cells, effector molecules, and pathogenetic implications. J Neuropathol Exp Neurol 51(6):630–643

Powers J et al (2005) Adreno-leukodystrophy: oxidative stress of mice and men. J Neuropathol Exp Neurol 64(12):1067–1079

Pujol A (2016) Novel therapeutic targets and drug candidates for modifying disease progression in adrenoleukodystrophy. Endocr Dev 30:147–160. https://doi.org/10.1159/000439340

Pujol A et al (2002) Late onset neurological phenotype of the X-ALD gene inactivation in mice: a mouse model for adrenomyeloneuropathy. Hum Mol Genet 11(5):499–505

Pujol A et al (2004) Functional overlap between ABCD1 (ALD) and ABCD2 (ALDR) transporters: a therapeutic target for X-adrenoleukodystrophy. Hum Mol Genet 13(23):2997–3006. https://doi.org/10.1093/hmg/ddh323

Raas Q et al (2019) CRISPR/Cas9-mediated knockout of Abcd1 and Abcd2 genes in BV-2 cells: novel microglial models for X-linked adrenoleukodystrophy. Biochim Biophys Acta Mol Cell Biol Lipids. https://doi.org/10.1016/j.bbalip.2019.02.006

Ranea Robles P et al (2018) Aberrant regulation of the GSK-3β/NRF2 axis unveils a novel therapy for adrenoleukodystrophy. EMBO Mol Med 10(8):e8604

Raymond GV et al (2010) Head trauma can initiate the onset of adreno-leukodystrophy. J Neurol Sci 290(1–2):70–74. https://doi.org/10.1016/j.jns.2009.11.005

Rizzo WB et al (1989) Dietary erucic acid therapy for X-linked adrenoleukodystrophy. Neurology 39(11):1415–1422

Sassa T et al (2014) Lorenzo's oil inhibits ELOVL1 and lowers the level of sphingomyelin with a saturated very long-chain fatty acid. J Lipid Res 55(3):524–530

Schlüter A et al (2018) Epigenomic signature of adrenoleukodystrophy predicts compromised oligodendrocyte differentiation. Brain Pathol 28(6):902–919

Semmler A et al (2008) Therapy of X-linked adrenoleukodystrophy. Expert Rev Neurother 8(9):1367–1379. https://doi.org/10.1586/14737175.8.9.1367

Singh J, Giri S (2014) Loss of AMP-activated protein kinase in X-linked adrenoleukodystrophy patient-derived fibroblasts and lymphocytes. Biochem Biophys Res Commun 445(1):126–131. https://doi.org/10.1016/j.bbrc.2014.01.126

Singh I et al (1998a) Lovastatin for X-linked adrenoleukodystrophy. N Engl J Med 339(10):702–703. https://doi.org/10.1056/NEJM199809033391012

Singh I et al (1998b) Lovastatin and sodium phenylacetate normalize the levels of very long chain fatty acids in skin fibroblasts of X-adrenoleukodystrophy. FEBS Lett 426(3):342–346

Strachan L et al (2017) A zebrafish model of X-linked adrenoleukodystrophy recapitulates key disease features and demonstrates a developmental requirement for abcd1 in oligodendrocyte patterning and myelination. Hum Mol Genet 26(18):3600–3614

Tagawa Y et al (2002) Anti-ganglioside antibodies bind with enhanced affinity to gangliosides containing very long chain fatty acids. Neurochem Res 27(7–8):847–855

Takahashi N et al (2007) Adrenoleukodystrophy: subcellular localization and degradation of adrenoleukodystrophy protein (ALDP/ABCD1) with naturally occurring missense mutations. J Neurochem 101(6):1632–1643. https://doi.org/10.1111/j.1471-4159.2007.04457.x

Takemoto Y et al (2002) Epidemiology of X-linked adrenoleukodystrophy in Japan. J Hum Genet 47(11):590–593. https://doi.org/10.1007/s100380200090

Taylor JL, Lee S (2019) Lessons learned from newborn screening in pilot studies. N C Med J 80(1):54–58. https://doi.org/10.18043/ncm.80.1.54

Thibert KA et al (2012) Cerebrospinal fluid matrix metalloproteinases are elevated in cerebral adrenoleukodystrophy and correlate with MRI severity and neurologic dysfunction. PLoS One 7(11):e50430. https://doi.org/10.1371/journal.pone.0050430

Tran C et al (2017) Long-term outcome of patients with X-linked adrenoleukodystrophy: a retrospective cohort study. Eur J Paediatr Neurol 21(4):600–609

Troffer Charlier N et al (1998) Mirror expression of adrenoleukodystrophy and adrenoleukodystrophy related genes in mouse tissues and human cell lines. Eur J Cell Biol 75(3):254–264

Tsuji S et al (1984) Fatty acid elongation activity in fibroblasts from patients with adrenoleukodystrophy (ALD). J Biochem 96(4):1241–1247

Turk B et al (2017a) Antioxidant capacity and superoxide dismutase activity in adrenoleukodystrophy. JAMA Neurol 74(5):519–524

Turk BR et al (2017b) Therapeutic strategies in adrenoleukodystrophy. Wien Med Wochenschr 167(9–10):219–226. https://doi.org/10.1007/s10354-016-0534-2

Turk B et al (2018) Dendrimer-N-acetyl-L-cysteine modulates monophagocytic response in adrenoleukodystrophy. Ann Neurol 84(3):452–462

van den Broek BTA et al (2018) Early and late outcomes after cord blood transplantation for pediatric patients with inherited leukodystrophies. Blood Adv 2(1):49–60

van der Voorn JP et al (2011) Correlating quantitative MR imaging with histopathology in X-linked adrenoleukodystrophy. AJNR Am J Neuroradiol 32(3):481–489. https://doi.org/10.3174/ajnr.A2327

van Engen CE et al (2016) CYP4F2 affects phenotypic outcome in adrenoleukodystrophy by modulating the clearance of very long-chain fatty acids. Biochim Biophys Acta 1862(10):1861–1870. https://doi.org/10.1016/j.bbadis.2016.07.006

van Geel BM et al (2001) Evolution of phenotypes in adult male patients with X-linked adrenoleukodystrophy. Ann Neurol 49(2):186–194

van Geel BM et al (2015) Hematopoietic cell transplantation does not prevent myelopathy in X-linked adrenoleukodystrophy: a retrospective study. J Inherit Metab Dis 38(2):359–361. https://doi.org/10.1007/s10545-014-9797-1

van Roermund CW et al (2008) The human peroxisomal ABC half transporter ALDP functions as a homodimer and accepts acyl-CoA esters. FASEB J 22(12):4201–4208. https://doi.org/10.1096/fj.08-110866

van Roermund CW et al (2011) Differential substrate specificities of human ABCD1 and ABCD2 in peroxisomal fatty acid beta-oxidation. Biochim Biophys Acta 1811(3):148–152. https://doi.org/10.1016/j.bbalip.2010.11.010

Vogel BH et al (2015) Newborn screening for X-linked adrenoleukodystrophy in New York State: diagnostic protocol, surveillance protocol and treatment guidelines. Mol Genet Metab 114(4):599–603

Weber F et al (2014) X-linked adrenoleukodystrophy: very long-chain fatty acid metabolism is severely impaired in monocytes but not in lymphocytes. Hum Mol Genet 23(10):2542–2550

Weinhofer I et al (2005) Liver X receptor alpha interferes with SREBP1c-mediated Abcd2 expression. Novel cross-talk in gene regulation. J Biol Chem 280(50):41243–41251. https://doi.org/10.1074/jbc.M509450200

Weinhofer I et al (2018) Impaired plasticity of macrophages in X-linked adrenoleukodystrophy. Brain 141(8):2329–2342

Wiesinger C et al (2013) Impaired very long-chain acyl-CoA beta-oxidation in human X-linked adrenoleukodystrophy fibroblasts is a direct consequence of ABCD1 transporter dysfunction. J Biol Chem 288(26):19269–19279. https://doi.org/10.1074/jbc.M112.445445

Wiesinger C et al (2015) The genetic landscape of X-linked adrenoleukodystrophy: inheritance, mutations, modifier genes, and diagnosis. Appl Clin Genet 8:109–121. https://doi.org/10.2147/TACG.S49590

Yamada T et al (2004) Therapeutic effects of normal cells on ABCD1 deficient cells in vitro and hematopoietic cell transplantation in the X-ALD mouse model. J Neurol Sci 218(1–2):91–97. https://doi.org/10.1016/j.jns.2003.11.006

Zhang X et al (2011) Conservation of targeting but divergence in function and quality control of peroxisomal ABC transporters: an analysis using cross-kingdom expression. Biochem J 436(3):547–557. https://doi.org/10.1042/BJ20110249

Part III
Topics in Peroxisome Research

Chapter 9
The Isolation of Peroxisomes

Tsuneo Imanaka

Abstract Peroxisomes are subcellular organelles bound by a single membrane. They are involved in a variety of metabolic processes such as β-oxidation of very long chain fatty acids. Isolated and purified peroxisomes are necessary to obtain a detailed understanding of their structure and function. Here, our three protocols for the isolation of peroxisomes from rat liver, rat hepatoma H4IIE cells and yeast *Komagataella phaffii* (previously called *Pichia pastoris*) are presented. Highly pure peroxisomes can be prepared from rat liver by differential centrifugation followed by Nycodenz gradient centrifugation. It is difficult to prepare highly purified peroxisomes from cultured mammalian cells, but the subcellular fractionation of peroxisomes is still a powerful tool to analyze whether a certain protein is localized to peroxisomes. Peroxisomes are potently induced in methylotrophic yeast such as *K. phaffii*. A large amount of mammalian peroxisomal proteins can be expressed in cells under the control of the alcohol oxidase gene promoter in a medium containing methanol as the only carbon source. Therefore, purified peroxisomes are useful for characterizing mammalian peroxisomal proteins. The protocols for the isolation of peroxisomes by immuno-beads and peroxisomal membranes by a sodium carbonate procedure are also presented.

Keywords Peroxisome · Peroxisomal membrane · Subcellular fractionation · Density gradient centrifugation · Rat liver · *Komagataella phaffii*

Abbreviations

ER	Endoplasmic reticulum
PMP	Peroxisomal membrane protein

T. Imanaka (✉)
Faculty of Pharmaceutical Sciences, Hiroshima International University, Kure, Hiroshima, Japan

University of Toyama, Toyama, Japan
e-mail: imanaka@hirokoku-u.ac.jp

T. Imanaka, N. Shimozawa (eds.), *Peroxisomes: Biogenesis, Function, and Role in Human Disease*, https://doi.org/10.1007/978-981-15-1169-1_9

9.1 Introduction

The isolation of peroxisomes is an essential technology for characterizing the constituents in peroxisomes along with their function. Peroxisomes play a critical role in a number of important metabolic pathways, including the β-oxidation of very long chain fatty acids (VLCFA) as well as the synthesis of ether-phospholipids and bile acid. Purified peroxisomes are needed to address the following questions. How are newly synthesized peroxisomal proteins imported into peroxisomes, and how are metabolites transported across peroxisomal membranes?

In mammals, peroxisomes are small organelles comprising only a few percent of the volume of the liver and kidney cells in which they are most abundant. However, highly purified peroxisomes can be prepared from the rat liver by mild homogenization of the tissue, differential centrifugation, and Nycodenz density gradient centrifugation. Nycodenz (*N*,*N′*-bis (2,3-dyhydroxypropyl)-5-[*N*-(2,3-dihydroxyprophyl)acetamido]-2,4,6-triiodo-isophthalmide; MW 821) is a nonionic iodinated density gradient medium of low viscosity and osmolality. The lower osmotic pressure and partial permeability of peroxisomal membranes should lead to improved separation of peroxisomes from other organelles. I present our protocol for the isolation of peroxisomes from rat liver based on certain published reports (Imanaka et al. 1987, 1991; Morita et al. 2000). I also touch on how we evaluate the purity of peroxisomes in the discussion section.

Highly purified peroxisomes are also prepared by immuno-beads (Luers et al. 1998). Rat liver peroxisomes are further purified from the peroxisomal fraction prepared by Nycodenz gradient centrifugation by using magnetic-beads covalently coupled to an anti-PMP70/ABCD3 antibody (Kikuchi et al. 2004).

Peroxisomal membranes lacking peripheral membrane proteins were prepared by a sodium carbonate procedure (Fujiki et al. 1982a, b) in which peroxisomes are disrupted and peroxisomal proteins denatured by the alkaline pH of sodium carbonate (0.1 M, pH ~11.0). As a result, peripheral membranes proteins detach from the peroxisomal membranes. Matrix proteins and core proteins such as urate oxidase are also released from the peroxisomes. After the sodium carbonate treatment, peroxisomal membranes with integral membrane proteins are recovered as membrane sheets by high speed centrifugation.

Peroxisomes are dynamic organelles and change their number and content in different environments. Cultured cells constitute an ideal tool for manipulating metabolic conditions, e.g. by incubation with selected compounds or modifying the expression of peroxisomal proteins by knockout or overexpression. Generally, the size of culture cell peroxisomes is considerably smaller and their amount less abundant compared to those in the rat liver and kidney. It is difficult to prepare highly purified peroxisomes. Nonetheless, the subcellular fractionation of peroxisomes provides useful information about peroxisomes. The subcellular fractionation procedure from rat hepatoma HIIE cells is presented (Imanaka et al. 1996).

Yeast cells contain only a few peroxisomes when glucose is used as the only carbon source. However, the peroxisome volume as well as number greatly increases when the carbon source is changed to methanol in methylotrophic yeast such as *Ogataea polymorpha* and *Komagataella phaffii* (formerly called *Hansenula polymorpha* and *Pichia pastoris,* respectively) (Ahmad et al. 2014). In addition, a large amount of mammalian peroxisomal proteins can be expressed in peroxisomes of yeast under control of a powerful methanol-inducible alcohol oxidase gene promoter. Therefore, the preparation of peroxisomes from yeast cells is useful for the characterization of mammalian peroxisomal proteins expressed in peroxisomes. Here I present our protocol for the isolation of peroxisomes from *K. phaffii*.

9.2 Isolation of Peroxisomes from the Rat Liver

As peroxisomes are most abundant in the rat liver in mammals, the basic procedure for isolating them has been developed in hepatic tissue. The isolation of peroxisomes from the rat liver is based on the procedures of Leighton et al. (1968) and Imanaka et al. (1987) with certain simplifications. As peroxisomes are fragile, mild conditions are employed during their isolation. The homogenate is normally prepared in an isotonic solution (0.25 M sucrose for mammalian tissues) with 0.1% (v/v) ethanol, 1 mM EDTA and a low concentration of buffer (pH 7.4). Ethanol is known to prevent the inactivation of catalase. The addition of a chelator reduces the aggregation of peroxisomes. The protease inhibitors leupeptin, pepstatin, chymostatin and antipain are added at a concentration of 10 μg/ml to prevent the degradation of peroxisomal membrane proteins (PMPs). Calculating the recovery of peroxisomes and their purity based on an assay of their marker enzymes are important for evaluating the quality of the preparation and reproducibility of the procedure (Leighton et al. 1968).

9.2.1 Procedure

1. Starve the rats overnight. Starvation reduces the glycogen particles in the liver that disturb the separation of the organelles.
2. Weigh the rats. After anesthesia and cutting the cervical aorta for bleeding them to death and removing the blood from the liver as soon as possible, rinse the liver with ice-cold 0.25 M sucrose, 1 mM EDTA and 0.1% ethanol at pH 7.4 (0.25 M SVE).
3. Weigh the liver and cut it into small pieces (~5 mm^3).
4. Homogenize the liver with one down-and-up stroke of a loose-fitting pestle (Potter-Elvehjem homogenizer) rotating at 1000 rpm. Adjust the volume of the homogenate to 5 volume of the tissue weight (5 ml/g tissue). All procedures are carried out at a temperature of 4 °C.

5. Centrifuge the homogenate at 2500 × *g* for 13 min[1] to pellet the cellular debris, nuclei and large population of mitochondria. (The pellet fraction is called the ν fraction, as in Leighton et al. (1968))
6. Retain the supernatant avoiding the lipid layer at the top, and remove the pink fluffy layer on the pellet carefully by pipetting and combine it with the supernatant (the ε fraction). Then suspend the pellet that has been detached by a glass rod in half of the original volume of 0.25 M SVE. Re-homogenize the pellet as before and centrifuge at 2500 × *g* for 13 min.
7. Pool the supernatants from both homogenization steps (including the pink fluffy layer), add the protease inhibitors leupeptin, pepstatin, chymostatin and antipain at a final concentration of 10 μg/ml, and centrifuge at 17,000 × *g* for 22 min to pellet the peroxisomes. The pellet includes mitochondria, lysosomes and endoplasmic reticulum (ER) and is called the λ fraction.
8. Remove the supernatant (the ϕ fraction) and then carefully remove the pink fluffy layer as thoroughly as possible by pipetting. Then suspend the pellet detached by the glass rod in a small volume of 0.25 M SVE and homogenize gently by two to three strokes with a Dounce homogenizer type B.
9. Centrifuge the suspension at 17,000 × *g* for 22 min and remove the supernatant with the pink fluffy layer, homogenize the pellet with Dounce homogenizer, and adjust the volume to 4 ml by 0.25 M SVE.
10. Apply 2 ml of each to the top of 20 ml of the Nycodenz linear gradient ($\rho = 1.15$–1.25) on 2 ml of Nycodenz cushion ($\rho = 1.30$). Nycodenz solutions are prepared in 0.25 M SVE plus 5 mM HEPES, pH 7.4 (0.25 M SVEH). Apply 1.0 ml of SVEH carefully on the sample. Centrifuge at 193,000 × *g* for 90 min with a vertical rotor using slow acceleration and deceleration. Peroxisomes form as a sediment of green color in the centrifuge tube (Fig. 9.1). Instead of the vertical rotor, a similar purity is obtained using an angle or swing rotor under the same conditions.
11. After centrifugation, make a small hole in the bottom of the centrifuge tube with an 18 gauge needle and recover the solution from the bottom drop by drop in 1 ml tubes. Measure refractive index to determine the density of the solution and determine the marker enzymes in each organelle.
12. To concentrate the peroxisomes in peroxisomal fractions, deliver the peroxisomal fraction drop by drop into 5 volumes of 0.25 M SVEH, mix gently and centrifuge at 17,000 × *g* for 22 min. Detach the pellet with a glass rod and

[1] Centrifugation conditions are express in integrate form; g-min, rather than in the conventional terms of at $a \times$ g for b min in the original paper (Leighton et al. 1968). The article considers the force during acceleration and deceleration, and determines the centrifugation condition (g-min) using the following formula; g-min = $\int_0^t \omega^2 dt \times R_{av}/981 \times 60$. R_{av} is the distance to the middle of the tube of the fluid during centrifugation. In this paper, conventionally, the determination of the centrifugation conditions is made from the plateau speed and the centrifugation time from the starting point as "$a \times g_{av}$ for b min". In this paper, g_{av} is expressed simply as g. The flowchart for the isolation of peroxisomes is shown in Fig. 9.1.

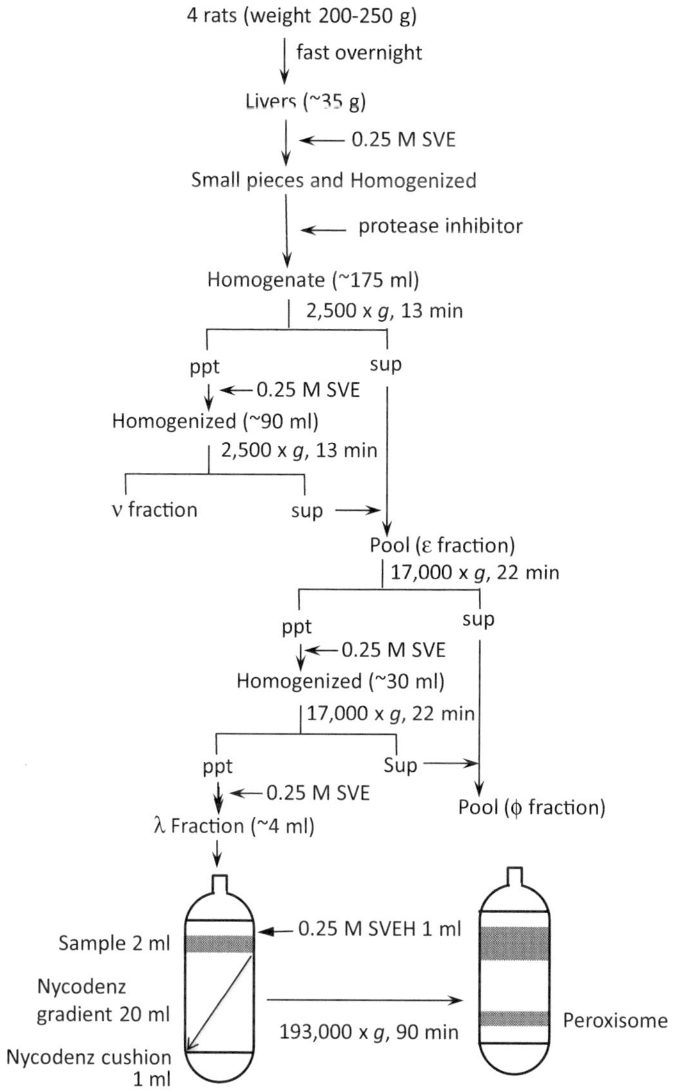

Fig. 9.1 Flowchart of the isolation of peroxisomes from the rat liver

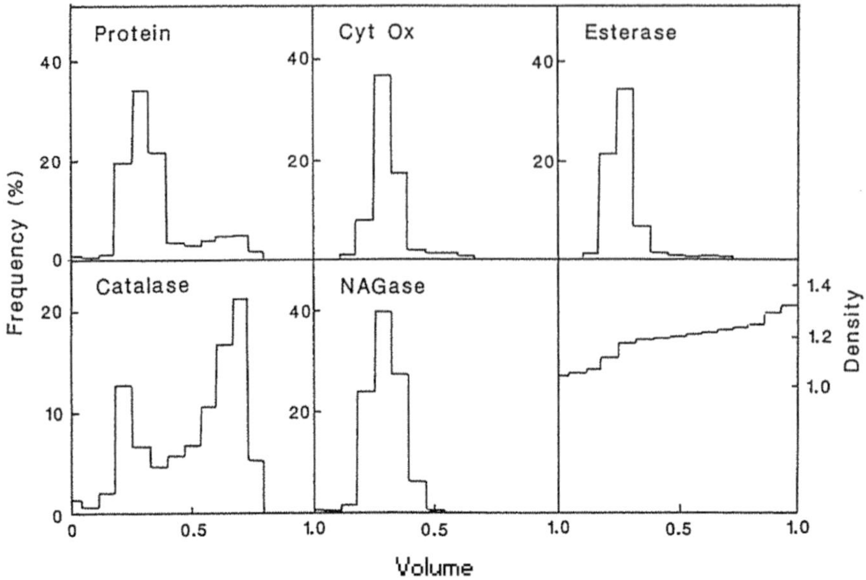

Fig. 9.2 Subcellular fractionation on a Nycodenz gradient. A light mitochondrial fraction prepared from the rat liver was fractionated by equilibrium density centrifugation on Nycodenz. Catalase, *N* -acetyl-β-D-glucosaminidase (NAGase), cytochrome *c* oxidase (Cyt Ox), esterase are measured as marker enzymes of peroxisome, lysosome, mitochondria and microsome (ER), respectively. The density of each fraction was calculated based on the refractive index

suspend it in several ml of SVEH, then homogenize it with the Dounce homogenizer as done previously.

9.2.2 Results and Discussion

The light mitochondrial fraction that had been prepared by differential centrifugation was further fractionated by equilibrium density centrifugation on Nycodenz. The distribution of marker enzymes is plotted in Fig. 9.2. Peroxisomes were well separated from the other organelles and formed a sediment at a density of approximately 1.25 in fractions 10–12.

The purity of the peroxisomal fraction is estimated by the relative specific activities (specific activity in the peroxisomal fraction/specific activity in the homogenate) of the marker enzymes. Generally, the relative specific activity of catalase increases to approximately 35, while that of mitochondria, ER and lysosomes decreases to approximately 0.1. The protein contents of peroxisomes, mitochondria, ER and lysosomes in rat liver cells are estimated to be approximately 2.5%, 20%, 20% and 2.0%, respectively (Leighton et al. 1968). Therefore, the purity of peroxisomes is estimated to be approximately 90%, based on the relative specific activity of catalase. Contamination by mitochondria and ER in the peroxisomal fraction is estimated as approximately 2% from relative specific activities of cytochrome *c* oxidase and esterase.

The isolation of peroxisomes from other tissues requires certain modification of the procedure, i.e. how to optimally homogenize the tissues and how long to centrifuge the sample. The brain contains myelin and a large amount of lipids, so other procedures are recommended (Schonenberger and Kovacs 2017).

9.3 Immuno-Isolation of Peroxisomes from the Rat Liver

Immuno-isolation is a powerful technique for the isolation of cells as well as subcellular organelles using the antigenic proteins on their surfaces (Luers et al. 1998). Here, I present our protocol for the further isolation of peroxisomes prepared by Nycodenz gradient centrifugation in order to eliminate any contaminant organelles such as mitochondria from the peroxisomal fraction (Kikuchi et al. 2004). This protocol can be used to isolate peroxisomes from crude subcellular fractions and cultured cell homogenates.

A polyclonal antibody raised against the cytoplasmic C-terminal amino acids of rat PMP70 (ABCD3) was covalently bound to magnetic beads (Dynabeads M-450). The coated beads were incubated with the purified peroxisomes. The peroxisomes were separated by magnetic beads that were bound to the antibody against PMP70. The polypeptides of these isolated peroxisomes were analyzed by SDS-PAGE.

9.3.1 Procedure

1. An anti-PMP70 specific antibody is bound to magnetic beads (1×10^8) using an anti-rabbit Fc fragment as a linker antibody according to the manufacturer's instructions. Peroxisomes are prepared by Nycodenz density gradient centrifugation.
2. Incubate the peroxisomes with the magnetic beads suspended in phosphate-buffered saline (PBS) containing 2 mM EDTA at 4 °C for 6 h. All procedures are carried out at a temperature of 4 °C.
3. After incubation, collect the bead complexes by placing the reaction tube in a magnet stand.
4. Wash the complex three times with PBS containing 2 mM EDTA and collect the bead complexes.
5. Elute the peroxisomes by boiling in 200 μl of 2× SDS sample buffer for SDS-PAGE. Separate the supernatant by placing the sample in a magnet stand and subject it to SDS-PAGE.

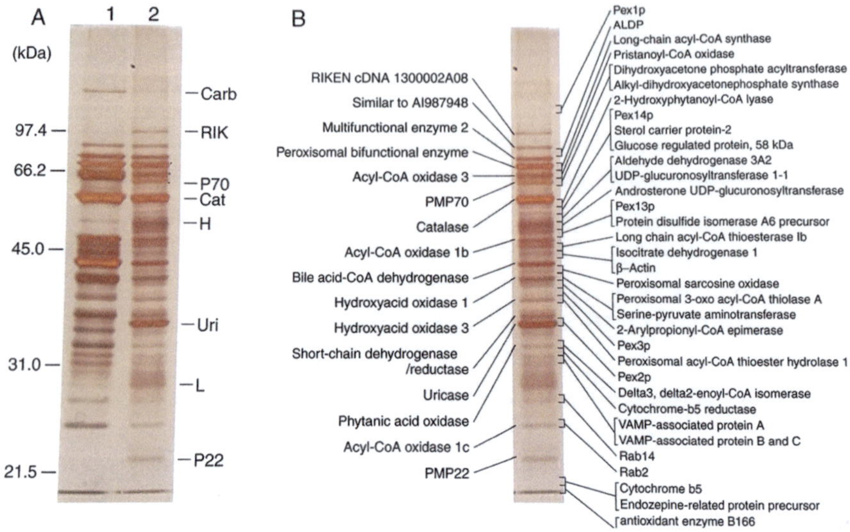

Fig. 9.3 Silver-stained gels of peroxisomal fraction. (**a**) Peroxisomal fraction before (line 1) or after immune-isolation (line 2) was subjected to SDS-PAGE and silver staining. Carb, carboamylphosphate synthase; RIK, RIKEN cDNA 1300002A08 (lon protease); P70, PMP70/ABCD3; Cat, catalase; H, IgG heavy chain; L, IgG light chain; Uri, urate oxidase; P22, PMP22. (**b**) Summary of the identified proteins. The major proteins corresponding to the prominent bands are listed on the left side of the silver-stained gel, while the less abundant proteins are shown on the right side

9.3.2 Results and Discussion

Immunomagnetic separation constitutes a novel approach to the isolation of highly purified rat liver peroxisomes. The fraction obtained is practically free of contaminating mitochondria. In fact, mitochondrial carboamylphosphate synthase is absent from purified fraction (Fig. 9.3). Utilizing an analysis of the peroxisomes by SDS-PAGE combined with liquid chromatography/mass spectrometry, 34 known peroxisomal proteins were identified together with a newly identified peroxisome-specific isoform of Lon protease (RIKEN cDNA 1300002A08). This procedure can be used to isolate peroxisomes from a crude preparation of tissues and cultured cells.

9.4 Isolation of the Peroxisomal Membranes

The sodium carbonate procedure is a simple, one-step procedure for the isolation of organelle membranes (Fujiki et al. 1982b). Treatment of peroxisomes with 0.1 M Na_2CO_3 at 4 °C denatures matrix proteins, peripheral membrane proteins and core protein complexes such as urate oxidase. These proteins are recovered in the

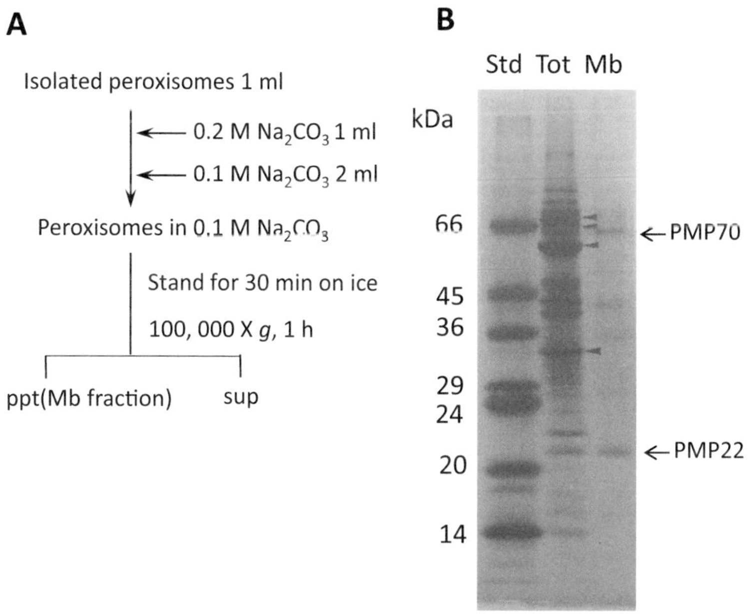

Fig. 9.4 Preparation of peroxisomal membranes by sodium carbonate procedure. (**a**) Flowchart of the isolation of peroxisomal membranes form rat liver peroxisomes. (**b**) The peroxisomal proteins (Tot) (50 μg) and peroxisomal membranes (Mb) prepared from 50 μg of peroxisomal proteins were subjected to SDS-PAGE and CBB staining. The arrowheads indicate hydratase-dehydrogenase, acyl-CoA oxidase, and catalase and urate oxidase, starting from the top

supernatant after centrifugation. Peroxisomal membranes with integral membrane proteins are recovered as membrane sheets at the bottom of the centrifuge tube (Fujiki et al. 1982a). It is simple to distinguish which protein is an integral membrane protein by SDS-PAGE followed by CBB staining or immunoblotting (Fig. 9.4).

9.4.1 *Procedure*

1. Mix the purified peroxisomal fraction with the same volume of 0.2 M Na_2CO_3 so as to make a final volume of approximately 4.0 ml by adding 0.1 M Na_2CO_3. Peroxisomal membranes are recovered quantitatively as membrane sheets at approximately 0.02–1.0 mg/ml of peroxisomal protein. Maintain the mixture on ice for 30 min. All procedures are carried out at a temperature of 4 °C.
2. Centrifuge the solution at 159,000 × *g* for 60 min. Any type of rotor is acceptable. At least 10,000 × *g* for 60 min is required to pellet the peroxisomal membranes.

3. Rinse the pellet gently with approximately 1.0 ml of ice-cold water.
4. Re-suspend the pellet in 100–200 μl of appropriate solution (sample buffer for SDS-PAGE).

9.4.2 Results and Discussion

As shown in Fig. 9.4, PMPs such as PMP70 and PMP22 recover quantitatively in the membrane fraction. On the other hand, major matrix proteins such as hydratase-dehydrogenase, acyl-CoA oxidase and catalase, as well as urate oxidase, comprise the core complex are not recovered in the membrane fraction. As the peroxisomal membranes form aggregates due to the alkaline pH of Na_2CO_3, the membranes are recovered as a pellet by centrifugation. However, a certain specific amount of the membranes is required to make the pellet after centrifugation. A reasonable peroxisomal membrane pellet is obtained without contamination of soluble proteins in a range between 20 μg/ml and 1 mg/ml of peroxisomal proteins. If the amount of peroxisomes is less than 20 μg/ml, erythrocyte membranes or *E. coli* cell membranes must be added as a carrier to precipitate the peroxisomal membranes (Santos et al. 1988).

9.5 Isolation of Peroxisomes from Cultured Mammalian Cells

On average, the peroxisomes in culture cells are considerably smaller and fewer in number than those in the rat liver. The peroxisome sediments occur at densities that are close to those in the ER and mitochondria. Although it is difficult to prepare highly purified peroxisomes, the distribution of the peroxisome marker enzymes can be distinguished from those of the ER and mitochondria. Culture cells have the advantage peroxisome manipulation by incubation with selected compounds, or by the knockout or overexpression of certain proteins. Isolated peroxisomes are useful for the characterization of these organelles. Rat hepatoma H4IIE cells possess a relatively large number of peroxisomes and express the proteins involved in the peroxisomal metabolic pathway. Here I present our protocol for the isolation of peroxisomes from rat hepatoma H4IIE cells (Imanaka et al. 1996) (Fig. 9.5).

9.5.1 Procedure

1. Wash confluent cells in 75-cm^2 culture flasks three times with PBS and harvest the cells with the aid of a rubber policeman in PBS.
2. Centrifuge the cell suspension at 800 × *g* for 10 min at room temperature. Suspend the cell pellet in 0.25 SVE, pH 7.4 containing 3 mM imidazole and

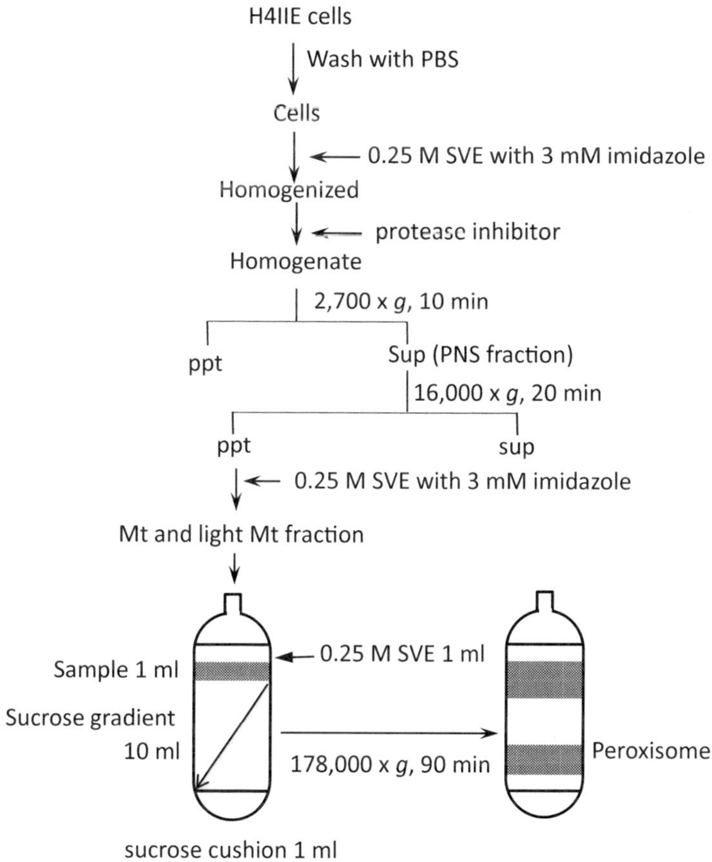

Fig. 9.5 Flowchart of the isolation of peroxisomes from rat hepatoma H4IIE cells

centrifuge at 1000 × g for 10 min at 4 °C. Re-suspend in the above solution at approximately 1 × 10^8 cells/ml and add leupeptin, antipain, chymostatin, and pepstatin A to a final concentration of 10 μg/ml each.

3. Homogenize the cells were with a Teflon/glass Potter-Elvehjem homogenizer and centrifuge at 2700 × g for 10 min to obtain a postnuclear supernatant (PNS) fraction. All procedures are carried out at a temperature of 4 °C.
4. Centrifuge the PNS fraction at 16,000 × g for 20 min to obtain a mitochondrial and light mitochondrial fraction.
5. Remove the supernatant and detach the pellet with a glass rod and suspend in ~1 ml of 0.25 M SVE containing 3 mM imidazole, then re-suspend gently using two to three strokes with a Dounce homogenizer type B.
6. Apply 1 ml of the suspended sample to the top of 10 ml of sucrose linear gradient (ρ = 1.15–1.27) with 1 ml of sucrose cushion (ρ = 1.30). Centrifuge at 178,000 × g for 90 min with an angle rotor. In addition to an angle rotor, a similar purity of peroxisomes (is prepared by vertical rotor) may be prepared with a vertical or swing rotor.

7. After centrifugation, make a small hole in the bottom of the centrifuge tube with an 18 gauge needle and recover the solution from the bottom drop by drop in tubes (1 ml each). Measure the refractive index to determine the density of the solution and the marker enzymes in each organelle.

9.5.2 *Results and Discussion*

Peroxisomes were separated from other organelles such as mitochondria, ER and lysosomes (Fig. 9.6). The LDH located in cytosol was recovered at the top fraction. The density of peroxisomes was approximately 1.17 and the peak of catalase differed from that of other organelles although the distribution patterns of the organelles were overlapped. A similar separation of peroxisomes was obtained in CHO cells (Imanaka et al. 1999).

We used sucrose as the gradient centrifugation medium since the separation of peroxisomes from other organelles was not improved by Nycodenz gradient in our hands. The reason for this is unknown at present. Because of the small size of peroxisomes, the penetration of Nycodenz into peroxisomes might not exert much influence on sedimentation. A longer period of centrifugation might improve the separation of peroxisomes if damage did not occur.

9.6 Isolation of Peroxisomes from *Komagataella phaffii*

Yeast is an excellent model organism for the study of peroxisomes. Yeast cells contain only a few peroxisomes when glucose is used as the sole carbon source. However, the peroxisome size and number markedly increase when the carbon source is changed to methanol in methylotrophic yeast such as *K. phaffii*. Furthermore, a large amount of mammalian peroxisomal proteins can be expressed in yeast peroxisomes under the control of the alcohol oxidase promoter (Ahmad et al. 2014). The peroxisomes, therefore, are useful to characterize the function of the proteins.

Yeast cells are surrounded by a cell wall that must be digested before cell lysis. Typically, yeast cells are enzymatically converted to spheroplasts using zymolyase. After cell wall digestion, yeast spheroplasts can be lysed easily under hypoosmotic conditions in a short period. Unbroken cells as well as nuclei are removed by low-speed centrifugation, and the resulting PNS fraction is used as the starting material for the isolation of peroxisomes. The organelle pellet is obtained by differential centrifugation of the PNS fraction. To separate peroxisomes from other organelles, the organelle fraction is subjected to Nycodenz gradient centrifugation. Here I present our protocol for the isolation of peroxisomes from *K. phaffii* (Fig. 9.7).

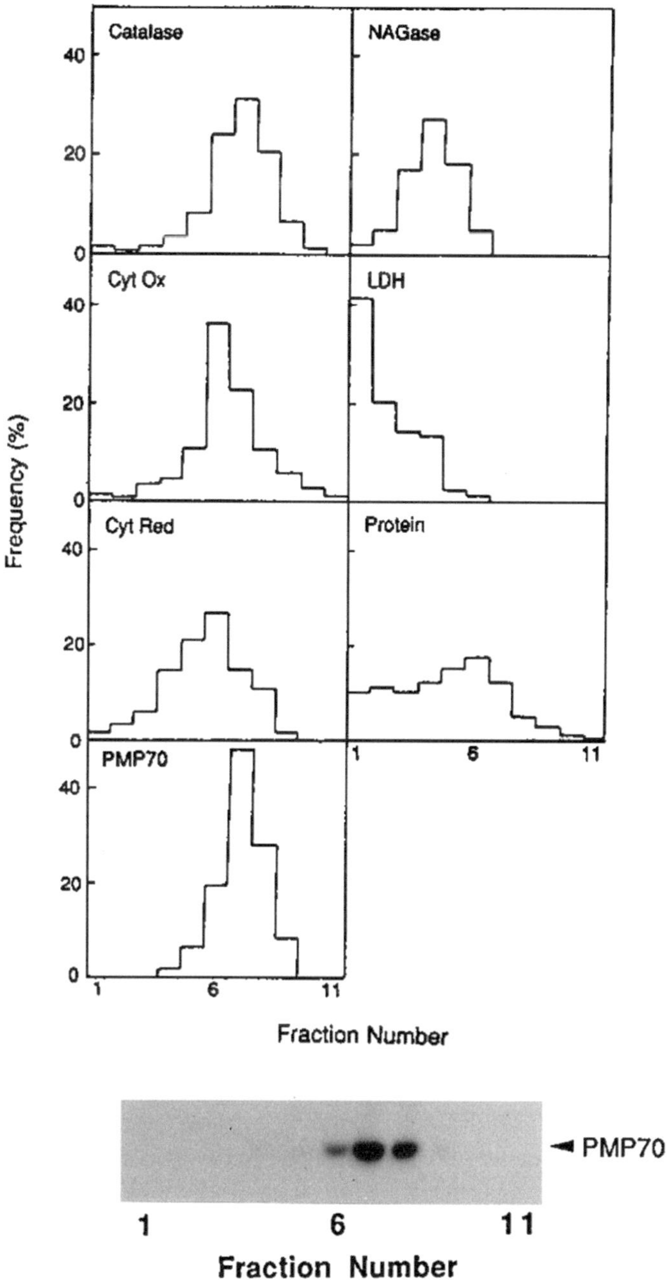

Fig. 9.6 Subcellular fractionation on a sucrose gradient. A light mitochondrial fraction prepared from rat hepatoma H4II cells was fractionated by equilibrium density centrifugation on sucrose. Catalase, *N*-acetyl-β-D-glucosaminidase (NAGase), cytochrome *c* oxidase (Cyt Ox), cytochrome *c* reductase (Cyt Red) and lactate dehydrogenase (LDH) are measured as marker enzymes of peroxisomes, lysosomes, mitochondria, microsomes and the cytosol, respectively. The distribution of PMP70 was determined by immunoblot analysis of PMP70

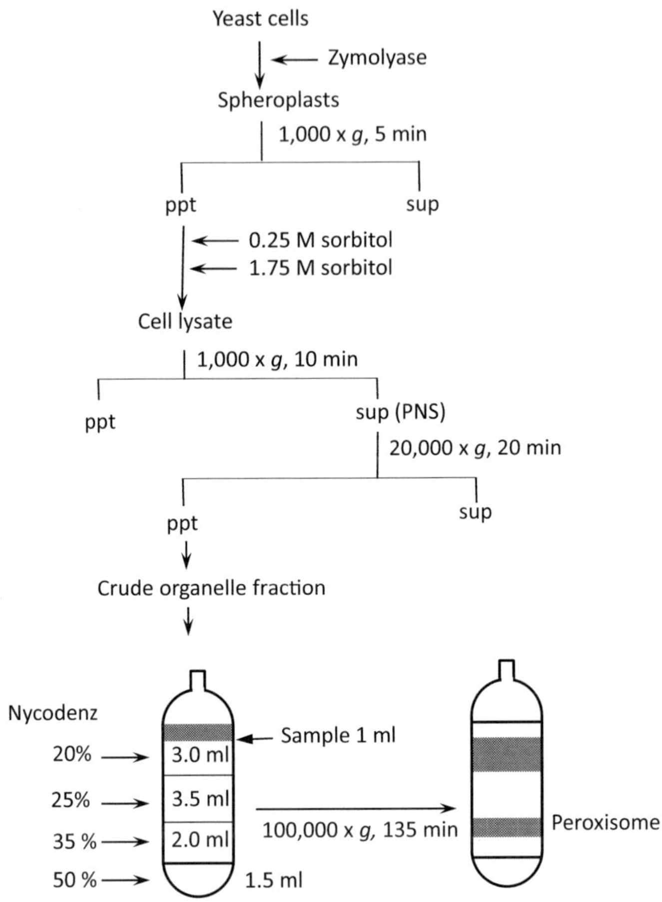

Fig. 9.7 Flowchart of the isolation of peroxisomes from *Komagataella phaffii*

9.6.1 Procedure

Cultivating Yeast Cells

1. Incubate the yeast cells overnight in 1% yeast extract, 2% peptone and 2% dextrose (YPD) medium under reciprocal shaking (200 rpm) at 30 °C.
2. Transfer the cells in YPD medium at an OD_{610} of 0.6 and culture for 6 h at 30 °C.
3. Harvest the cells by centrifugation at 600 × *g* for 5 min and wash the cells twice with sterile H_2O.
4. Harvest the cells by centrifugation at 600 × *g* for 5 min, then suspend the cells in 0.5% yeast extract and 0.5% methanol (BM) medium as briefly as possible (~5 ml).

5. Transfer the cell suspension to BM medium and adjust the OD_{610} to 0.6, then incubate the cells under reciprocal shaking (200 rpm) for 14–16 h at 30 °C.

Generating Spheroplasts

1. Harvest the cells by centrifugation at 1000 × *g* for 10 min and wash the cells with sterile H_2O 2–3 times.
2. Suspend the cells pellet in 20 ml of 0.1 M Tris-HCl, pH 7.5 and 50 mM EDTA. Add 14 μl of β-mercaptoethanol and gently invert for 20 min at room temperature.
3. Harvest the cell by centrifugation at 1000 × *g* for 10 min and wash the cells twice with 20 ml of 1 M sorbitol and 20 mM K_2PO_4, pH 7.5, and suspend in 20 ml of 1.4 M sorbitol and 20 mM K_2PO_4, pH 7.5.
4. Add Zymolyase at a concentration of 1 mg/g of cells and incubate at 30 °C.
5. Monitor the digestion of the yeast cell wall photometrically at OD_{610} in 0.1% SDS. The burst of spheroplasts can also be observed microscopically by adding 0.1% SDS from the slide to the cells under the coverslip.
6. Harvest the cell by centrifugation at 1000 × *g* for 10 min and wash the cells twice with 1 M sorbitol and 5 mM MES, pH 5.5. Harvest the cell by centrifugation at 1000 × *g* for 10 min. Suspend in 3 ml of 1 M sorbitol and 5 mM MES, pH 5.5 and let them stand on ice.

Obtaining the PNS and Subcellular Fractionation

1. Add 15 ml of 0.25 M sorbitol, 5 mM MES and 1 mM PMSF to the spheroplast suspension (add 1 ml each and pipette gently) and check the lysis of the spheroplasts microscopically. All procedures are carried out at a temperature of 4 °C.
2. Add 15 ml of 1.75 M sorbitol, 5 mM MES, pH 5.5 and 1 mM PMSF (add 1 ml each and pipette gently) to adjust it back osmotically to 1.0 M sorbitol. Centrifuge at 1000 × *g* for 10 min to remove unlysed cells, nuclei and other cell debris. Repeat several times until no pellet is observed after centrifugation.
3. Transfer the PNS to a new tube and centrifuge at 20,000 × *g* for 20 min to obtain a crude pellet containing peroxisomes and mitochondria.
4. Remove the supernatant and detach the pellet with a glass rod and suspend in approximately 1 ml of 1 M sorbitol and 5 mM MES, pH 5.5 and re-suspend gently using two to three strokes with Dounce homogenizer type B.
5. To prepare a continuous Nycodenz gradient solution, a step gradient (1.5 ml of 50% Nycodenz, 2.0 ml of 35% Nycodenz, 3.5 ml of 25% Nycodenz and 3.0 ml of 20% Nycodenz in 5 mM MES, pH 6.0, 1 mM KCl, 1 mM EDTA and 0.1% ethanol) was frozen once in liquid nitrogen and then thawed.
6. Apply 1 ml of sample to the top of the Nycodenz gradient and centrifuge at 100,000 × g for 135 min.
7. After centrifugation, make a small hole in the bottom of the centrifuge tube by a needle with an 18 gauge and recover the solution from the bottom drop by drop in tubes (1 ml each). Measure the refractive index to determine the density of the solution and marker enzymes in each organelle.

9.6.2 Results and Discussion

Culturing *K. phaffii* in the medium with methanol as the only carbon source results in peroxisomes being highly induced, and the peroxisome volume of peroxisomes comes to occupy more than 50% of the cells. Therefore, highly purified peroxisomes can be readily obtained. The preparation of spheroplasts is a critical step in preparing peroxisomes. The digestion of the yeast cell wall may be observed microscopically by adding 0.1% SDS. Then the pellet looks like that of the rat cerebrum after centrifugation of the spheroplasts.

Peroxisomes were well separated from mitochondria after Nycodenz gradient centrifugation (Fig. 9.8). When human ABCD1 was expressed in *K. phaffii* under the control of the alcohol oxidase gene promoter, ABCD1 was mainly recovered in fraction 3–4 on the Nycodenz gradient, which corresponded to the position of the peroxisomes shown by the catalase distribution (Okamoto et al. 2018). ABCD1 was

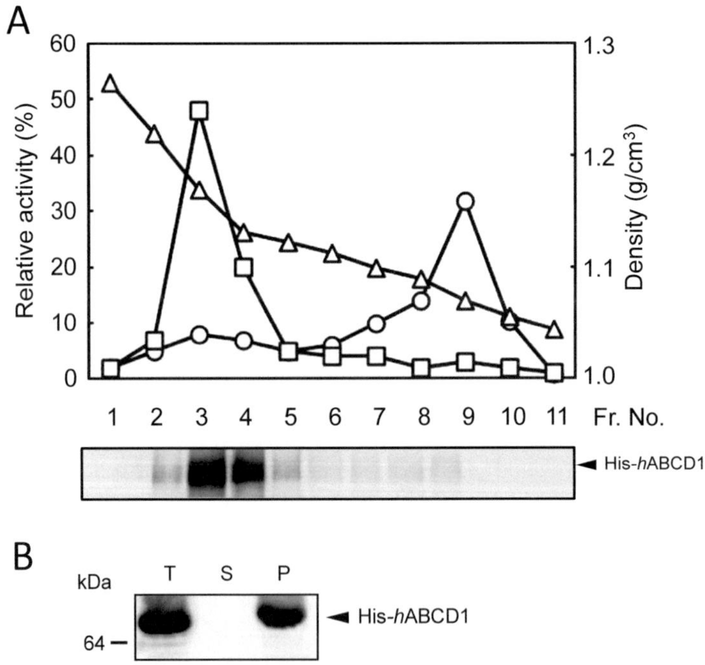

Fig. 9.8 (**a**) Subcellular fractionation on a Nycodenz gradient. The organelle pellet fraction from methanol-induced cells was fractionated by Nycodenz density gradient centrifugation. The density (open triangle) of each fraction was calculated based on the refractive index. Catalase activity (open square) and cytochrome *c* oxidase activity (open circle) were used as markers of peroxisomes and mitochondria. The peroxisomal localization of His-ABCD1 was confirmed by immunoblotting using an anti-His antibody. (**b**) The peroxisomal fraction (T) was separated into soluble (S) and membrane (P) fractions by Na_2CO_3 treatment, and an equal portion of each fraction was subjected to immunoblot analysis

recovered in the membrane fraction after sodium carbonate treatment, suggesting the ABCD1 expressed in *K. phaffii* was properly targeted to the peroxisome and inserted in the peroxisomal membrane. Highly purified peroxisomes may be prepared from yeast *Saccharomyces cerevisiae* in which peroxisomes were induced by oleic acid as the only carbon source (Cramer et al. 2015).

Acknowledgments The publication is supported in part by a Grant-in-Aid for Intractable Diseases from the Ministry of Health, Labour and Welfare of Japan. Pacific Edit reviewed the manuscript prior to submission.

References

Ahmad M et al (2014) Protein expression in *Pichia pastoris*: recent achievements and perspectives for heterologous protein production. Appl Microbiol Biotechnol 98(12):5301–5317

Cramer J et al (2015) Large-scale purification of peroxisomes for preparative applications. Cold Spring Harb Protoc 2015(9):pdb.prot083725

Fujiki Y et al (1982a) Polypeptide and phospholipid composition of the membrane of rat liver peroxisomes: comparison with endoplasmic reticulum and mitochondrial membranes. J Cell Biol 93(1):103–110

Fujiki Y et al (1982b) Isolation of intracellular membranes by means of sodium carbonate treatment: application to endoplasmic reticulum. J Cell Biol 93(1):97–102

Imanaka T et al (1987) Translocation of acyl-CoA oxidase into peroxisomes requires ATP hydrolysis but not a membrane potential. J Cell Biol 105(6 Pt 2):2915–2922

Imanaka T et al (1991) A novel 57 kDa peroxisomal membrane polypeptide detected by monoclonal antibody (PXM1a/207B). Biochim Biophys Acta 1062(2):264–270

Imanaka T et al (1996) Insertion of the 70-kDa peroxisomal membrane protein into peroxisomal membranes *in vivo* and *in vitro*. J Biol Chem 271(7):3706–3713

Imanaka T et al (1999) Characterization of the 70-kDa peroxisomal membrane protein, an ATP binding cassette transporter. J Biol Chem 274(17):11968–11976

Kikuchi M et al (2004) Proteomic analysis of rat liver peroxisome: presence of peroxisome-specific isozyme of Lon protease. J Biol Chem 279(1):421–428

Leighton F et al (1968) The large-scale separation of peroxisomes, mitochondria, and lysosomes from the livers of rats injected with triton WR-1339. Improved isolation procedures, automated analysis, biochemical and morphological properties of fractions. J Cell Biol 37(2):482–513

Luers GH et al (1998) Immuno-isolation of highly purified peroxisomes using magnetic beads and continuous immunomagnetic sorting. Electrophoresis 19(7):1205–1210

Morita M et al (2000) Insulin-degrading enzyme exists inside of rat liver peroxisomes and degrades oxidized proteins. Cell Struct Funct 25(5):309–315

Okamoto T et al (2018) Characterization of human ATP-binding cassette protein subfamily D reconstituted into proteoliposomes. Biochem Biophys Res Commun 496(4):1122–1127

Santos MJ et al (1988) Peroxisomal integral membrane proteins in control and Zellweger fibroblasts. J Biol Chem 263(21):10502–10509

Schonenberger MJ, Kovacs WJ (2017) Isolation of peroxisomes from mouse brain using a continuous Nycodenz gradient: a comparison to the isolation of liver and kidney peroxisomes. Methods Mol Biol 1595:13–26

Chapter 10
Structure Biology of Peroxisomal Proteins, Peroxins

Hiroaki Kato

Abstract Peroxins participate in importing necessary proteins into the peroxisomes as well as producing the membrane that separates the peroxisome from the rest of the cell. One of the most powerful ways to elucidate the mechanisms underlying biological phenomena is to determine the three-dimensional structure of the macromolecules and their complexes. This review summarizes recent findings on available structural data of peroxins and their complexes. It provides structure-based mechanistic interpretation of the processes they are involved in, especially, two cases of structural investigations for PTS2 recognition complex that consists of Pex7p, Pex21p and an artificial PTS2-cargo protein, and for the complex of PMP carrier Pex19p and its receptor Pex3p. Recent breakthroughs in cryo-electron microscopy and X-ray free electron lasers have accelerated structural studies of difficult-to-crystallize proteins and have opened up new opportunities in understanding conformational dynamics and visualizing the process of biological actions. Their current results related to peroxins and future prospects are also explained to non-structural biologists.

Keywords Pex3 · Pex7 · Pex19 · Pex21 · PTS2 · X-ray crystallography · Intrinsically disordered proteins · X-ray free electron laser (XFEL) · Cryo-electron microscopy (Cryo-EM) · AAA+ ATPase

10.1 Introduction: Peroxisome Biogenesis and Structural Biology

Three-dimensional knowledge of the structures of macromolecules in the cell is essential for understanding how they work. Investigation of peroxisome biogenesis has much progressed by cellular and genetic biological studies. However, structural insight into the underlying molecular mechanisms at atomic resolution is still at an

H. Kato (✉)
Department of Structural Biology, Graduate School of Pharmaceutical Sciences, Kyoto University, Kyoto, Japan
e-mail: katohiro@pharm.kyoto-u.ac.jp

T. Imanaka, N. Shimozawa (eds.), *Peroxisomes: Biogenesis, Function, and Role in Human Disease*, https://doi.org/10.1007/978-981-15-1169-1_10

early and incomplete stage. This review summarizes recent findings on available structural data of proteins and their complexes that are participated in the peroxisome biogenesis and provides structure-based mechanistic interpretation of the processes they are involved in. Peroxisome biogenesis requires peroxins, which are encoded by *PEX* genes (Fujiki et al. 2014). To date, there are 16 known human peroxins and 36 altogether in all species (Smith and Aitchison 2013; Yofe et al. 2017; Farre et al. 2017).

Peroxins participate in importing necessary proteins into the peroxisomes as well as producing the membrane that separates the peroxisome from the rest of the cell. Currently available molecular structures for peroxins are summarized in Table 10.1. The functional roles of those peroxins are interpreted as follows. In the import of the peroxisomal matrix proteins, Pex5p recognizes the peroxisomal targeting signal 1 (PTS1), as the PTS1-cargo-receptor. There is another targeting signal, PTS2 and PTS2-cargo is recognized by Pex7p together with Pex21p (or Pex18p) in yeast, or with Pex5L in mammals. The PTS receptor Pex5p interacts with the PTS1 of cargo in the cytoplasm, docks at the docking complex, Pex13p and Pex14p, and is integrated into the membrane to form the transport channel with Pex14p. Pex1p/Pex6p are involved in the export (recycling) of the PTS receptors. In yeast Pex15p possesses a role in anchoring Pex6 onto the peroxisomal membrane. Pex4 belongs to the ubiquitin conjugating enzymes required to the receptor export. Monoubiquitination of Pex5p (or perhaps Pex7p) by ubiquitin-conjugating and ligase enzymes, Pex4p, enables Pex5p recycling back to the cytosol for another round of import. Pex4p binds the peroxisomal membrane protein Pex22p, a co-activator of E2 enzyme (Williams et al. 2012). Pex22p is an integral peroxisomal membrane protein with its NH_2 terminus in the matrix and its COOH terminus in the cytosol and anchors Pex4p at the peroxisomal membrane (Koller et al. 1999). On the other hand, Pex3p and Pex19p are involved in the proper integration of peroxisomal membrane proteins (PMPs) in yeast as well as mammal. PMPs that are direct targeting to peroxisomes (class I PMPs) are recognized by the shuttling receptor Pex19p though the membrane PTS (mPTS) in the cytosol. Pex19p-cargo complexes are then recruited to peroxisomes by membrane-bound Pex3p and cargo PMPs are inserted into membrane and assembled into PMP complexes by mechanisms that seems to be mediated by the chaperone activity of Pex19p.

Considerable amount of the structural information has been apparently available, but the analyzed structures are restricted for 12 peroxins in the 36 totals (Table 10.1). They belong to the soluble proteins or soluble domain of PMPs (Pex3p, Pex13p and Pex14p). There are three peroxins, Pex3p, Pex4p, and Pex7p, of which structures determined are of whole molecule (Pex4p and Pex7p) or that without the membrane anchor peptide region (Pex3p). The other structures are of a folded domain or a short peptide region bound to another peroxin.

Many peroxins possess molecular properties of the proteins called intrinsically disordered proteins (IDPs; also known as intrinsically unstructured proteins), which lack tertiary contacts and typically do not exhibit stable secondary structure. Some proteins are predicted to be entirely disordered, whereas others contain disordered regions (IDRs), as well as structured globular domains. For example, Pex5p has

Table 10.1 Peroxin structures

Structural state of pexoxins	PDB ID	Species	Method/resolution	Reference
Pex1 N-terminal domain (13–179)	1WLF	Mouse	X-ray/2.05 Å	Shiozawa et al. (2004)
Pex1–Pex6 complex	–	*S. cerevisiae*	CryoEM/6.2–7.3 Å	Blok et al. (2015)
Pex1–Pex6 complex	–	*S. cerevisiae*	CryoEM/21–23 Å	Ciniawsky et al. (2015)
Pex1–Pex6 complex	–	*S. cerevisiae*	CryoEM/23 Å	Gardner et al. (2018)
Pex3 (40–368)–Pex19 peptide (14–30)	3MK4	Human	X-ray/2.42 Å	Schmidt et al. (2010)
Pex3 (52–368)–Pex19 peptide (15–40)	3AJB	Human	X-ray/2.5 Å	Sato et al. (2010)
Pex4 (54–180)–Pex22 (57–180)	2Y9M	*S. cerevisiae*	X-ray/2.6 Å	Williams et al. (2012)
Pex4–Pex22 mutant	2Y9P	*S. cerevisiae*	X-ray/3.25 Å	Unpublished
Pex4–Pex22 SS bond mutant	4BWF	*S. cerevisiae*	X-ray/3.23 Å	Williams et al. (2013)
Pex4 (1–183)–Pex22 (50–154)	5NKZ	*O. polymorpha*	X-ray/2.85 Å	Groves et al. (2018)
Pex4–Pex22 (50–154)	5NL8	*O. polymorpha*	X-ray/2.85 Å	Groves et al. (2018)
Pex5 TPR domain (280–602)–PTS1	1FCH	Human	X-ray/2.2 Å	Gatto Jr. et al. (2000)
Pex5 TPR domain (335–639)–MSCP2 (22–143)	2C0L	Human	X-ray/2.3 Å	Stanley et al. (2006)
Pex5 TPR domain apo (284–602)	2C0M	Human	X-ray/2.5 Å	Stanley et al. (2006)
Pex5 TPR another conformation	2J9Q	Human	X-ray/2.65 Å	Stanley et al. (2007)
Pex5–PTS1	3CV0	*T. brucei*	X-ray/2 Å	Sampathkumar et al. (2008)
Pex5 TRP domain–cargo protein	3R9A	Human	X-ray/2.6 Å	Fodor et al. (2012)
Pex5 TRP domain–cargo protein K390/K391A	4KXK	Human	X-ray/2.9 Å	Fodor et al. (2015)
Pex5 TRP domain–cargo protein K390A	4KYO	Human	X-ray/2.2 Å	Fodor et al. (2015)
Pex7 (1–374)–Pex21 (197–288)–PTS2	3W15	*S. cerevisiae*	X-ray/1.8 Å	Pan et al. (2013)
Pex13 SH3 domain	1WXU	Mouse	NMR	Unpublished
Pex13 (301–386) SH3 domain	1JQQ	*S. cerevisiae*	X-ray/2.65 Å	Douangamath et al. (2002)
Pex13 SH3 domain–Pex14 peptide	1N5Z	*S. cerevisiae*	X-ray/2.7 Å	Douangamath et al. (2002)
Pex13 SH3 domain	1NM7	*S. cerevisiae*	NMR	Pires et al. (2003)

(continued)

Table 10.1 (continued)

Structural state of pexoxins	PDB ID	Species	Method/resolution	Reference
Pex13 SH3 domain–Pex14 peptide	2V1R	*S. cerevisiae*	X-ray/2.1 Å	Unpublished
Pex14 N-terminal domain	3FF5	Rat	X-ray/1.8 Å	Su et al. (2009)
Pex14 (20–76)–Pex5 (93–112)	2W84	Human	NMR	Neufeld et al. (2009)
Pex14 (20–76)–Pex19 (101–112)	2W85	Human	NMR	Neufeld et al. (2009)
Pex14 (12–80)–Pex5 peptide (57–71)	4BXU	Human	NMR	Neuhaus et al. (2014)
Pex14 N-terminal domain (23–71)	5AON	*T. brucei*	X-ray/1.65 Å	Watanabe et al. (2016)
Pex14 N-terminal domain (15–84)	5MMC	*T. brucei*	NMR	Dawidowski et al. (2017)
Pex14 (5–66)–inhibitor 6RD	5L87	*T. brucei*	X-ray/0.87 Å	Dawidowski et al. (2017)
Pex14 (5–66)–inhibitor 6RB	5L8A	*T. brucei*	X-ray/1.57 Å	Dawidowski et al. (2017)
Pex14 (5–66)–inhibitor KZZ	5N8V	*T. brucei*	X-ray/1.55 Å	Dawidowski et al. (2017)
Pex15 (43–259)	5VXV	*S. cerevisiae*	X-ray/1.55 Å	Gardner et al. (2018)
Pex19 (171–280)	2WL8	Human	X-ray/2.05 Å	Schueller et al. (2010)
Pex19 farnesylated C-terminal domain (161–299)	5LNF	Human	NMR	Emmanouilidis et al. (2017)

S. cerevisiae, *Saccharomyces cerevisiae*; *O. polymorpha*, (formerly called *Hancenula polymorpha*); *T. brucei*, *Trypanosoma brucei*; CryoEM, cryoelectron microscopy; X-ray, X-ray crystallography; PDB, Protein Data Bank; SH3, SRC homology 3

IDR in the N-terminal half (1–279) as well as the structured three tetratricopeptide repeats, call TPR domain, in the C-terminal half (280–602) (Barros-Barbosa et al. 2019). Pex19p also has IDR in the N-terminal half (1–155) and alpha-helical C-terminal half (156–296). Interestingly, purified Pex19 exhibits quite high solubility, with concentrations of more than 200 mg/ml remaining in solution (Shibata et al. 2004). They are characterized by their biased amino acid composition and low sequence complexity, as well as by their low proportions of bulky hydrophobic amino acids and high proportions of charged and hydrophilic amino acids. In general, IDPs frequently interact with, or function as hubs in, protein-interaction networks. Atomic structures of IDRs in their bound states can be obtained by X-ray crystallography, if they are well ordered. However, there is a difficulty in the investigation to find the correctly folded conditions of molecules or the bound partner peptide regions to fix the tertiary structure of molecules. Thus, I introduce two case studies of structural investigations for PTS2 recognition complex that consists of

Pex7p, Pex21p, and an artificial PTS2-cargo protein, and for the complex of PMP carrier Pex19p and its receptor Pex3p. The readers could obtain experimental strategy or tactics that facilitate the other structural investigation of peroxins.

10.2 Structural Basis for the Recognition Mechanism of the Second Peroxisomal Targeting Signal, PTS2 by Its Receptor Complex Pex7p and Pex21p

There are two types of targeting signals for peroxisomal matrix proteins: PTS1 at C-terminal, and PTS2 near N-terminal (Platta and Erdmann 2007). PTS1 is recognized by the peroxin Pex5p, whereas PTS2 is recognized by Pex7p together with a co-receptor peroxin. Crystal structures of the PTS1-binding domain of Pex5p have been solved in apo-form (Stanley et al. 2006), with a PTS1 peptide (Gatto Jr. et al. 2000) or with a PTS1-cargo (mSCP2) (Stanley et al. 2006). On the other hand, crystal structure of Pex7p or the PTS2-interaction domain of the co-receptor peroxin, ligated with or without PTS2, had been reported (Pan et al. 2013). The first Pex7p identified was *S. cerevisiae* Pex7p (also known as Peb1p or Pas7p), which was shown to be the receptor of Fox3p (also known as Pot1p) (Marzioch et al. 1994; Zhang and Lazarow 1995; Rehling et al. 1996). Fox3p contains a functional PTS2 at its N-terminus (Rehling et al. 1996; Glover et al. 1994) and functions as a 3-ketoacyl-CoA thiolase (EC 2.3.1.16) at the final step of the β-oxidation (Wanders and Waterham 2006). Human Pex7p was identified as the homolog of *S. cerevisiae* Pex7p and linked to a fatal peroxisome biogenesis disorder (PBD), rhizomelic chondrodysplasia punctata type 1 (RCDP1) (Braverman et al. 1997, 2002; Motley et al. 1997, 2002; Purdue et al. 1997).

The PTS2 sequence was initially identified in rat peroxisomal thiolase (Swinkels et al. 1991; Osumi et al. 1991). Later, the sequence motif of PTS2 was refined to R–[L/V/I/Q]–xx–[L/V/I/H]–[L/S/G/A]–x–[H/Q]–[L/A] (x = any residue) by bioinformatics analysis of available functional PTS2 sequences (Petriv et al. 2004). Positions 2, 5, and 9 of this motif are mostly limited to hydrophobic residues whereas positions 1 and 8 are limited to hydrophilic residues. In order to explain these sequence constraints, the detailed interactions between PTS2 and its receptors must be determined based on the three-dimensional structure.

An intriguing question regarded how the co-receptor peroxin assists Pex7p in PTS2 recognition. The co-receptor peroxin for Pex7p is species-specific. Paralogous peroxins, Pex18p and Pex21p, act as the co-receptor of Pex7p in *S. cerevisiae* (Purdue et al. 1998), and their homolog Pex20p exists in other yeast and fungi (Sichting et al. 2003; Einwachter et al. 2001; Leon et al. 2006). The PTS1 receptor, Pex5p, also acts as the co-receptor of Pex7p in *Arabidopsis thaliana* (Nito et al. 2002; Khan and Zolman 2010) and in mammals (Braverman et al. 1998; Otera et al. 1998; Matsumura et al. 2000). The domain structures of these co-receptor peroxins share common features: First, their N-terminal domains contain a conserved motif

(Einwachter et al. 2001; Dodt et al. 2001) that is ubiquitinated during the recycling of receptors from peroxisomal membrane to cytosol (Leon and Subramani 2007; Hensel et al. 2011). Second, the Pex14p-binding motif (Wxxx[F/Y]) interacts with Pex14p (Neufeld et al. 2009), a peroxin that forms a docking complex on the peroxisomal membrane. Third, the Pex7p-binding domain (Pex7pBD) is essential for the interaction with Pex7p and the import of PTS2-cargos into peroxisomes (Einwachter et al. 2001; Matsumura et al. 2000; Dodt et al. 2001). In Pex18p, Pex21p, and Pex20p, their Pex7pBD resided close to the C-terminal, whereas in Pex5p, an additional PTS1-binding domain flanked the C-terminal side of Pex7pBD (Einwachter et al. 2001). Pex7pBD seemed to be the most important region for interaction with Pex7p, but where does Pex7pBD bind on Pex7p, and how does it support PTS2-cargo transport? The structure of Pex7p–Pex7pBD complex unveiled their interactions (Pan et al. 2013).

10.2.1 Preparation of Pex7p–Pex21pC–Fox3pN-MBP Complex

To crystallize the ternary complex of Pex7p–co-receptor–PTS2-cargo, Pex7p, Pex21p and Fox3p from *S. cerevisiae* were chosen. It is noted that since Pex7p was difficult to obtain in soluble form using an *Escherichia coli* expression system, it was expressed in *Komagataella phaffii* (formerly called *Pichia pastoris*) as GST-fusion protein. Well-diffracting plate crystals grew when a Pex7p mutant with a nine-residue deletion (Δ257–265) was used instead of wild-type Pex7p. In the Pex21p preparation, the C-terminal region from Lys190 to Asp288, called Pex21pC was sufficient to form the ternary complex and was more resistant to degradation than the homologous region of Pex18p. Since the full length Fox3p was unsuitable to the crystallization, Fox3pN-MBP in which the N-terminal 15 residues of Fox3p (hereafter referred to as Fox3pN) was attached to the N-terminal of the maltose-binding protein (MBP) from *E. coli* by a two-residue linker (Arg-Ser) was used for PTS2–cargo (Fig. 10.1). Finally, plate like crystals diffracted to a 1.8 Å resolution were obtained reproducibly using polyethylene glycol 2000 as a precipitant (Pan et al. 2013).

10.2.2 The Structure of the Pex7p–Pex21pC–Fox3pN-MBP Complex

The hetero-trimeric structure of Pex7p–Pex21pC–Fox3pN-MBP possesses the characteristic assembly of the three proteins (Fig. 10.2). Pex7p forms a ring structure with a seven-bladed β-propeller fold (Fig. 10.3a). The co-receptor, Pex21pC consists of three parts: a cluster of three α-helices (α1–3), a small three-stranded

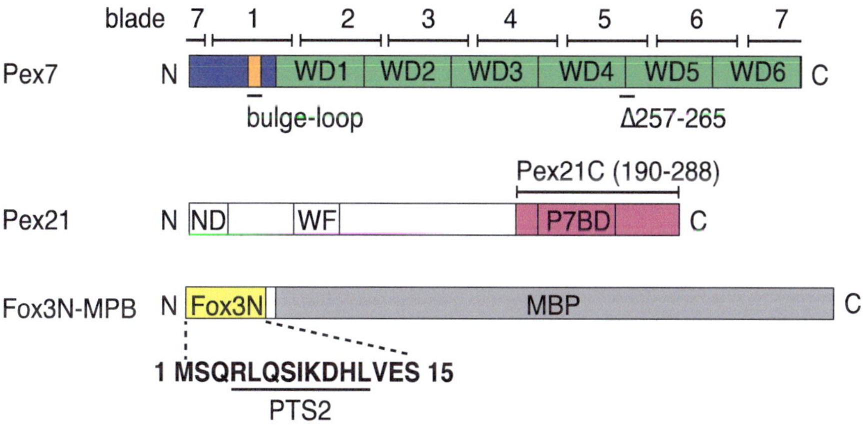

Fig. 10.1 Schematic diagrams of the constructs used for crystallization. The color code here is used for all figures, unless otherwise noted. WD1–6 indicate WD40 motifs. ND and WF indicate N-terminal conserved domain and WxxxF/Y Pex14 binding motif, respectively. (This is adapted from Fig. 1a in Pan et al. (2013))

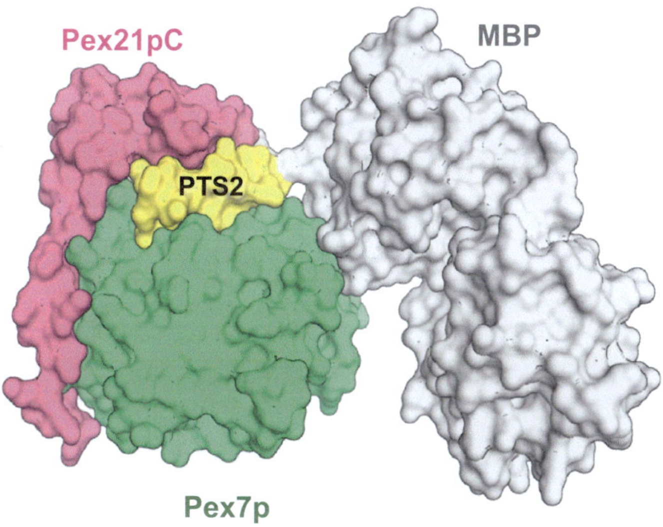

Fig. 10.2 Skeleton representation of the hetero-trimer of Pex7p–Pex21pC–Fox3pN(PTS2)-MBP in the crystallographic asymmetric unit

β-sheet (β1–3), and a stretched C-terminal loop (Fig. 10.3a). The helical cluster and the β-sheet of Pex21pC bind on the rim of the top surface of Pex7p, and they cooperatively form a binding pocket for PTS2 (Fox3pN). The C-terminal loop of Pex21pC extends along the side of Pex7p and does not interact with Fox3pN. Fox3pN adopts an α-helical conformation and stretches from the globular region of MBP into the binding pocket prepared by Pex7p and Pex21pC.

10.2.3 Special Motif of Pex7p and Pex7p–Pex21p Complex Formation

The crystal structure revealed structural features important for Pex7p–Pex21p complex formation. Although the framework of Pex7p is the same as that of typical WD40 proteins (Smith et al. 1999), it contains a special region at its N-terminus. The N-terminal 44 residues of Pex7p have low sequence similarity to other WD40

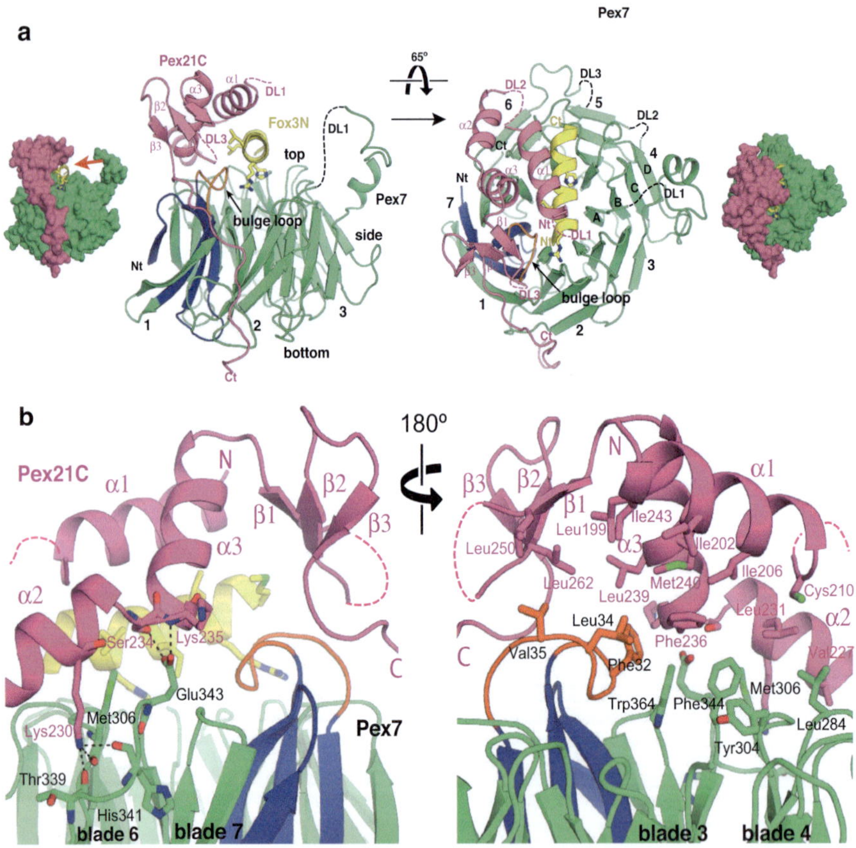

Fig. 10.3 (**a**) Ribbon and surface representations of the side view (left) and the top view (right) of the hetero-trimer Pex7p–Pex21pC–Fox3pN. Side chains of Arg4, Leu5, His11, and Leu12 in Fox3pN(PTS2) are shown as stick models. The disordered loops (DLs) in Pex7p and Pex21pC are presented as dashed curves. The red arrow in the leftmost panel represents the viewpoint of the right panel in **b** and the viewpoint of Fig. 10.4a. Nt, N-terminus; Ct, C-terminus. (This is adapted from Fig. 1c in Pan et al. (2013)). (**b**) Interface residues which determine the specific binding of the complex. Residues forming hydrogen bonds between Pex7 and Pex21C (left panel) and forming hydrophobic interfaces between Pex7 and Pex21C (right panel). The helix of Fox3N is omitted in the right panel for clarity. In this figure, dashed black lines indicate possible hydrogen bonds. (This is adapted from Fig. 3b in Pan et al. (2013))

motifs (Zhang and Lazarow 1995), and the loop between β-strand 1B and 1C (Asn31-Asn37) forms a bulge (Fig. 10.3a), referred as to the bulge loop. The bulge loop contains a six-residue consensus motif, [N/H]-[F/Y]-G-[L/I]-[V/L/I/S/A]-G, whose length and composition are conserved in Pex7p homologs. The bulge loop exposes the hydrophobic residues Phe32, Leu34, and Val35 on its surface (Fig. 10.3b). When Pex7p forms a complex with Pex21pC, this protruding hydrophobic bulge loop is covered by the hydrophobic pocket of Pex21pC formed at the hinge region between α3 and β1 (Fig. 10.3b).

There is a hydrophobic ridge on the top surface of Pex7p (Fig. 10.3a) that consists of residues from blade 1 (the bulge loop), blade 6 (Leu284, Tyr304, Met306), and blade 7 (Phe344, Trp364) (Fig. 10.3b). The hydrophobic ridge contacts the hydrophobic surface of Pex21p, which contains nine hydrophobic residues: Cys210 on α1; Val227 and Leu231 on α2; Phe236, Ile237, Leu239, and Met240 on α3; Leu250 on β1; and Leu262 on β2 (Fig. 10.3b). Four residues of Pex7p, Phe32, Leu34, Phe344, and Trp364, surround Pex21p Phe236 (Fig. 10.3b), which is highly conserved among the co-receptor peroxins (Pan et al. 2013).

Pex7p has another characteristic region involved in binding with Pex21pC, called the acidic patch at the side of the hydrophobic ridge on blades 6 and 7 (Fig. 10.3b). The acidic patch consists of two components: the side chain of Pex7p Glu343 and the clustered main-chain C=O groups of Pex7p Met306, Thr339, and His341 (Fig. 10.3b). The side chain of Pex7p Glu343 interacts electrostatically with the side-chain OH group of Pex21p Ser234 and the main-chain NH group of Pex21p Lys235 (Fig. 10.3b). Because Pex21p Ser234 and Lys235 are located in the N-terminal portion of α3, they are positively charged by the dipole effect of the α-helix. The clustered main-chain C=O groups of Pex7p Met306, Thr339, and His341 are in range to form hydrogen bonds with Pex21p Lys230 (Fig. 10.3b). Most of the residues forming the hydrophobic and electrostatic interfaces between Pex7p and Pex21pC are conserved among the corresponding homologs.

10.2.4 PTS2 Recognition by Cooperation of Pex7p and Pex21p

The binding site of PTS2 is a large cleft that holds the α-helix of Fox3pN and contains two minor grooves that accept side chains protruding from the α-helix (Figs. 10.1 and 10.4a). One groove is hydrophilic and the other is hydrophobic. The α-helix of Fox3pN is amphipathic. Fox3p Met1, Leu5, Ile8, Leu12, and Val13 form the hydrophobic surface, whereas the remaining residues form the hydrophilic surface (Fig. 10.4b). Thus, the key residues of PTS2 are separated in two groups: the hydrophilic key residues (Fox3p Arg4, His11) and the hydrophobic key residues (Fox3p Leu5, Ile8, Leu12). The hydrophilic groove at the center of the top surface of Pex7p contains two pockets that accept the hydrophilic key residues of PTS2 (Fox3p Arg4 and His11) (Fig. 10.4a). Fox3p Arg4 is surrounded by the residues of Pex7p blades 1–3, and its guanidine group is fixed by formation of salt bridges with

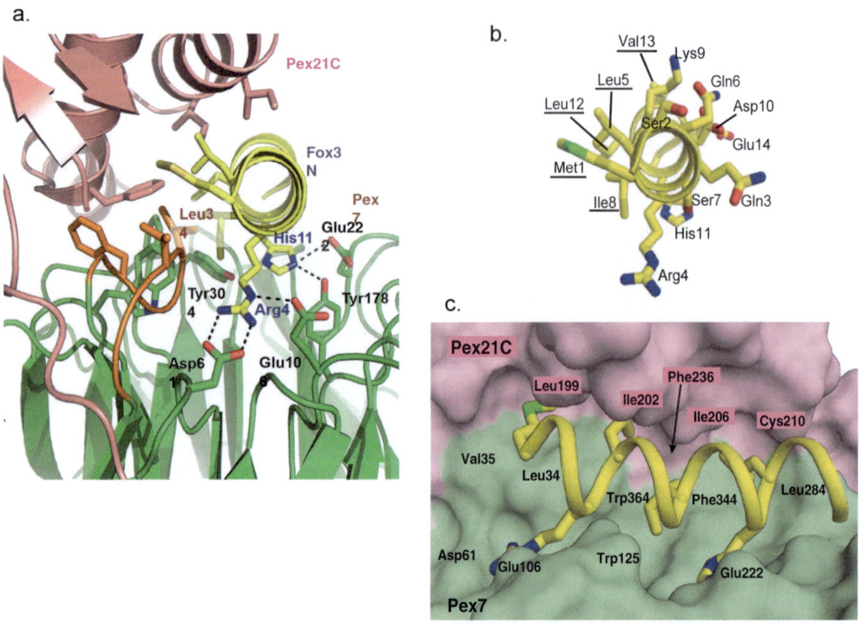

Fig. 10.4 Molecular recognition of key residues of PTS2. (**a**) The PTS2 binding pocket. The side chains of Met1, Arg4, Leu5, Ile8, His11, and Leu12 of Fox3pN are shown (see **b**). Dashed black lines indicate possible hydrogen bonds. (**b**) The amphipathic α-helix of Fox3pN. Hydrophobic residues are highlighted by underlines. (This is adapted from Fig. 4b in Pan et al. (2013)). (**c**) The surface of the PTS2-binding pocket is shown and colored according to the peptide chain (Pex21pC, light pink; Pex7p, pale green). The viewpoint is indicated in Fig. 10.3a by a red arrow. (This is adapted from Fig. 4c (left) in Pan et al. (2013))

the carboxyl groups of two acidic residues, Pex7p Asp61 and Glu106 (Fig. 10.4a, c). Fox3p His11 is surrounded by the residues of Pex7p blades 4–6, and its imidazole ring is fixed by formation of hydrogen bonds with Pex7p Tyr178, Glu222, and Tyr304 (Fig. 10.4a). When Fox3p Arg4 and His11 bind to the top surface of Pex7p in the correct orientation, the hydrophobic residues of Fox3pN consequently fit into the hydrophobic groove (Fig. 10.4c), which is formed in between the hydrophobic interfaces of Pex7p and Pex21pC (Fig. 10.3a). All three hydrophobic key residues (Fox3p Leu5, Ile8, Leu12) of PTS2 are hidden inside the hydrophobic groove and form van der Waals contacts with Pex21p Phe236 (Fig. 10.4b, c). The residues of Pex7p and Pex21pC that form the binding site for PTS2 are well conserved among the corresponding homologs. In particular, the α1 helix of Pex21pC, which forms a lid to cover PTS2, is also a conserved region among the co-receptor peroxins as with Pex21p α2 and α3, which compose Pex7pBD in other species (Schliebs and Kunau 2006).

10.2.5 In Vitro *and* In Vivo *Assays of Pex7p and Pex21p Mutants Support Structural Interpretations*

The functional importance of the interaction modules of Pex21p and Pex7p in ternary complex formation and PTS2-cargo transport is examined by, *in vitro* and *in vivo* functional assays (Pan et al. 2013). The *in vitro* assays were conducted by pull-down experiments using amylose resin with purified MBP-Pex21pC (or its variants), Pex7p (or its variants), and Fox3p, whereas the *in vivo* assays by growth defects of *S. cerevisiae* gene deletion strains on oleic-acid medium (SCOT agar plate). As described previously (Zhang and Lazarow 1995; Purdue et al. 1998), *Δpex7* and *Δpex18Δpex21* exhibited growth defects on SCOT plates due to the inability to import sufficient Fox3p into peroxisomes, whereas *Δpex7* expressing wild-type Pex7p and *Δpex18Δpex21* expressing Pex21p exhibited normal growth. Simultaneously, Fox3pN-EGFP and mCherry-PTS1 (PTS1 = Ser-Lys-Leu) were co-expressed in all strains, and their localizations were observed using a confocal microscope.

The mutants containing Pex21p Phe236 or Pex7p Leu34 to Asp (Pex21p F236D, Pex7p L34D), or simultaneously mutated two nearby residues of Pex7p to Ala (Pex7p F32A L34A, F344A W364A), were exhibited clear defects in complex formation by pull-down assays (Pan et al. 2013). These mutants also failed to restore growth of *Δpex7* or *Δpex18Δpex21* on SCOT plates and were unable to restore localization of Fox3pN-EGFP to peroxisomes. Pex21p Ile206, located on the α1 helix, is also a component of the hydrophobic core. The Pex21p I206D mutant exhibited lower affinity for Pex7p and Fox3p and was unable to restore growth of *Δpex18Δpex21* on SCOT plates. The results of Pex21p I206D were similar to those of the Pex21p Δα1 mutant (Δ197–211). Mutations of residues outside the hydrophobic core (Pex21p I202D, K230E) did not affect complex formation, and these mutants were able to restore growth of *Δpex18Δpex21* on SCOT plates.

The role of Pex7p residues that form electrostatic interactions with the hydrophilic key residues of PTS2 (Fox3p Arg4, His11) were also examined (Pan et al. 2013). The mutated Pex7p containing D61R E106H lost the ability to import wild-type Fox3p into peroxisomes and failed to restore the growth of *Δpex7* on SCOT plate. However, co-expression of the Pex7p D61R E106H mutant with the Fox3p R4E mutant enabled to grow the transformants on SCOT plate as rapidly as a strain expressing wild-type Pex7p.

The structure of the bulge loop is essential not only to form a landmark for Pex21p but also to constitute a part of the binding site for PTS2 as described above. Among the single-amino acid substitutions in human Pex7p that cause RCDP1 (Braverman et al. 2002; Motley et al. 2002), substitutions of His39 (Asn31 in *S. cerevisiae*) to Pro, and Gly41 (Gly33 in *S. cerevisiae*) to Val, are mutations on the bulge loop. This fact, together with the results of our mutation analyses of Pex7p Phe32 and Leu34, imply that the conserved shape and the surface property of the bulge loop are required for Pex7p's function.

10.2.6 Comparison with Mitochondrial Targeting Signal

Other helical N-terminal targeting signals, such as the signal peptide (Janda et al. 2010) and mitochondrial targeting signal (MTS) (Abe et al. 2000), also contain hydrophobic residues in their sequence motifs. However, the receptor proteins of the signal peptide and MTS (SRP54 and Tom20, respectively) do not recognize any hydrophilic residue as specifically as Pex7p–Pex21p does. The PTS2 system shares this characteristic with the PTS1 system, in which basic key residue of the PTS1 sequence motif (-[S/A/C]-[K/H/R]-[L/M]-CO_2^-) is specifically recognized by Pex5p (Stanley et al. 2006; Gatto Jr. et al. 2000). It is reported that a mutated PTS1 peptide (Leu-Gln-Ser-Glu-Leu) has an affinity for Pex5p two orders of magnitude lower than that of the wild-type PTS1 peptide (Tyr-Gln-Ser-Lys-Leu) (Gatto Jr. et al. 2000). Consequently, both hydrophobic and hydrophilic key residues in PTS1 and PTS2 achieve their specific complementarity with the receptors.

10.3 Structural Basis for Interactions Between Pex3p and Pex19p in Peroxisomes

At least two peroxin proteins, Pex3p and Pex19p, play central roles in the generation of peroxisome membrane structure (Fujiki et al. 2006). Disruption of the genes encoding Pex3p or Pex19p gives rise to no detectable peroxisome membrane structures (Matsuzono et al. 1999; Muntau et al. 2000; Shimozawa et al. 2000). In the cycle of peroxisome formation, Pex3p and Pex19p play an important role in the insertion of newly synthesized PMPs into pre-existing peroxisomes (Geuze et al. 2003; Kim et al. 2006). Furthermore, it is suggested that in yeast, interaction of Pex3p with Pex19p is also required for the exit of pre-mature peroxisomes from the ER (van der Zand et al. 2010). Thus, in order to understand the peroxisomal membrane biogenesis, it is essential to analyze the molecular architecture in which Pex3p and Pex19p participate.

Pex19p functions as a cytosolic carrier for the PMPs, while Pex3p acts as Pex19p's docking factor on peroxisomes (Fang et al. 2004; Jones et al. 2004). Pex19p distinguishes PMPs from other membrane proteins, such as mitochondrial ones, which are imported with the assistance of general chaperones, Hsp70 and Hsp90 (Young et al. 2003). Pex19p is an intrinsically disordered protein (Dyson and Wright 2005) that exhibits a characteristic domain organization (Fig. 10.5), including a random coil in the N-terminal half and a rigid core in the C-terminal half (Shibata et al. 2004). The N-terminal half of Pex19p seems to contain two binding sites for Pex3p: one is on the N-terminal end and has a strong affinity; the other is near the PMP-binding site and has weak affinity (Fransen et al. 2005; Matsuzono et al. 2006). The main Pex3p-binding site in Pex19p has been localized by two independent studies; it lies within both peptide Met1-Gly56 (Fang et al. 2004) and peptide Ala12-Glu73 (Matsuzono et al. 2006) (Fig. 10.5). The sequence alignment indicates that Glu17-Lys31 in human Pex19p is highly conserved among animals and plants, whereas yeasts have some exceptions. This region is occupied by a unique amino acid sequence containing five hydrophobic residues and seven acidic residues.

On the other hand, Pex3p is a membrane-anchored protein; the N-terminal 33-residue region binds to the peroxisomal membrane, whereas the C-terminal remainder is exposed to the cytosol (Ghaedi et al. 2000; Kammerer et al. 1998;

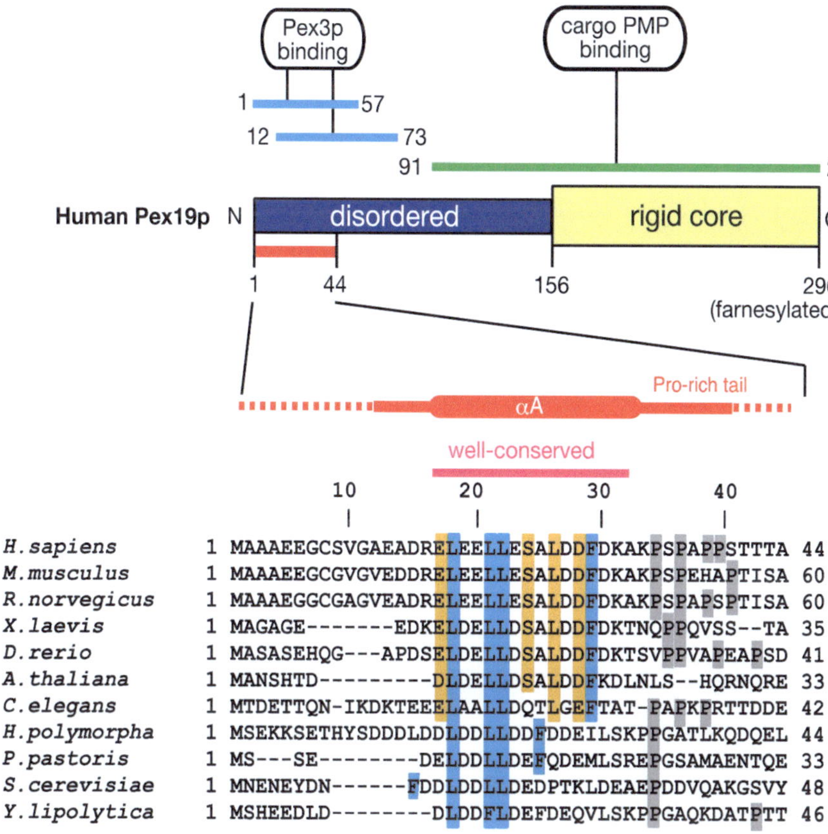

Fig. 10.5 Features of the N-terminal region of Pex19p. Domain organization of human Pex19p and N-terminal sequence alignment of vertebrate Pex19p proteins. In the upper panel, the N-terminal disordered half and the C-terminal rigid core of Pex19p are shown in blue and yellow, respectively. The regions containing the previously reported binding sites for Pex3p and cargo PMP are colored in cyan and green, respectively. In the lower panel with the alignment of human Pex19p (Met1-Ala44), the helix and the subsequent tail are illustrated in red. Amino acid residues significant for the interaction with Pex3p are shown in cyan. Residues involved in the Pex3p binding are shown in orange. Proline residues in the tail region are shown in gray. The yeast Pex19p and Pex3p interaction motif may exhibit exceptional behavior. Thus, based on the alignment, hydrophobic residues, which are estimated to be parts of Pex3p-binding motif, are shown in cyan. (This is adapted from Fig. 1A in Sato et al. (2010))

Soukupova et al. 1999). Even though the hydrophobic segment was removed, Pex3p is prone to aggregate formation. Sato et al. found that a low temperature and a high salt concentration were required to improve its solubility (Sato et al. 2008). For example, during the dialysis of His10–Pex3p (34–373) at 0.6 mg/ml in a high salt buffer (10 mM HEPES-Na, pH 7.5, 0.5 M NaCl, 10% glycerol) against a buffer without salt (10 mM HEPES-Na, pH 7.5, 10% glycerol), the protein formed precipitates and finally achieved a 0.1 mg/ml concentration in a 20,000 × *g* supernatant of the dialyzed solution. This cytosolic domain of Pex3p docks with Pex19p in a specific manner, with a dissociation constant (K_D) of 3.4 nM (Sato et al. 2008). The W104A mutant of Pex3p exhibits significantly diminished binding affinity for Pex19p, as well as lower peroxisomal restoration activity (Sato et al. 2008). Lipid molecules can bind to Pex3p in competition with Pex19p (Pinto et al. 2009), implying the existence of a hydrophobic interaction between Pex3p and Pex19p. Thus, the crystal structure of Pex3p–Pex19p complex was expected to reveal how the Pex3p cytosolic domain and its docking site for Pex19p are arranged on the peroxisome membrane, and how this complex achieves the PMP translocation.

10.3.1 Definition of the Pex3p-Binding Region in Pex19p, for Use in Crystallization

To unify the conformational differences in the Pex3p-Pex19p complexes, an N-terminal Pex19p fragment sufficient for binding to Pex3p is identified. Pex19p fragments of the N-terminal end (Met1-Ala90, Met1-Ala44, and Met1-Asp28) were purified as GST-fused proteins, and examined their affinity for Pex3p by pull-down assay and by surface plasmon resonance (SPR). The experiment showed that at least the N-terminal 44 residues of Pex19p are required for the binding, whereas the Met1-Asp28 fragment on its own has no detectable affinity for Pex3p. The apparent K_D of the Pex19p peptide for Pex3p was 40.8 nM; this affinity is ten times lower than that of the full-length Pex19p (3.4 nM) (Sato et al. 2008), but still significant for a crystallographic analysis. In contrast, the remaining region from Pro45 to the C-terminal end has the K_D 26 μM, 8000-fold higher than that of the wild type. Thus, the binding between Pex3p and Pex19p can be principally attributed to the N-terminal 44-residue segment of Pex19p, although the other C-terminal region may play an auxiliary role in the binding. This N-terminal Pex19p fragment and the cytosolic region of Pex3p (Ile49-Lys373) were independently purified, and the complex was crystallized in the presence of polyethylene glycol 3350.

10.3.2 Overall Structure of the Pex3p Cytosolic Domain in Complex with the N-Terminal Pex19p Peptide

The crystal structure of the complex containing the cytosolic region of Pex3p (Ile49-Lys373) and the N-terminal fragment of Pex19p (Met1-Ala44) was determined at 2.5 Å resolution. The final model of the structure includes residues Ala52-Pro368 of Pex3p and Asp15-Ser40 of Pex19p. Pex3p has a prolate spheroidal shape with approximate dimensions of 80 × 35 × 30 Å. This spheroid exhibits a novel antiparallel helical fold with six α-helical units: α1 (Arg53-Asn95), α2 (Lys100-Ala146), α3 (Pro158-Gly192), α4 (Leu202-Gln249), α5 (Ile258-Arg300), and α6 (Leu322-Ser366), with five inter-helical loops (IHLs). The helical units are twisted up and down roughly along the major axis (Fig. 10.6a, b). Based on the structural characteristics, we named this novel fold a "twisted six-helix bundle".

The most striking feature of the Pex3p structure is that the central longest helical unit α2 runs through the major axis, forming a gentle curve (Fig. 10.6a). Around α2, the other five helical units are assembled in the order α1, α3, α4, α5, and α6. Thus, a pair of spatially adjacent helical units, α1 and α3, make contacts in a parallel fashion, whereas the other adjacent pairs (α3–α4, α4–α5, α5–α6, and α6–α1) make contacts in an antiparallel fashion. Except for α2, each helical unit is bent in one or two places near its middle. These bent regions are designated as helix-bending loops (HBLs) (Fig. 10.6b). The separated helices are defined, applying alphabetical suffixes to the name of the helical unit. Such bending of the surrounding helical units causes tight entwinement around α2. Notably, the middle of α2, Ser115-Leu128 (the thick bar region assigned in Fig. 10.6a right panel) is almost shielded from the solvent and is closely packed against the surrounding helices (Fig. 10.6c). The kinked helices, α3b–α3c, α4a–α4b, α5a–α5b, and α6a–α6b enclose the middle of α2. The side-chain interactions between α2 and the surrounding helical units mainly involve hydrophobic residues. These residues are widely conserved among eukaryotes, suggesting the importance of the hydrophobic packing around the centrally located α2. This shielded region of α2 overlaps a transmembrane helix, spanning from Ile109 to Val131, previously predicted based on hydrophobicity analysis (Kammerer et al. 1998).

The G138E mutation of Pex3p, is isolated as a complementary group in a peroxisome-deficient CHO mutants (Ghaedi et al. 2000). In the crystal structure, Gly138 is positioned near the C-terminal end of α2, which is distal to the binding site for Pex19p (Fig. 10.6a). The methylene group of Gly138 is associated with the main chain of Phe365 and Ser366 on α6c. In the G138E mutant, the bulky side chain of Glu138 seems to push out the contacting α6c and to generate a local structural alteration (Sato et al. 2010).

The N-terminal Pex19p peptide is bound to one of the apexes of the spheroidal Pex3p, and is oblique to α2 (Fig. 10.6a). The Pex3p-binding site of the Pex19p peptide is spans from Glu17 to Ala32, which is widely conserved among multicellular eukaryotes (Fig. 10.5). This short segment of Pex19p forms an α-helix, and is in contact with the Pex3p surface (composed of α1b, α2, and α6a) forming a left-handed

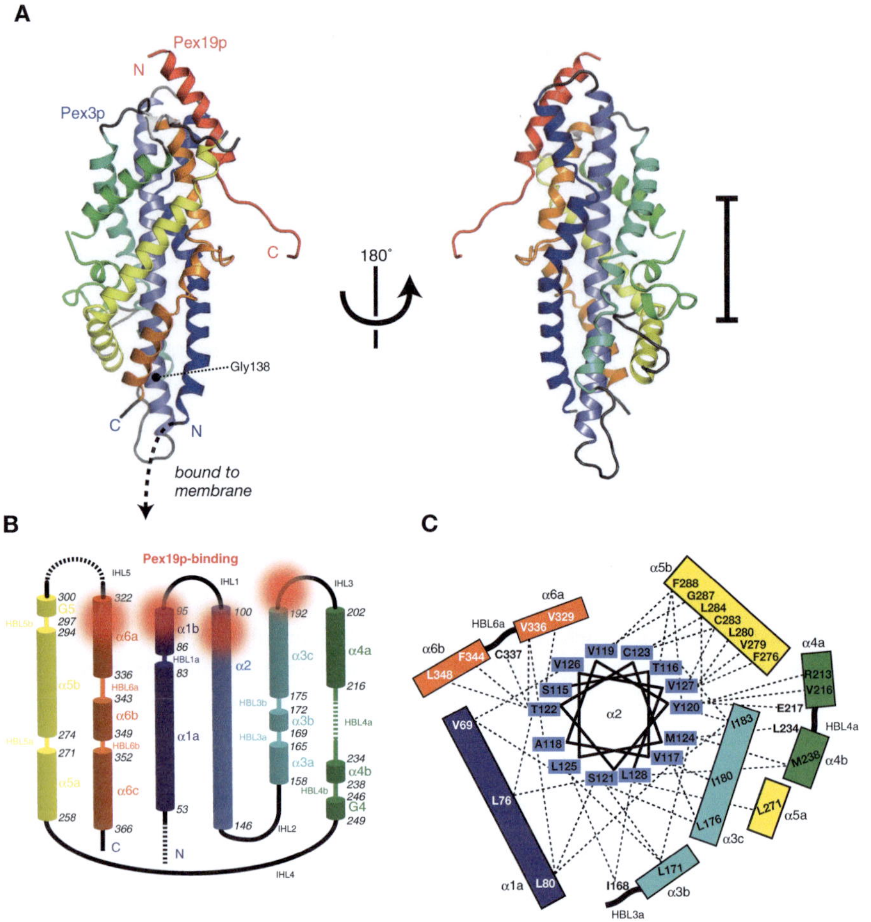

Fig. 10.6 Three-dimensional structure of Pex3p in complex with the Pex19p fragment. (**a**) Overall structure of the complex of the cytosolic domain of Pex3p (Ile49-Lys373) with the N-terminal Pex19p peptide (Met1-Ala44). The Pex19p peptide is colored in red. Six helical units in Pex3p are labeled with the following scheme: α1, navy; α2, light blue; α3, cyan; α4, green; α5, yellow; and α6, orange. The middle of α2 region, buried within the molecule, is shown by a thick bar on the right panel. Inter-helical loops (IHLs) are colored in black, and helix-bending loops (HBLs) are shown in the color of the associated helical unit. Gly138, a mutational site in peroxisome-deficient CHO cell line ZPG208, is pointed out in black on α2. (**b**) Topology of Pex3p. Helices are colored in the same way as in **a** (α1a, Arg53-Gln83; α1b, Ser86-Asn95; α2, Lys100-Ala146; α3a, Pro158-Leu165; α3b, Gln169-Leu172; α3c, Gly175-Gly192; α4a, Leu202-Val216; α4b, Leu234-Met238; G4 (3_{10} helix), Leu246-Gln249; α5a, Ile258-Leu271; α5b, Pro274-Asn294; G5, Glu297-Arg300; α6a, Leu322-Val336; α6b, His343-Leu349; and α6c, Glu352-Ser366). IHLs are shown in black (IHL1, Arg96-Asn99; IHL2, Val147-Pro157; IHL3, Ser193-Ser201; IHL4, Ala250-Asp257, and IHL5, Pro301-Pro321). HBLs are shown in the same way as in **a** (HBL1a, Leu84-Asn85; HBL3a, Ser166-Ile168; HBL3b, Gly173-Asp174; HBL4a, Glu217-Leu233; HBL4b, Met239-Pro245; HBL5a, Glu272-Ser273; HBL5b, Met295-Ala296; HBL6a, Cys337-Ser342; and HBL6b, Thr350-Met351). The Pex19p-binding regions are shown as red shading. Dotted lines are the unassigned

four-helix bundle. Since α1b and α6a are connected to HBL1a and HBL6a, respectively, these helices are likely to be flexible relative to α2. This raises the possibility that a small induced-fit conformational change of 1b and/or α6a could occur upon Pex19p helix binding. Although the Pex3p-bound Pex19p peptide includes the α-helix, we have previously demonstrated that the N-terminal half of Pex19p (Met1-Ala156) is disordered in the absence of Pex3p binding (Shibata et al. 2004). The circular dichroism spectrum for this Pex19p fragment indicates that the peptide Met1-Ala44 also adopts a random coil structure in an aqueous buffer solution (Sato et al. 2010). Interestingly, the α-helical content was increased with an increasing the methanol concentration in the buffer. Therefore, in hydrophobic environment, the Pex19p peptide (Met1-Ala44) tends to acquire secondary structure, including α-helix, from a random coil. The crystal structure also shows that the Pex19p tail region containing Pro33-Ser40, extends from the helix to the solvent, without intermolecular interactions between Pex19p and Pex3p (Figs. 10.5 and 10.6a). It was suggested that the helix formation in this region of Pex19p could be coupled with hydrophobic binding to the Pex3p surface (Sato et al. 2010). If this Pex19p amphipathic helix were maintained without Pex3p binding, the hydrophobic side of the helix could interact with various cytosolic proteins in a non-specific manner. Therefore, during the cytosolic translocation of a cargo PMP, the Pex19p helix would assume an unstructured conformation in order to avoid such non-specific binding. A similar behavior has been discovered in mitochondrial protein targeting. Tom20, a constituent of the translocon of the outer mitochondrial membrane (TOM) complex, and recognizes classical N-terminal mitochondrial-targeting presequences existing on mitochondrial-precursor proteins (Wiedemann et al. 2009). The each presequence also forms α-helix upon interaction with the hydrophobic groove of Tom20, though presequences alone exhibit very little secondary structure in aqueous solutions (Abe et al. 2000; Saitoh et al. 2007). Thus, the helical formation in the amphiphilic and specific regions of target proteins coupled with hydrophobic binding sites of acceptor proteins could represent a widely applicable mechanism for selective targeting of organelle proteins by posttranslational import systems.

A crystal structure of residues 41–368 of the C235S mutant Pex3p in complex with a synthetic peptide corresponding to residues 14–30 of Pex19p has also been solved (Schmidt et al. 2010). Although the two complexes produce crystals with different space groups ($P6_522$ or $P2_1$), and the structures were solved by different methods, (Multi-wavelength anomalous diffraction or molecular replacement), both structures are very similar. This consistency supports a reliability of the Pex3p–Pex19p peptide structure.

Fig. 10.6 (continued) regions; His219-Ser230 (green) in HBL4a and Thr302-Asn313 (black) in IHL5. (**c**) Packing of α2 by the surrounding helices in Pex3p. Ser115-Leu128 on α2, the buried region shown by a thick bar on the right panel in **a**, is illustrated using a wheel diagram. The side chains on surrounding helices located in 2.5–4.0 Å from Ser115-Leu128 on α2 are drawn around the wheel. (Figure is adapted from Fig. 1 in Sato et al. (2010))

10.3.3 Architecture of the Pex3p–Pex19p Interface

The Pex19p binding/docking surface on Pex3p presents a large tapered cavity. The side chain of Trp104 of Pex3p protrudes from this cavity and divides it into a wider surface area near the surface and a narrower area deeper inside (Fig. 10.7a). This cavity is complementary to the Pex19p helix structure in which two types of "protruding tooth" are formed from the characteristic side chains: one is composed of

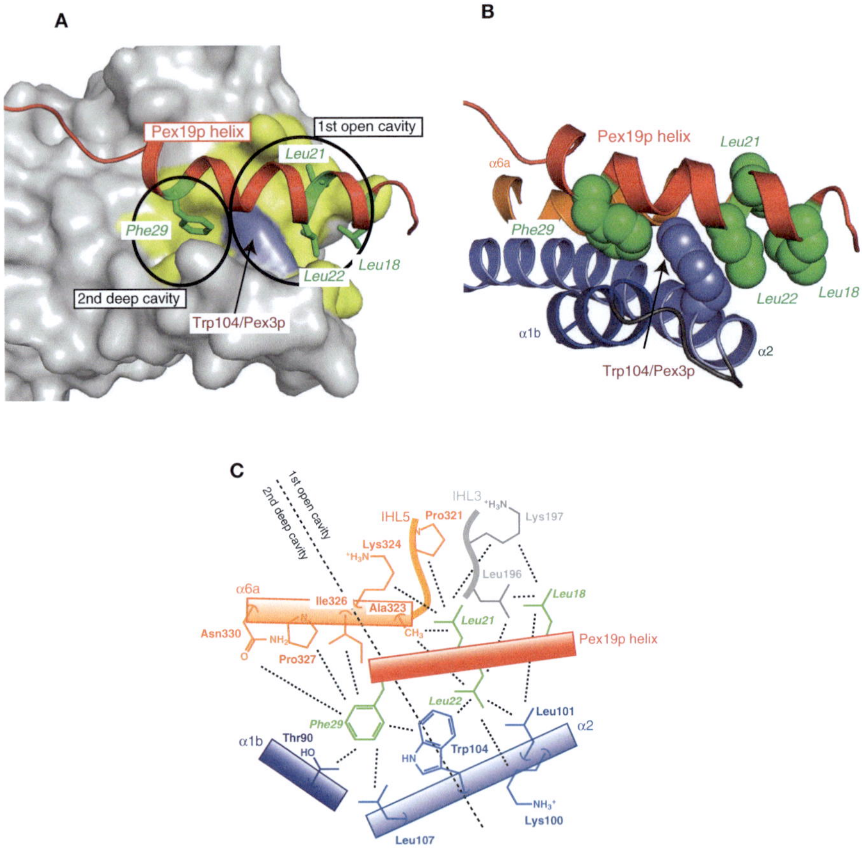

Fig. 10.7 Binding of Pex19p on Pex3p. (**a**) The Pex19p-binding region on Pex3p. Pex19p is shown in red, and the side chains essential for the binding to Pex3p are highlighted in green. The binding cavities on Pex3p are colored in yellow, and Trp104 of Pex3p is highlighted in light blue. Two cavities are shown in black circles. (**b**) Hydrophobic packing between two types of protuberance of Pex19p and Trp104 of Pex3p. Leu18, Leu21, Leu22, and Phe29 of Pex19p are shown in green, and Trp104 of Pex3p is shown in light blue. The ribbon model is colored in the same way as in Fig. 10.6a. (**c**) Schematic drawing of the interaction between Pex3p and Pex19p. The colors are coded in the same way as Fig. 10.6a. (Figure is adapted from Fig. 4 in Sato et al. (2010))

three leucine residues, Leu18, Leu21, and Leu22, thus termed the "leucine triad"; and the other is composed of the phenyl side chain of Phe29 (Fig. 10.7a). The leucine triad is surrounded by the wider surface portion of the cavity (Fig. 10.7a), which composed of Lys100, Leu101, and Trp104 at the N-terminus of α2; Leu196 and Lys197 on the middle of IHL3; Pro321 at the C-terminal end of IHL5; and, Ala323 and Lys324 on the subsequent α6a (Fig. 10.7c). The other "protruding tooth", Phe29, is surrounded by the narrower deeper part of the cavity (Fig. 10.7a), which is composed of Thr90 on α1b; Trp104 and Leu107 on α2; and Ile326, Pro327, and Asn330 on α6a (Fig. 10.7c). The indole ring of Trp104 of Pex3p is inserted into the notch on the Pex19p helix between the two "protruding teeth" (Fig. 10.7b). These features indicate that the interaction between Pex3p and Pex19p is supported mainly by van der Waals forces and/or hydrophobic effects. The lack of the side chain at residue 104 (Trp in the wild-type protein) is expected to destabilize the binding; indeed, this destabilization has been demonstrated (Sato et al. 2008). The structure also suggests that several other residues could participate in the interaction between Pex3p and Pex19p. Glu17 and Leu26 on the Pex19p helix seem to interact with Lys197 on IHL3 and Leu93 on α1b of Pex3p, respectively, and both Ser 24 and Asp28 on the Pex19p helix seem to bind to Lys324 on α6a of Pex3p.

In the case of mitochondrial protein targeting, the portion of presequence peptide that binds Tom 20 contains a diverse consensus motif (ϕχχϕϕ, where ϕ is hydrophobic and χ is any amino acid) and adopts an amphiphilic helical conformation (Wiedemann et al. 2009). The diverse consensus motif is likely to guarantee the broadly selective specificity of the Tom20 receptor toward diverse mitochondrial presequences (Abe et al. 2000; Saitoh et al. 2007). In the case of the peroxisome, the Pex19p–Pex3p interaction requires more complex motif (Leu-X-X-Leu-Leu-X_6-Phe on the animal Pex19p) to confer both high specificity and tight-binding. It is suggested that the sequence should be designated as the Pex3p-binding motif with an intrinsically disordered manner, one that is tolerant of conformational changes according to its intracellular environment.

10.3.4 Mutational Analysis of Functional Residues in Pex19p Required for Binding to Pex3p and for Peroxisome Biogenesis

The structural assignment of the residues involved in the binding between Pex3p and Pex19p is revealed by mutations (Sato et al. 2010). The Pex19p mutants L18A, L21A, L22A, or F29A, in which a part of the Pex3p-binding motif is disrupted, clearly diminished binding affinity for Pex3p in the pull-down assays and SPR analysis. It is revealed that their K_D values for binding of Pex3p were increased more than 100 times relative to the wild type. On the other hand, L26A, another mutant of a hydrophobic residue on the Pex19p helix bound to Pex3p with weaker affinity; the K_D value for Pex3p binding was 46.3 nM. A mutants of one polar residues S24A, had a K_D of 4.7 nM, essentially the same affinity as the wild type; however, two other polar residue mutants E17A and D28A, had K_D of 17.3 nM and 21.8 nM, respectively, in both cases slightly weaker affinity than the wild type. The functional defect of the mutants in peroxisome biogenesis is also revealed by peroxisome-restoring experiments using the PEX19 deficient fibroblasts (Sato et al. 2010).

10.3.5 Proposed Mechanism for the PMP Translocation Process

Based on the structural and functional findings, Pex3p and Pex19p participate in a proposed mechanism in the PMP translocation process as follows. Pex3p is anchored on the membrane by the N-terminal hydrophobic region attached to one of its apexes. In contrast, the binding site for Pex19p is located on the opposite apex, which is most distal from the membrane. Therefore, Pex19p could be effectively captured by the apex of Pex3p, "standing" on the membrane like a tower. After Pex3p captures the N-terminal Pex19p helix, the remainder of Pex19p (Pro45-farnesylated Cys296), containing the PMP-binding region, appears to hang down toward the membrane from the top of Pex3p, along the spheroid. The molecular surface of the human Pex3p exhibits interhelical grooves rich in hydrophobic residues, which are reasonable candidates for mediators of docking with other protein molecules involved in the translocation process. Future research should aim to identify other binding partners for Pex3p and the structural basis for their interaction; this knowledge would facilitate our understanding of their biological roles in peroxisome biogenesis.

10.4 Future Prospect in Structural Biology Research of Peroxins (Outlook for Structural Biology of Peroxins)

A major challenge will be the structural characterization of full-length peroxins containing both folded and disordered domains. Usually, peroxins have been analyzed as fragments, and their interactions with physiological partners examined by atomic-scale structural approaches, such as X-ray crystallography and NMR. Many peroxins are large modular proteins and do not function as isolated domains, but instead, their component regions, both ordered and disordered, act synergistically in performing their functions. Pex7p and Pex3p are few cases in which their whole molecule or at least all soluble part of structures have been determined as described above. In the case of the Pex7p study, the multi-protein complex instead of the binary complex is chosen to order the structure and to improve crystallizability of the target molecules. Moreover, *Komagataella phaffii* is used for the protein expression of Pex7p since *E. coli* did not produce Pex7p with properly folded molecules.

The selection of an appropriate host of protein production is suggested to be important for structural biology approach. However, still the crystallization is the bottleneck of the structural study of peroxins at an atomic level.

Recently, cryo-electron microscopy (cryo-EM) is evolutionally improved their resolution. The advent of direct electron detectors with improved quantum efficiencies allowed for the correction of beam-induced motions of the specimen in vitrified ice (Li et al. 2013). Thus, structures at near-atomic resolution are no longer the prerogative of X-ray crystallography or nuclear magnetic resonance (NMR) spectroscopy (Kuhlbrandt 2014). Cryo-EM requires only small amounts of material. Samples that cannot be isolated in large enough quantities for X-ray crystallography or NMR can now be analyzed and yield high-resolution structures. The same holds for heterogeneous samples or flexible complexes that do not crystallize readily, because cryo-EM images of different particles or conformations are easily separated at the image processing stage.

The hexameric structures of Pex1/Pex6 in the presence of nucleotides have been determined by cryo-EM at 7.2–23 Å resolution (Blok et al. 2015; Ciniawsky et al. 2015; Gardner et al. 2015). These structures agree in the overall architecture of the complex, but a common mode of action cannot be deduced from these works (Saffert et al. 2017). While some studies observe domain movements in the whole complex (Ciniawsky et al. 2015; Gardner et al. 2015), others detect no movements whatever upon ATP hydrolysis (Blok et al. 2015). In particular domain movements in the N-terminal domains and the nucleotide binding domain, NBD2 are reminiscent of nucleotide dependent movements observed for the homologous protein p97 (Saffert et al. 2017). P97 is by far the best studied and best characterized ATPase among the exciting AAA+ protein family (Saffert et al. 2017; Banerjee et al. 2016; Schuller et al. 2016; Huang et al. 2016; Ripstein et al. 2017) and the cryo-EM structure of p97 with 540 kDa has been determined at 2.3 Å resolution (Banerjee et al. 2016). Another cryo-EM structure of a peroxisomal enzyme, alcohol oxidase (AOX) with a 600-kDa homo-octamer from *Pichia pastoris* has been determined at 3.4 Å resolution (Vonck et al. 2016).

In parallel to the ground-breaking cryo-EM developments, crystallography has been also revolutionized using X-ray free-electron lasers (XFELs). XFELs generate extremely intense X-ray pulses of tens of femtosecond duration with nine-to-ten orders of magnitude higher peak brilliance than third-generation synchrotrons, like SPring-8. Such unique characteristics of XFELs prompted a new approach for crystallographic data collection called serial femtosecond crystallography (SFX) (Chapman et al. 2011). Unlike traditional crystallography, where a complete dataset is collected from a single large (or a few small) crystals, SFX data are acquired from tens to hundreds of thousands of very small crystals intersecting the XFEL beam in random orientations. Although each crystal is destroyed by the beam, the short pulse duration allows outrunning radiation damage and obtaining structural information from intact molecules at room temperature without the necessity of cryocooling (Neutze et al. 2000). The extremely high brightness of each XFEL pulse provides sufficient signal for the detection of high-resolution diffraction patterns from micrometer (Liu et al. 2013) and even sub-micrometer-sized crystals

(Gati et al. 2017). We determined the structure of enzyme that is related to the luciferin–luciferase reaction using SFX with de novo phasing (Yamashita et al. 2015, 2017) at the Japanese XFEL facility, SACLA (Mizohata et al. 2018). The crystals we used are with a micrometer size of suspension in a precipitant solution of several tens of microliters. Near future, XFELs offer the possibility of stepping beyond X-ray crystallography, to extend structural studies to single, non-crystalline particles or molecules. The method has been applied to an imaging of large viruses (Seibert et al. 2011). Therefore, cryo-EM and XFELs will allow us to determine the structures of peroxins without conventional crystals with either the static states or dynamic ones in the biological actions with respect to the peroxisomal biogenesis.

References

Abe Y, Shodai T, Muto T, Mihara K, Torii H, Nishikawa S, Endo T, Kohda D (2000) Structural basis of presequence recognition by the mitochondrial protein import receptor Tom20. Cell 100(5):551–560

Banerjee S, Bartesaghi A, Merk A, Rao P, Bulfer SL, Yan Y, Green N, Mroczkowski B, Neitz RJ, Wipf P, Falconieri V, Deshaies RJ, Milne JL, Huryn D, Arkin M, Subramaniam S (2016) 2.3 A resolution cryo-EM structure of human p97 and mechanism of allosteric inhibition. Science 351(6275):871–875. https://doi.org/10.1126/science.aad7974

Barros-Barbosa A, Rodrigues TA, Ferreira MJ, Pedrosa AG, Teixeira NR, Francisco T, Azevedo JE (2019) The intrinsically disordered nature of the peroxisomal protein translocation machinery. FEBS J 286(1):24–38. https://doi.org/10.1111/febs.14704

Blok NB, Tan D, Wang RY, Penczek PA, Baker D, DiMaio F, Rapoport TA, Walz T (2015) Unique double-ring structure of the peroxisomal Pex1/Pex6 ATPase complex revealed by cryo-electron microscopy. Proc Natl Acad Sci U S A 112(30):E4017–E4025. https://doi.org/10.1073/pnas.1500257112

Braverman N, Steel G, Obie C, Moser A, Moser H, Gould SJ, Valle D (1997) Human PEX7 encodes the peroxisomal PTS2 receptor and is responsible for rhizomelic chondrodysplasia punctata. Nat Genet 15(4):369–376. https://doi.org/10.1038/ng0497-369

Braverman N, Dodt G, Gould SJ, Valle D (1998) An isoform of pex5p, the human PTS1 receptor, is required for the import of PTS2 proteins into peroxisomes. Hum Mol Genet 7(8):1195–1205

Braverman N, Chen L, Lin P, Obie C, Steel G, Douglas P, Chakraborty PK, Clarke JT, Boneh A, Moser A, Moser H, Valle D (2002) Mutation analysis of PEX7 in 60 probands with rhizomelic chondrodysplasia punctata and functional correlations of genotype with phenotype. Hum Mutat 20(4):284–297. https://doi.org/10.1002/humu.10124

Chapman HN, Fromme P, Barty A, White TA, Kirian RA, Aquila A, Hunter MS, Schulz J, DePonte DP, Weierstall U, Doak RB, Maia FR, Martin AV, Schlichting I, Lomb L, Coppola N, Shoeman RL, Epp SW, Hartmann R, Rolles D, Rudenko A, Foucar L, Kimmel N, Weidenspointner G, Holl P, Liang M, Barthelmess M, Caleman C, Boutet S, Bogan MJ, Krzywinski J, Bostedt C, Bajt S, Gumprecht L, Rudek B, Erk B, Schmidt C, Homke A, Reich C, Pietschner D, Struder L, Hauser G, Gorke H, Ullrich J, Herrmann S, Schaller G, Schopper F, Soltau H, Kuhnel KU, Messerschmidt M, Bozek JD, Hau-Riege SP, Frank M, Hampton CY, Sierra RG, Starodub D, Williams GJ, Hajdu J, Timneanu N, Seibert MM, Andreasson J, Rocker A, Jonsson O, Svenda M, Stern S, Nass K, Andritschke R, Schroter CD, Krasniqi F, Bott M, Schmidt KE, Wang X, Grotjohann I, Holton JM, Barends TR, Neutze R, Marchesini S, Fromme R, Schorb S, Rupp D, Adolph M, Gorkhover T, Andersson I, Hirsemann H, Potdevin G, Graafsma H, Nilsson B, Spence JC (2011) Femtosecond X-ray protein nanocrystallography. Nature 470(7332):73–77. https://doi.org/10.1038/nature09750

Ciniawsky S, Grimm I, Saffian D, Girzalsky W, Erdmann R, Wendler P (2015) Molecular snapshots of the Pex1/6 AAA+ complex in action. Nat Commun 6:7331. https://doi.org/10.1038/ncomms8331

Dawidowski M, Emmanouilidis L, Kalel VC, Tripsianes K, Schorpp K, Hadian K, Kaiser M, Maser P, Kolonko M, Tanghe S, Rodriguez A, Schliebs W, Erdmann R, Sattler M, Popowicz GM (2017) Inhibitors of PEX14 disrupt protein import into glycosomes and kill Trypanosoma parasites. Science 355(6332):1416–1420. https://doi.org/10.1126/science.aal1807

Dodt G, Warren D, Becker E, Rehling P, Gould SJ (2001) Domain mapping of human PEX5 reveals functional and structural similarities to Saccharomyces cerevisiae Pex18p and Pex21p. J Biol Chem 276(45):41769–41781. https://doi.org/10.1074/jbc.M106932200

Douangamath A, Filipp FV, Klein AT, Barnett P, Zou P, Voorn-Brouwer T, Vega MC, Mayans OM, Sattler M, Distel B, Wilmanns M (2002) Topography for independent binding of alpha-helical and PPII-helical ligands to a peroxisomal SH3 domain. Mol Cell 10(5):1007–1017

Dyson HJ, Wright PE (2005) Intrinsically unstructured proteins and their functions. Nat Rev Mol Cell Biol 6(3):197–208. https://doi.org/10.1038/nrm1589. nrm1589 [pii]

Einwachter H, Sowinski S, Kunau WH, Schliebs W (2001) Yarrowia lipolytica Pex20p, Saccharomyces cerevisiae Pex18p/Pex21p and mammalian Pex5pL fulfil a common function in the early steps of the peroxisomal PTS2 import pathway. EMBO Rep 2(11):1035–1039. https://doi.org/10.1093/embo-reports/kve228

Emmanouilidis L, Schutz U, Tripsianes K, Madl T, Radke J, Rucktaschel R, Wilmanns M, Schliebs W, Erdmann R, Sattler M (2017) Allosteric modulation of peroxisomal membrane protein recognition by farnesylation of the peroxisomal import receptor PEX19. Nat Commun 8:14635. https://doi.org/10.1038/ncomms14635

Fang Y, Morrell JC, Jones JM, Gould SJ (2004) PEX3 functions as a PEX19 docking factor in the import of class I peroxisomal membrane proteins. J Cell Biol 164(6):863–875. https://doi.org/10.1083/jcb.200311131. jcb.200311131 [pii]

Farre JC, Carolino K, Stasyk OV, Stasyk OG, Hodzic Z, Agrawal G, Till A, Proietto M, Cregg J, Sibirny AA, Subramani S (2017) A new yeast peroxin, Pex36, a functional homolog of mammalian PEX16, functions in the ER-to-peroxisome traffic of peroxisomal membrane proteins. J Mol Biol 429(23):3743–3762. https://doi.org/10.1016/j.jmb.2017.10.009

Fodor K, Wolf J, Erdmann R, Schliebs W, Wilmanns M (2012) Molecular requirements for peroxisomal targeting of alanine-glyoxylate aminotransferase as an essential determinant in primary hyperoxaluria type 1. PLoS Biol 10(4):e1001309. https://doi.org/10.1371/journal.pbio.1001309. PBIOLOGY-D-11-04741 [pii]

Fodor K, Wolf J, Reglinski K, Passon DM, Lou Y, Schliebs W, Erdmann R, Wilmanns M (2015) Ligand-induced compaction of the PEX5 receptor-binding cavity impacts protein import efficiency into peroxisomes. Traffic 16(1):85–98. https://doi.org/10.1111/tra.12238

Fransen M, Vastiau I, Brees C, Brys V, Mannaerts GP, Van Veldhoven PP (2005) Analysis of human Pex19p's domain structure by pentapeptide scanning mutagenesis. J Mol Biol 346(5):1275–1286. https://doi.org/10.1016/j.jmb.2005.01.013. S0022-2836(05)00035-5 [pii]

Fujiki Y, Matsuzono Y, Matsuzaki T, Fransen M (2006) Import of peroxisomal membrane proteins: the interplay of Pex3p- and Pex19p-mediated interactions. Biochim Biophys Acta 1763(12):1639–1646. https://doi.org/10.1016/j.bbamcr.2006.09.030. S0167-4889(06)00304-1 [pii]

Fujiki Y, Okumoto K, Mukai S, Honsho M, Tamura S (2014) Peroxisome biogenesis in mammalian cells. Front Physiol 5:307. https://doi.org/10.3389/fphys.2014.00307

Gardner BM, Chowdhury S, Lander GC, Martin A (2015) The Pex1/Pex6 complex is a heterohexameric AAA+ motor with alternating and highly coordinated subunits. J Mol Biol 427(6 Pt B):1375–1388. https://doi.org/10.1016/j.jmb.2015.01.019

Gardner BM, Castanzo DT, Chowdhury S, Stjepanovic G, Stefely MS, Hurley JH, Lander GC, Martin A (2018) The peroxisomal AAA-ATPase Pex1/Pex6 unfolds substrates by processive threading. Nat Commun 9(1):135. https://doi.org/10.1038/s41467-017-02474-4

Gati C, Oberthuer D, Yefanov O, Bunker RD, Stellato F, Chiu E, Yeh SM, Aquila A, Basu S, Bean R, Beyerlein KR, Botha S, Boutet S, DePonte DP, Doak RB, Fromme R, Galli L, Grotjohann I, James DR, Kupitz C, Lomb L, Messerschmidt M, Nass K, Rendek K, Shoeman RL, Wang D, Weierstall U, White TA, Williams GJ, Zatsepin NA, Fromme P, Spence JC, Goldie KN, Jehle JA, Metcalf P, Barty A, Chapman HN (2017) Atomic structure of granulin determined from native nanocrystalline granulovirus using an X-ray free-electron laser. Proc Natl Acad Sci U S A 114(9):2247–2252. https://doi.org/10.1073/pnas.1609243114

Gatto GJ Jr, Geisbrecht BV, Gould SJ, Berg JM (2000) Peroxisomal targeting signal-1 recognition by the TPR domains of human PEX5. Nat Struct Biol 7(12):1091–1095. https://doi.org/10.1038/81930

Geuze HJ, Murk JL, Stroobants AK, Griffith JM, Kleijmeer MJ, Koster AJ, Verkleij AJ, Distel B, Tabak HF (2003) Involvement of the endoplasmic reticulum in peroxisome formation. Mol Biol Cell 14(7):2900–2907. https://doi.org/10.1091/mbc.E02-11-0734. E02-11-0734 [pii]

Ghaedi K, Tamura S, Okumoto K, Matsuzono Y, Fujiki Y (2000) The peroxin pex3p initiates membrane assembly in peroxisome biogenesis. Mol Biol Cell 11(6):2085–2102

Glover JR, Andrews DW, Subramani S, Rachubinski RA (1994) Mutagenesis of the amino targeting signal of Saccharomyces cerevisiae 3-ketoacyl-CoA thiolase reveals conserved amino acids required for import into peroxisomes in vivo. J Biol Chem 269(10):7558–7563

Groves MR, Schroer CFE, Middleton AJ, Lunev S, Danda N, Ali AM, Marrink SJ, Williams C (2018) Structural insights into K48-linked ubiquitin chain formation by the Pex4p-Pex22p complex. Biochem Biophys Res Commun 496(2):562–567. https://doi.org/10.1016/j.bbrc.2017.12.150

Hensel A, Beck S, El Magraoui F, Platta HW, Girzalsky W, Erdmann R (2011) Cysteine-dependent ubiquitination of Pex18p is linked to cargo translocation across the peroxisomal membrane. J Biol Chem 286(50):43495–43505. https://doi.org/10.1074/jbc.M111.286104

Huang R, Ripstein ZA, Augustyniak R, Lazniewski M, Ginalski K, Kay LE, Rubinstein JL (2016) Unfolding the mechanism of the AAA+ unfoldase VAT by a combined cryo-EM, solution NMR study. Proc Natl Acad Sci U S A 113(29):E4190–E4199. https://doi.org/10.1073/pnas.1603980113

Janda CY, Li J, Oubridge C, Hernandez H, Robinson CV, Nagai K (2010) Recognition of a signal peptide by the signal recognition particle. Nature 465(7297):507–510. https://doi.org/10.1038/nature08870

Jones JM, Morrell JC, Gould SJ (2004) PEX19 is a predominantly cytosolic chaperone and import receptor for class 1 peroxisomal membrane proteins. J Cell Biol 164(1):57–67. https://doi.org/10.1083/jcb.200304111. jcb.200304111 [pii]

Kammerer S, Holzinger A, Welsch U, Roscher AA (1998) Cloning and characterization of the gene encoding the human peroxisomal assembly protein Pex3p. FEBS Lett 429(1):53–60. S0014-5793(98)00557-2 [pii]

Khan BR, Zolman BK (2010) pex5 Mutants that differentially disrupt PTS1 and PTS2 peroxisomal matrix protein import in Arabidopsis. Plant Physiol 154(4):1602–1615. https://doi.org/10.1104/pp.110.162479. 110.162479 [pii]

Kim PK, Mullen RT, Schumann U, Lippincott-Schwartz J (2006) The origin and maintenance of mammalian peroxisomes involves a de novo PEX16-dependent pathway from the ER. J Cell Biol 173(4):521–532. https://doi.org/10.1083/jcb.200601036. jcb.200601036 [pii]

Koller A, Snyder WB, Faber KN, Wenzel TJ, Rangell L, Keller GA, Subramani S (1999) Pex22p of Pichia pastoris, essential for peroxisomal matrix protein import, anchors the ubiquitin-conjugating enzyme, Pex4p, on the peroxisomal membrane. J Cell Biol 146(1):99–112. https://doi.org/10.1083/jcb.146.1.99

Kuhlbrandt W (2014) Biochemistry. The resolution revolution. Science 343(6178):1443–1444. https://doi.org/10.1126/science.1251652

Leon S, Subramani S (2007) A conserved cysteine residue of Pichia pastoris Pex20p is essential for its recycling from the peroxisome to the cytosol. J Biol Chem 282(10):7424–7430. https://doi.org/10.1074/jbc.M611627200

Leon S, Zhang L, McDonald WH, Yates J III, Cregg JM, Subramani S (2006) Dynamics of the peroxisomal import cycle of PpPex20p: ubiquitin-dependent localization and regulation. J Cell Biol 172(1):67–78. https://doi.org/10.1083/jcb.200508096

Li X, Mooney P, Zheng S, Booth CR, Braunfeld MB, Gubbens S, Agard DA, Cheng Y (2013) Electron counting and beam-induced motion correction enable near-atomic-resolution single-particle cryo-EM. Nat Methods 10(6):584–590. https://doi.org/10.1038/nmeth.2472

Liu W, Wacker D, Gati C, Han GW, James D, Wang D, Nelson G, Weierstall U, Katritch V, Barty A, Zatsepin NA, Li D, Messerschmidt M, Boutet S, Williams GJ, Koglin JE, Seibert MM, Wang C, Shah ST, Basu S, Fromme R, Kupitz C, Rendek KN, Grotjohann I, Fromme P, Kirian RA, Beyerlein KR, White TA, Chapman HN, Caffrey M, Spence JC, Stevens RC, Cherezov V (2013) Serial femtosecond crystallography of G protein-coupled receptors. Science 342(6165):1521–1524. https://doi.org/10.1126/science.1244142

Marzioch M, Erdmann R, Veenhuis M, Kunau WH (1994) PAS7 encodes a novel yeast member of the WD-40 protein family essential for import of 3-oxoacyl-CoA thiolase, a PTS2-containing protein, into peroxisomes. EMBO J 13(20):4908–4918

Matsumura T, Otera H, Fujiki Y (2000) Disruption of the interaction of the longer isoform of Pex5p, Pex5pL, with Pex7p abolishes peroxisome targeting signal type 2 protein import in mammals. Study with a novel Pex5-impaired Chinese hamster ovary cell mutant. J Biol Chem 275(28):21715–21721. https://doi.org/10.1074/jbc.M000721200. M000721200 [pii]

Matsuzono Y, Kinoshita N, Tamura S, Shimozawa N, Hamasaki M, Ghaedi K, Wanders RJ, Suzuki Y, Kondo N, Fujiki Y (1999) Human PEX19: cDNA cloning by functional complementation, mutation analysis in a patient with Zellweger syndrome, and potential role in peroxisomal membrane assembly. Proc Natl Acad Sci U S A 96(5):2116–2121

Matsuzono Y, Matsuzaki T, Fujiki Y (2006) Functional domain mapping of peroxin Pex19p: interaction with Pex3p is essential for function and translocation. J Cell Sci 119(Pt 17):3539–3550. https://doi.org/10.1242/jcs.03100. jcs.03100 [pii]

Mizohata E, Nakane T, Fukuda Y, Nango E, Iwata S (2018) Serial femtosecond crystallography at the SACLA: breakthrough to dynamic structural biology. Biophys Rev 10(2):209–218. https://doi.org/10.1007/s12551-017-0344-9

Motley AM, Hettema EH, Hogenhout EM, Brites P, ten Asbroek AL, Wijburg FA, Baas F, Heijmans HS, Tabak HF, Wanders RJ, Distel B (1997) Rhizomelic chondrodysplasia punctata is a peroxisomal protein targeting disease caused by a non-functional PTS2 receptor. Nat Genet 15(4):377–380. https://doi.org/10.1038/ng0497-377

Motley AM, Brites P, Gerez L, Hogenhout E, Haasjes J, Benne R, Tabak HF, Wanders RJ, Waterham HR (2002) Mutational spectrum in the PEX7 gene and functional analysis of mutant alleles in 78 patients with rhizomelic chondrodysplasia punctata type 1. Am J Hum Genet 70(3):612–624. https://doi.org/10.1086/338998. S0002-9297(07)60265-1 [pii]

Muntau AC, Mayerhofer PU, Paton BC, Kammerer S, Roscher AA (2000) Defective peroxisome membrane synthesis due to mutations in human PEX3 causes Zellweger syndrome, complementation group G. Am J Hum Genet 67(4):967–975. https://doi.org/10.1086/303071. S0002-9297(07)63288-1 [pii]

Neufeld C, Filipp FV, Simon B, Neuhaus A, Schuller N, David C, Kooshapur H, Madl T, Erdmann R, Schliebs W, Wilmanns M, Sattler M (2009) Structural basis for competitive interactions of Pex14 with the import receptors Pex5 and Pex19. EMBO J 28(6):745–754. https://doi.org/10.1038/emboj.2009.7. emboj20097 [pii]

Neuhaus A, Kooshapur H, Wolf J, Meyer NH, Madl T, Saidowsky J, Hambruch E, Lazam A, Jung M, Sattler M, Schliebs W, Erdmann R (2014) A novel Pex14 protein-interacting site of human Pex5 is critical for matrix protein import into peroxisomes. J Biol Chem 289(1):437–448. https://doi.org/10.1074/jbc.M113.499707

Neutze R, Wouts R, van der Spoel D, Weckert E, Hajdu J (2000) Potential for biomolecular imaging with femtosecond X-ray pulses. Nature 406(6797):752–757. https://doi.org/10.1038/35021099

Nito K, Hayashi M, Nishimura M (2002) Direct interaction and determination of binding domains among peroxisomal import factors in Arabidopsis thaliana. Plant Cell Physiol 43(4):355–366

Osumi T, Tsukamoto T, Hata S, Yokota S, Miura S, Fujiki Y, Hijikata M, Miyazawa S, Hashimoto T (1991) Amino-terminal presequence of the precursor of peroxisomal 3-ketoacyl-CoA thiolase is a cleavable signal peptide for peroxisomal targeting. Biochem Biophys Res Commun 181(3):947–954

Otera H, Okumoto K, Tateishi K, Ikoma Y, Matsuda E, Nishimura M, Tsukamoto T, Osumi T, Ohashi K, Higuchi O, Fujiki Y (1998) Peroxisome targeting signal type 1 (PTS1) receptor is involved in import of both PTS1 and PTS2: studies with PEX5-defective CHO cell mutants. Mol Cell Biol 18(1):388–399

Pan D, Nakatsu T, Kato H (2013) Crystal structure of peroxisomal terget signal-2 bound to its receptor complex Pex7p–Pex21p. Nat Struct Mol Biol 20(8):987–993. https://doi.org/10.1038/nsmb.2618. nsmb.2618 [pii]

Petriv OI, Tang L, Titorenko VI, Rachubinski RA (2004) A new definition for the consensus sequence of the peroxisome targeting signal type 2. J Mol Biol 341(1):119–134. https://doi.org/10.1016/j.jmb.2004.05.064

Pinto MP, Grou CP, Fransen M, Sa-Miranda C, Azevedo JE (2009) The cytosolic domain of PEX3, a protein involved in the biogenesis of peroxisomes, binds membrane lipids. Biochim Biophys Acta 1793(11):1669–1675. https://doi.org/10.1016/j.bbamcr.2009.08.007. S0167-4889(09)00211-0 [pii]

Pires JR, Hong X, Brockmann C, Volkmer-Engert R, Schneider-Mergener J, Oschkinat H, Erdmann R (2003) The ScPex13p SH3 domain exposes two distinct binding sites for Pex5p and Pex14p. J Mol Biol 326(5):1427–1435

Platta HW, Erdmann R (2007) Peroxisomal dynamics. Trends Cell Biol 17(10):474–484. https://doi.org/10.1016/j.tcb.2007.06.009

Purdue PE, Zhang JW, Skoneczny M, Lazarow PB (1997) Rhizomelic chondrodysplasia punctata is caused by deficiency of human PEX7, a homologue of the yeast PTS2 receptor. Nat Genet 15(4):381–384. https://doi.org/10.1038/ng0497-381

Purdue PE, Yang X, Lazarow PB (1998) Pex18p and Pex21p, a novel pair of related peroxins essential for peroxisomal targeting by the PTS2 pathway. J Cell Biol 143(7):1859–1869

Rehling P, Marzioch M, Niesen F, Wittke E, Veenhuis M, Kunau WH (1996) The import receptor for the peroxisomal targeting signal 2 (PTS2) in Saccharomyces cerevisiae is encoded by the PAS7 gene. EMBO J 15(12):2901–2913

Ripstein ZA, Huang R, Augustyniak R, Kay LE, Rubinstein JL (2017) Structure of a AAA+ unfoldase in the process of unfolding substrate. eLife 6. https://doi.org/10.7554/eLife.25754

Saffert P, Enenkel C, Wendler P (2017) Structure and function of p97 and Pex1/6 type II AAA+ complexes. Front Mol Biosci 4:33. https://doi.org/10.3389/fmolb.2017.00033

Saitoh T, Igura M, Obita T, Ose T, Kojima R, Maenaka K, Endo T, Kohda D (2007) Tom20 recognizes mitochondrial presequences through dynamic equilibrium among multiple bound states. EMBO J 26(22):4777–4787. https://doi.org/10.1038/sj.emboj.7601888. 7601888 [pii]

Sampathkumar P, Roach C, Michels PA, Hol WG (2008) Structural insights into the recognition of peroxisomal targeting signal 1 by Trypanosoma brucei peroxin 5. J Mol Biol 381(4):867–880. https://doi.org/10.1016/j.jmb.2008.05.089. S0022-2836(08)00760-2 [pii]

Sato Y, Shibata H, Nakano H, Matsuzono Y, Kashiwayama Y, Kobayashi Y, Fujiki Y, Imanaka T, Kato H (2008) Characterization of the interaction between recombinant human peroxin Pex3p and Pex19p: identification of TRP-104 IN Pex3p as a critical residue for the interaction. J Biol Chem 283(10):6136–6144. https://doi.org/10.1074/jbc.M706139200

Sato Y, Shibata H, Nakatsu T, Nakano H, Kashiwayama Y, Imanaka T, Kato H (2010) Structural basis for docking of peroxisomal membrane protein carrier Pex19p onto its receptor Pex3p. EMBO J 29(24):4083–4093. https://doi.org/10.1038/emboj.2010.293

Schliebs W, Kunau WH (2006) PTS2 co-receptors: diverse proteins with common features. Biochim Biophys Acta 1763(12):1605–1612. https://doi.org/10.1016/j.bbamcr.2006.08.051

Schmidt F, Treiber N, Zocher G, Bjelic S, Steinmetz MO, Kalbacher H, Stehle T, Dodt G (2010) Insights into peroxisome function from the structure of PEX3 in complex with a soluble fragment of PEX19. J Biol Chem 285(33):25410–25417. https://doi.org/10.1074/jbc.M110.138503. M110.138503 [pii]

Schueller N, Holton SJ, Fodor K, Milewski M, Konarev P, Stanley WA, Wolf J, Erdmann R, Schliebs W, Song YH, Wilmanns M (2010) The peroxisomal receptor Pex19p forms a helical mPTS recognition domain. EMBO J 29(15):2491–2500. https://doi.org/10.1038/emboj.2010.115. emboj2010115 [pii]

Schuller JM, Beck F, Lossl P, Heck AJ, Forster F (2016) Nucleotide-dependent conformational changes of the AAA+ ATPase p97 revisited. FEBS Lett 590(5):595–604. https://doi.org/10.1002/1873-3468.12091

Seibert MM, Ekeberg T, Maia FR, Svenda M, Andreasson J, Jonsson O, Odic D, Iwan B, Rocker A, Westphal D, Hantke M, DePonte DP, Barty A, Schulz J, Gumprecht L, Coppola N, Aquila A, Liang M, White TA, Martin A, Caleman C, Stern S, Abergel C, Seltzer V, Claverie JM, Bostedt C, Bozek JD, Boutet S, Miahnahri AA, Messerschmidt M, Krzywinski J, Williams G, Hodgson KO, Bogan MJ, Hampton CY, Sierra RG, Starodub D, Andersson I, Bajt S, Barthelmess M, Spence JC, Fromme P, Weierstall U, Kirian R, Hunter M, Doak RB, Marchesini S, Hau-Riege SP, Frank M, Shoeman RL, Lomb L, Epp SW, Hartmann R, Rolles D, Rudenko A, Schmidt C, Foucar L, Kimmel N, Holl P, Rudek B, Erk B, Homke A, Reich C, Pietschner D, Weidenspointner G, Struder L, Hauser G, Gorke H, Ullrich J, Schlichting I, Herrmann S, Schaller G, Schopper F, Soltau H, Kuhnel KU, Andritschke R, Schroter CD, Krasniqi F, Bott M, Schorb S, Rupp D, Adolph M, Gorkhover T, Hirsemann H, Potdevin G, Graafsma H, Nilsson B, Chapman HN, Hajdu J (2011) Single mimivirus particles intercepted and imaged with an X-ray laser. Nature 470(7332):78–81. https://doi.org/10.1038/nature09748

Shibata H, Kashiwayama Y, Imanaka T, Kato H (2004) Domain architecture and activity of human Pex19p, a chaperone-like protein for intracellular trafficking of peroxisomal membrane proteins. J Biol Chem 279(37):38486–38494. https://doi.org/10.1074/jbc.M402204200. M402204200 [pii]

Shimozawa N, Suzuki Y, Zhang Z, Imamura A, Ghaedi K, Fujiki Y, Kondo N (2000) Identification of PEX3 as the gene mutated in a Zellweger syndrome patient lacking peroxisomal remnant structures. Hum Mol Genet 9(13):1995–1999

Shiozawa K, Maita N, Tomii K, Seto A, Goda N, Akiyama Y, Shimizu T, Shirakawa M, Hiroaki H (2004) Structure of the N-terminal domain of PEX1 AAA-ATPase. Characterization of a putative adaptor-binding domain. J Biol Chem 279(48):50060–50068. https://doi.org/10.1074/jbc.M407837200

Sichting M, Schell-Steven A, Prokisch H, Erdmann R, Rottensteiner H (2003) Pex7p and Pex20p of Neurospora crassa function together in PTS2-dependent protein import into peroxisomes. Mol Biol Cell 14(2):810–821. https://doi.org/10.1091/mbc.E02-08-0539

Smith JJ, Aitchison JD (2013) Peroxisomes take shape. Nat Rev Mol Cell Biol 14(12):803–817. https://doi.org/10.1038/nrm3700

Smith TF, Gaitatzes C, Saxena K, Neer EJ (1999) The WD repeat: a common architecture for diverse functions. Trends Biochem Sci 24(5):181–185

Soukupova M, Sprenger C, Gorgas K, Kunau WH, Dodt G (1999) Identification and characterization of the human peroxin PEX3. Eur J Cell Biol 78(6):357–374

Stanley WA, Filipp FV, Kursula P, Schuller N, Erdmann R, Schliebs W, Sattler M, Wilmanns M (2006) Recognition of a functional peroxisome type 1 target by the dynamic import receptor pex5p. Mol Cell 24(5):653–663. https://doi.org/10.1016/j.molcel.2006.10.024. S1097-2765(06)00730-1 [pii]

Stanley WA, Pursiainen NV, Garman EF, Juffer AH, Wilmanns M, Kursula P (2007) A previously unobserved conformation for the human Pex5p receptor suggests roles for intrinsic flexibility and rigid domain motions in ligand binding. BMC Struct Biol 7:24. https://doi.org/10.1186/1472-6807-7-24. 1472-6807-7-24 [pii]

Su JR, Takeda K, Tamura S, Fujiki Y, Miki K (2009) Crystal structure of the conserved N-terminal domain of the peroxisomal matrix protein import receptor, Pex14p. Proc Natl Acad Sci U S A 106(2):417–421. https://doi.org/10.1073/pnas.0808681106. 0808681106 [pii]

Swinkels BW, Gould SJ, Bodnar AG, Rachubinski RA, Subramani S (1991) A novel, cleavable peroxisomal targeting signal at the amino-terminus of the rat 3-ketoacyl-CoA thiolase. EMBO J 10(11):3255–3262

van der Zand A, Braakman I, Tabak HF (2010) Peroxisomal membrane proteins insert into the endoplasmic reticulum. Mol Biol Cell 21(12):2057–2065. https://doi.org/10.1091/mbc.E10-02-0082. E10-02-0082 [pii]

Vonck J, Parcej DN, Mills DJ (2016) Structure of alcohol oxidase from Pichia pastoris by cryo-electron microscopy. PLoS One 11(7):e0159476. https://doi.org/10.1371/journal.pone.0159476

Wanders RJ, Waterham HR (2006) Biochemistry of mammalian peroxisomes revisited. Annu Rev Biochem 75:295–332. https://doi.org/10.1146/annurev.biochem.74.082803.133329

Watanabe Y, Kawaguchi K, Okuyama N, Sugawara Y, Obita T, Mizuguchi M, Morita M, Imanaka T (2016) Characterization of the interaction between Trypanosoma brucei Pex5p and its receptor Pex14p. FEBS Lett 590(2):242–250. https://doi.org/10.1002/1873-3468.12044

Wiedemann N, van der Laan M, Pfanner N (2009) SnapShot: import and sorting of mitochondrial proteins. Cell 138(4):808–808.e801. https://doi.org/10.1016/j.cell.2009.08.004. S0092-8674(09)00966-0 [pii]

Williams C, van den Berg M, Panjikar S, Stanley WA, Distel B, Wilmanns M (2012) Insights into ubiquitin-conjugating enzyme/co-activator interactions from the structure of the Pex4p:Pex22p complex. EMBO J 31(2):391–402. https://doi.org/10.1038/emboj.2011.411. emboj2011411 [pii]

Williams C, van den Berg M, Stanley WA, Wilmanns M, Distel B (2013) A disulphide bond in the E2 enzyme Pex4p modulates ubiquitin-conjugating activity. Sci Rep 3:2212. https://doi.org/10.1038/srep02212. srep02212 [pii]

Yamashita K, Pan D, Okuda T, Sugahara M, Kodan A, Yamaguchi T, Murai T, Gomi K, Kajiyama N, Mizohata E, Suzuki M, Nango E, Tono K, Joti Y, Kameshima T, Park J, Song C, Hatsui T, Yabashi M, Iwata S, Kato H, Ago H, Yamamoto M, Nakatsu T (2015) An isomorphous replacement method for efficient de novo phasing for serial femtosecond crystallography. Sci Rep 5:14017. https://doi.org/10.1038/srep14017

Yamashita K, Kuwabara N, Nakane T, Murai T, Mizohata E, Sugahara M, Pan D, Masuda T, Suzuki M, Sato T, Kodan A, Yamaguchi T, Nango E, Tanaka T, Tono K, Joti Y, Kameshima T, Hatsui T, Yabashi M, Manya H, Endo T, Kato R, Senda T, Kato H, Iwata S, Ago H, Yamamoto M, Yumoto F, Nakatsu T (2017) Experimental phase determination with selenomethionine or mercury-derivatization in serial femtosecond crystallography. IUCrJ 4(Pt 5):639–647. https://doi.org/10.1107/s2052252517008557

Yofe I, Soliman K, Chuartzman SG, Morgan B, Weill U, Yifrach E, Dick TP, Cooper SJ, Ejsing CS, Schuldiner M, Zalckvar E, Thoms S (2017) Pex35 is a regulator of peroxisome abundance. J Cell Sci 130(4):791–804. https://doi.org/10.1242/jcs.187914

Young JC, Hoogenraad NJ, Hartl FU (2003) Molecular chaperones Hsp90 and Hsp70 deliver preproteins to the mitochondrial import receptor Tom70. Cell 112(1):41–50. S0092867402012503 [pii]

Zhang JW, Lazarow PB (1995) PEB1 (PAS7) in Saccharomyces cerevisiae encodes a hydrophilic, intra-peroxisomal protein that is a member of the WD repeat family and is essential for the import of thiolase into peroxisomes. J Cell Biol 129(1):65–80

Chapter 11
Lipidomics of Peroxisomal Disorders

Kotaro Hama, Yuko Fujiwara, and Kazuaki Yokoyama

Abstract Peroxisomes are cytoplasmic organelles that play a key role in metabolic and synthetic processes of various lipids. Peroxisomal disorders are caused by defects in peroxisome biogenesis or a peroxisomal single enzyme, and lead to multiple pathological features such as enlarged ventricles, demyelination, hearing loss and psychomotor retardation. Specific lipid profiles are observed for each peroxisomal disorder, which reflects the abnormal process in metabolism or synthesis of lipids. Therefore, lipid analysis of biological samples (e.g. blood plasma and fibroblasts) from patients is essential for the precise diagnosis and identification of responsible genes of peroxisomal disorders. Recent advances in mass spectrometry have enabled both identification and simultaneous quantification of a large number of lipid species. In this review, we introduce the principles of 'lipidomics' and the latest research using the lipidomic approach to study peroxisomal disorders.

Keywords Lipidomics · LC-ESI-MS · PBD · X-ALD

Abbreviations

ABCD1	ATP-binding cassette sub-family D1
AMACR	A-methylacyl-CoA racemase
AOX	Acyl-CoA oxidase
DBP	D-bifunctional protein
DHCA	Dihydroxycholestanoic acid
ESI	Electrospray ionization
GC-MS	Gas chromatography–mass spectrometry
IMS	Imaging mass spectrometry
IRD	Infantile Refsum's disease
LC	Liquid chromatography
MRM	Multiple reaction monitoring
MS	Mass spectrometry
MS/MS	Tandem mass

K. Hama (✉) · Y. Fujiwara · K. Yokoyama
Faculty of Pharma-Sciences, Teikyo University, Tokyo, Japan
e-mail: khama@pharm.teikyo-u.ac.jp; yfujiwar@pharm.teikyo-u.ac.jp; yokoyama@pharm.teikyo-u.ac.jp

T. Imanaka, N. Shimozawa (eds.), *Peroxisomes: Biogenesis, Function, and Role in Human Disease*, https://doi.org/10.1007/978-981-15-1169-1_11

NALD	Neonatal adrenoleukodystrophy
PBD	Peroxisome biogenesis disorders
PC	Phosphatidylcholine
PE	Phosphatidylethanolamine
Q	Quadrupole
RCDP	Rhizomelic chondrodysplasia punctata
SCPx	Sterol carrier protein X
THCA	Trihydroxycholestanoic acid
TOF	Time-of-Flight
VLCFA	Very long chain fatty acids
X-ALD	X-linked adrenoleukodystrophy
ZS	Zellweger syndrome

11.1 Introduction

Peroxisomes are subcellular organelles with a single phospholipid bilayer and are involved in various biological processes including lipid metabolism. In particular, peroxisomes contain a number of enzymes essential for β-oxidation of very long chain fatty acids (VLCFA) and bile acids, α-oxidation of phytanic acid, and the synthesis of plasmalogen and polyunsaturated fatty acids. Peroxisomal disorders are inherited defects that affect metabolic and synthetic processes and are classified primarily into two categories: (a) peroxisome biogenesis disorders (PBD) caused by mutations in peroxin (PEX) genes that are essential for the assembly of peroxisomes; and (b) disorders caused by defects in single enzymes or transporters in peroxisomes. Because most peroxisomal disorders display specific lipid profiles that reflect the abnormal metabolic or synthetic processes of lipids in peroxisomes, lipid analysis represents an accurate and rapid diagnostic approach to identify peroxisomal disorders, which is essential for appropriate and effective treatment. In addition, elucidating responsible lipids involved in the development of the specific peroxisomal disorders should aid the development of effective therapeutic strategies.

Lipidomics is the large-scale study of lipid metabolic pathways, such as synthesis, metabolism and transport in biological systems. The "Lipidome" describes the quality and quantity of total lipids in biological samples, and lipidomics uses the principles and techniques of analytical chemistry to comprehensively analyze whole lipid species in cell or biological samples. Among several techniques, mass spectrometry (MS) employing soft ionization techniques is one of the most fascinating methods used for current lipidomics.

In this review, we first present lipid species involved in peroxisomal disorders. We then introduce the principles of lipidomic analysis using MS and explain several lipidomic approaches conducted in the study of PBD and X-linked adrenoleukodystrophy (X-ALD).

11.2 Lipid Species Affected in Peroxisomal Disorders

11.2.1 Zellweger Syndrome (ZS), Neonatal Adrenoleukodystrophy (NALD) and Infantile Refsum's Disease (IRD)

ZS, NALD and IRD have similar symptoms and are categorized as Zellweger spectrum disorders. The formation of a functional peroxisome can be disrupted because of dysfunctional PEX genes, which results in aberrant metabolism of various lipid species. In particular, excessive accumulation of VLCFA, branched fatty acids and intermediate metabolites of bile acid occur and quantitation of these metabolites is clinically important for diagnosing a Zellweger spectrum disorder (Suzuki et al. 2001). The amount of each VLCFA species, branched fatty acids and intermediate metabolites of bile acid can be analyzed by gas chromatography (GC) or gas chromatography-mass spectrometry (GC-MS), and are reliable techniques for diagnosing PBD and other peroxisomal disorders (Takemoto et al. 2003). For example, Moser et al. carried out an intensive analysis of fatty acid species using GC and revealed that VLCFA species such as cerotic acid (C26:0 FA) (Fig. 11.1a) were significantly elevated in plasma and red blood cells of patients with PBD disorders (Moser et al. 1999). Schutgens et al. used GC-MS and revealed that age-dependent accumulation of phytanic acid (Fig. 11.1b) was observed in fibroblasts from patients with ZS (Schutgens et al. 1987). Excessive amounts trihydroxycholestanoic acid (THCA) (Fig. 11.1d) and dihydroxycholestanoic acid (DHCA) (Fig. 11.1e) were also observed in urine and serum from patients with ZS (Hanson et al. 1979; Parmentier et al. 1979).

11.2.2 X-ALD, Acyl-CoA Oxidase (AOX) Deficiency, D-bifunctional Protein (DBP) Deficiency and Acyl-CoA Binding Domain-Containing 5 (ACBD5) Deficiency

VLCFA are transported via ATP-binding cassette sub-family D1 (ABCD1) and degraded by β-oxidation catalyzed by AOX and DBP in the peroxisome. ABCD1, acyl-CoA oxidase 1a and DBP are the responsible genes of X-ALD, AOX deficiency and DBP deficiency, respectively, and VLCFA such as cerotic acid (hexacosanoic acid, C26:0 FA) (Fig. 11.1a) are abnormally accumulated in patients with these disorders (Mosser et al. 1993). In addition, recent observation that ACBD5 deficiency leads to accumulation of VLCFA indicates the important role of ACBD5 in peroxisomal lipid metabolism (Ferdinandusse et al. 2017; Yagita et al. 2017). The precise quantitation of VLCFA in samples such as blood plasma and fibroblasts is essential for the diagnosis of X-ALD (Kemp et al. 2016). Recently C26:0-lysophosphatidylcholine (lysoPC) (Fig. 11.1f) was identified as a candidate marker for diagnosis of X-ALD (Hubbard et al. 2006). C26:0-lysoPC is analyzed by

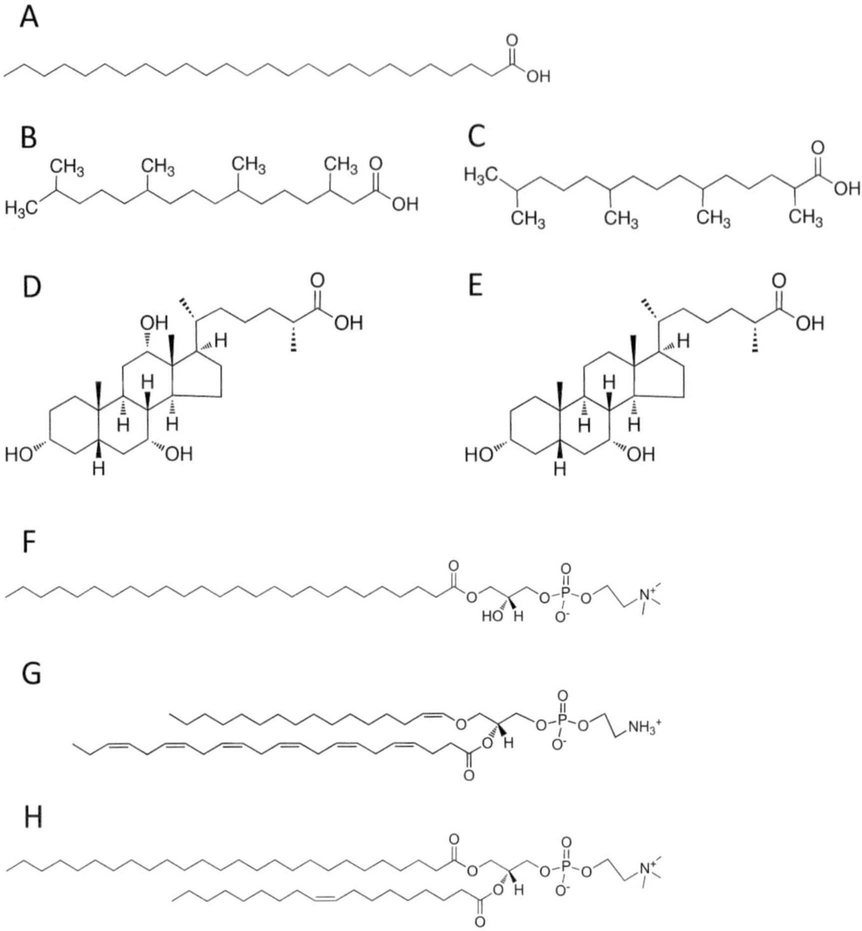

Fig. 11.1 The structure of lipid species affected in peroxisomal disorders. (**a**) The structure of C26:0 FA. VLCFA including C26:0 FA degraded mainly in peroxisome but not in mitochondria, and thus accumulated in a number of peroxisomal disorders. The structures of phytanic acid (**b**) and pristanic acid (**c**). Phytanic acid accumulates in patients with RCDP, SCPx deficiency and Refsum disease. Phytanic acid and pristanic acid accumulate in age- and diet-dependent manners. The structure of THCA (**d**) and DHCA (**e**). THCA and DHCA accumulate in patients with ZS, NALD, IRD, DBP deficiency, SCPx deficiency and AMACR deficiency. (**f**) The structure of C26:0-lysoPC. C26:0-lysoPC levels are conventionally analyzed in dried blood spot samples. The accumulation of C26:0-lysoPC can be a sensitive biomarker for peroxisomal disorders including X-ALD. (**g**) The structure of plasmalogen with docosahexaenoic acid (PlsEtn (16:0/22:6)). Plasmalogen species with oleic acid, linoleic acid or arachidonic acid linked to the *sn*-2 position of the glycerol backbone are also observed. (**h**) The structure of PC (26:0/18:1). Most VLCFA moieties are linked to the *sn*-1 position of the glycerol backbone in each phospholipid species

liquid chromatography-electrospray ionization-MS (LC-ESI-MS), in which a derivatization process is not required for sample preparation in LC-ESI-MS. Thus, the analysis of C26:0-lysoPC in dried blood spot samples is suitable for high-throughput screening of X-ALD (Hubbard et al. 2009; Theda et al. 2014). Screening of newborns for X-ALD is now conducted in several USA states and has successfully identified novel linages of X-ALD (Vogel et al. 2015). Intriguingly, it was found that C26:0-lysoPC was unexpectedly elevated in a sample from a newborn with Aicardi–Goutieres syndrome that was originally screened for X-ALD (Armangue et al. 2017), enhancing the significance of newborn screening using LC-ESI-MS for identifying novel diseases related to specific metabolic processes.

11.2.3 Rhizomelic Chondrodysplasia Punctata (RCDP) Type-1, RCDP Type-2 and RCDP Type-3

RCDP displays systemic shortening of proximal bones, congenital cataracts, distinctive facial features and intellectual disabilities, and is classified into three types according to the responsible gene: PEX7, dihydroxyacetone phosphate (DHAP) acyltransferase and alkyl-DHAP synthase cause RCDP type-1, type-2 and type-3, respectively (Braverman et al. 1997; Ofman et al. 1998; de Vet et al. 1998). RCDP is biochemically characterized by the aberrant synthesis of plasmalogen (e.g. PlsEtn (16:0/22:6)) (Fig. 11.1g). The total amount of plasmalogen can be analyzed by GC-MS and used as a diagnostic marker of RCDP (Takemoto et al. 2003). Recently, each plasmalogen species was analyzed by ESI-LC-MS and it was found that a specific ethanolamine plasmalogen species (PlsEtn (16:0/22:6)) was replete by the administration of a plasmalogen precursor analog (PPI-1011) (Wood et al. 2011).

11.2.4 Refsum Disease, Sterol Carrier Protein X (SCPx) Deficiency and A-Methylacyl-CoA Racemase (AMACR) Deficiency

Branched chain fatty acids are obtained mainly through diet and are degraded through α-oxidation in the peroxisome. Dysfunctional activity of SCPx, AMACR and phytanoyl-CoA hydroxylase are responsible for SCPx deficiency, AMACR deficiency and Refsum disease, respectively, and are involved in the degradation of branched chain fatty acids (Shimozawa 2007). SCPx and AMACR are also involved in the metabolism of bile acids (Shimozawa 2007). Thus, the precise quantitation of both branched chain fatty acids and intermediate metabolites of bile acids is important for the diagnosis of these three diseases. Phytanic acid and pristanic acid (Fig. 11.1c), as branched chain fatty acids, and THCA (Fig. 11.1d) and DHCA (Fig. 11.1e), as intermediate metabolites of bile acids, have been analyzed by GC-MS (Takemoto et al. 2003) and LC-MS (Ferdinandusse et al. 2006; Ferdinandusse et al.

2001). Moreover, lysoPC species with branched chain fatty acids have been identified as unique phospholipid species in plasma from patients with Refsum disease and AMACR deficiency in a recent lipidomic analysis (Herzog et al. 2018).

11.3 Recent Advances in MS-Based Methods for Lipid Analysis

MS-based lipidomic methods are listed in Table 11.1. Soft ionization techniques as innovative progress in MS and computer-based analysis that integrate metabolite databases have enabled lipidomic analysis that target a large number of lipid species. Electrospray ionization-MS (ESI-MS) coupled to high performance liquid chromatography systems (LC-ESI-MS) is the approach that is frequently used for lipid analysis. In LC-ESI-MS analysis, lipid species are separated by their physical properties in LC and then analyzed according to the molecular size of the ions (*m/z*) in ESI-MS. Imaging mass spectrometry (IMS) is a new MS technology and can be applied to lipidomics. In IMS, ions are generated in a very small area of the samples (e.g. organ sections) and analyzed by MS. By scanning the surface of the samples, the spatial distribution of compounds of interest is visualized. Reports using IMS in lipidomics and metabolomics continue to increase in number (Sugiura et al. 2009; Sun et al. 2018; Goto-Inoue et al. 2011). Other MS-based techniques such as ion mobility-mass spectrometry and SWATH (sequential window acquisition of all theoretical fragment-ion spectra) technology have also been developed recently and are applied in lipidomics (Cajka and Fiehn 2016).

Two strategies are conducted for lipidomics. The first strategy is targeted lipidomics, in which defined lipid classes are analyzed, whereas the other strategy is untargeted lipidomics, which focuses on the analysis of all detectable compounds including unknown chemicals in samples. In targeted lipidomics, multiple reaction monitoring (MRM) (also termed selected reaction monitoring, SRM) is frequently applied using triple quadrupole MS. The precursor ion of interest is selected in the first quadrupole (Q1) and processed to collision-activated dissociation in the second quadrupole (Q2). Then, one of the fragment ions generated in the Q2, which has characteristics to the precursor ion of interest, is selected in the third quadrupole (Q3) and detected by an ion detector (Fig. 11.2a, b). Metabolites including lipids were intensively analyzed by the precursor ion scan mode or neutral loss scan mode previously (Taguchi et al. 2005). Currently, very short dwell times (1–5 ms) can be applied in a triple quadrupole mass spectrometer and therefore each lipid species can be analyzed intensively with high sensitivity by monitoring multiple Q1/Q3 MRM transition channels. In contrast, untargeted lipidomics is usually conducted by high-resolution MS such as Time-of-Flight (TOF) MS (Fig. 11.2c), orbital ion trap MS and Fourier transform ion cyclotron resonance MS. The high-resolution properties of these instruments resolve co-eluting isobaric compounds, leading to both quantitative analysis with high sensitivity and determination of the composition formula of precursor ions. Each lipid species corresponding to the precursor ion analyzed is assigned by integrating the

Table 11.1 MS-based lipidomic methods

Methods	Principles	Advantages (+) and Disadvantages (−)
GC-MS	Each substance in test samples is separated by gas-chromatography, ionized and then analyzed by MS	+ Powerful method to identify the precise structure of lipids such as the location of double bonds in unsaturated fatty acids + Several methods for ionization are available according to the purpose of the analysis and the properties of substances of interest − Derivatization is required to analyze nonvolatile substances
LC-MS	Each substance in test samples is separated by LC before MS analysis	+ A number of substances of interest can be analyzed simultaneously. Thus, this approach is the most popular method for current lipidomics + Used for both quantitative and qualitative (structural) analysis − Derivatization is required to analyze substances with low ionization efficiency
Direct infusion-MS (shotgun)	Test samples are processed directly to ionization without separation by LC and analyzed by MS	+ The period required for each analysis is short when compared with that of LC-MS and thus suitable for high throughput analysis − The ionization efficiency of the substances of interest can be affected by ion suppression or ion enhancement due to the coexistence of other substances
MALDI-MS	Test samples are mixed with the matrix, ionized by the laser and then analyzed by MS	+ Useful to analyze high molecular compounds + The period required for each analysis is short and thus suitable for high throughput analysis
Imaging MS	Sections of tissues or organs on glass slides are sprayed by a matrix, ionized by the laser and then analyzed by MS	+ Precise distributions of target substances are visualized − Quantitative ranges are limited − Longer periods are required for each analysis. Thus, this method is not suitable for high throughput analysis
Ion-mobility spectrometry-MS	Before MS each substance is separated by ion-mobility spectrometry through interaction between charged molecules and the buffer gas	+ Substances can be separated according to differences in physical properties such as structure, chirality and hydrophobicity − Database for comprehensive analysis is under-developed

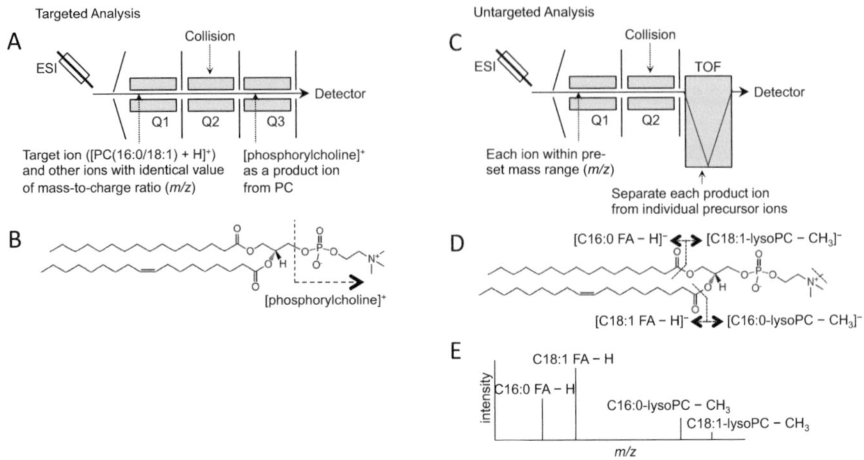

Fig. 11.2 The schematic diagram of targeted and untargeted analysis. The schematic diagram of ESI-triple quadrupole mass spectrometer (**a**) and quadrupole time-of-flight mass spectrometry (**c**) are presented as representative mass spectrometry instruments used for targeted and untargeted analysis, respectively. In targeted analysis, the precursor ions of target molecules generated at ESI source were selected in Q1, and processed to collision activated dissociation in the second quadrupole (Q2). Then, one of the fragment ions generated in Q2 is selected in the Q3 and detected by ion detector. In the case of PC (16:0/18:1) as a target molecule, for example, the precursor ion ($[M + H]^+$ (*m/z* 760)) and the product ion (phosphorylcholine (*m/z* 184)) is selected in Q1 and Q3, respectively (**b**) (Hama et al. 2018). In untargeted analysis, each precursor ion within the pre-set mass range is scanned with unbiased manner for quantitation. Simultaneously, all product ions from each precursor ion are scanned in TOF system and are processed for the assignment of each precursor ion (**c**). For example, product ions corresponding to fatty-acyl moieties and demethylated lysoPC species are generally produced from PC species (**d**). Thus the precursor ion detected with unbiased manner is assigned as PC (16:0/18:1) if the product ions corresponding to C16:0 FA, C18:1 FA, demethylated C16:0-lysoPC and demethylated C18:1-lysoPC are observed (**e**)

fragment ion spectra through data-dependent or data-independent tandem mass (MS/MS) analysis (Fig. 11.2d, e). In untargeted lipidomics, novel lipids or metabolic processes that are affected by genes or diseases of interest are identified. For example, fatty acid esters of hydroxyl fatty acids have been identified as novel biomarkers and as a therapeutic target against type 2 diabetes (Yore et al. 2014).

11.4 Application of MS-Based Lipidomics in Peroxisomal Disorders

Most fatty acids are not present in the free form but are associated with complex lipids such as phospholipids in biological samples (Moser et al. 2001), whereas fatty acyl moieties are liberated from complex lipids during the derivatizing process of sample preparation for GC or GC-MS, which leads to loss of information about the endogenous lipid species. In LC-MS, by contrast, a number of lipid species including phospholipids and neutral lipids can be analyzed directly without derivatization.

Thus, LC-MS is useful for analyzing the endogenous form of complex lipid species and the free fatty acids, and is frequently used in recent lipidomic analyses (Taguchi et al. 2005; Takashima et al. 2017). Our recent lipidomic analysis revealed that a large number of VLCFA-containing molecular species were detected in several phospholipid classes (PC, phosphatidylethanolamine (PE) and phosphatidylserine) as well as neutral lipids, including triacylglycerol and cholesteryl esters in fibroblasts from patients with ZS (Hama et al. 2013). In addition to C26:0 FA, MS/MS analysis revealed the presence of a variety of VLCFA moieties, including polyunsaturated VLCFA; the number of carbons and double bonds in VLCFA moieties range from 23 to 32 and 0 to 6, respectively (Hama et al. 2013). Interestingly, most VLCFA moieties are linked to the *sn*-1 position of the glycerol backbone in each phospholipid species such as PC (26:0/18:1) (Fig. 11.1h), providing insight into the machinery used for the synthesis of phospholipids with VLCFA moieties (Hama et al. 2013). Abe et al. conducted target lipidomic analysis of fibroblasts from ZS, AOX deficiency, DBP deficiency and X-ALD samples (Abe et al. 2014). This analysis found that plasmenyl ethanolamine species were reduced while PE species were elevated in fibroblasts from ZS (Abe et al. 2014). They also found that phospholipid species with docosahexaenoic acid were reduced, while PC species with polyunsaturated VLCFA moieties accumulate in fibroblasts from ZS, AOX deficiency and DBP deficiency samples (Abe et al. 2014). Interestingly, recent untargeted lipidomics offer the possibility to predict the severity of peroxisomal disorders. For example, Herzog et al. used high-resolution MS and showed that a set of specific phospholipid ratios correlate with the severity of patients with a Zellweger spectrum disorder (Herzog et al. 2016). In addition, Lee et al. proposed that ceramide and triacylglycerol species can function as distinct biomarkers for phenotypic variations and the severity of X-ALD (Lee et al. 2019). Lipidomic analysis was also conducted for the study of *Abcd1*-deficient mice as model animals for X-ALD (Hama et al. 2018). We developed a method composed of several hundreds of MRM transition channels and intensively analyzed each phospholipid species in the brain of *Abcd1*-deficient mice, revealing that the number of carbons and double bonds in VLCFA moieties ranged from 23 to 34 and mostly 0 to 2, respectively. As observed for results examining fibroblasts from patients with ZS, most VLCFA moieties are linked to the *sn*-1 position of the glycerol backbone in each phospholipid species such as PC (26:0/18:1) (Fig. 11.1h) in the brain of *Abcd1*-deficient mice, indicating that specific enzymes such as acyltransferases are involved in the synthesis of phospholipids with VLCFA moieties (Hama et al. 2018; Hama et al. 2013).

11.5 Concluding Remarks

In this review, we introduced concepts and the general procedure of lipidomics, and provided a general overview of recent lipidomic analyses using LC-MS in the field of peroxisomal disorders. Identification of the precise machinery responsible for the progression of peroxisomal disorders remains challenging because of the lack of appropriate mice models (e.g. X-ALD) and difficulties associated with obtaining samples from lesions (e.g. human brain). Nonetheless, several novel metabolites such as oxysterols and oxidized fatty acids have been analyzed by LC-MS and are identified as possible markers for X-ALD (Jang et al. 2016; Nury et al. 2017). In addition, technical developments in MS have underpinned the development of new MS-based lipidomic methods to analyze peroxisomal disorders. For example, imaging mass spectrometry was used to investigate X-ALD; we found that very long chain fatty acid-containing PC (PC (26:0/18:1), Fig. 11.1h) was enriched in the cerebral cortex and the molecular layer in the cerebellum but not in white matter or the cerebellum medulla of brain tissue from *Abcd1*-deficient mice (Hama et al. 2018). Further analysis of brain samples from X-ALD patients using IMS instruments with higher sensitivity is necessary to clarify the precise distribution of multiple lipid species and for investigating correlations between specific lipid species and pathological features such as demyelination. In conclusion, expanding MS-based lipidomic analyses will provide intriguing insights that aid our understanding of the pathogenesis of peroxisomal disorders and the development of novel biomarkers. Such advances should lead to effective therapy of peroxisomal diseases.

Conflict of Interest The authors declare that they have no conflict of interest.

References

Abe Y, Honsho M, Nakanishi H, Taguchi R, Fujiki Y (2014) Very-long-chain polyunsaturated fatty acids accumulate in phosphatidylcholine of fibroblasts from patients with Zellweger syndrome and acyl-CoA oxidase1 deficiency. Biochim Biophys Acta 1841:610–619

Armangue T, Orsini JJ, Takanohashi A, Gavazzi F, Conant A, Ulrick N, Morrissey MA, Nahhas N, Helman G, Gordish-Dressman H, Orcesi S, Tonduti D, Stutterd C, van Haren K, Toro C, Iglesias AD, van der Knaap MS, Goldbach Mansky R, Moser AB, Jones RO, Vanderver A (2017) Neonatal detection of Aicardi Goutieres syndrome by increased C26:0 lysophosphatidylcholine and interferon signature on newborn screening blood spots. Mol Genet Metab 122(3):134–139

Braverman N, Steel G, Obie C, Moser A, Moser H, Gould SJ, Valle D (1997) Human PEX7 encodes the peroxisomal PTS2 receptor and is responsible for rhizomelic chondrodysplasia punctata. Nat Genet 15:369–376

Cajka T, Fiehn O (2016) Toward merging untargeted and targeted methods in mass spectrometry-based metabolomics and lipidomics. Anal Chem 88:524–545

de Vet EC, Ijlst L, Oostheim W, Wanders RJ, van den Bosch H (1998) Alkyl-dihydroxyacetonephosphate synthase. Fate in peroxisome biogenesis disorders and identification of the point mutation underlying a single enzyme deficiency. J Biol Chem 273:10296–10301

Ferdinandusse S, Overmars H, Denis S, Waterham HR, Wanders RJ, Vreken P (2001) Plasma analysis of di- and trihydroxycholestanoic acid diastereoisomers in peroxisomal alpha-methylacyl-CoA racemase deficiency. J Lipid Res 42:137–141

Ferdinandusse S, Kostopoulos P, Denis S, Rusch H, Overmars H, Dillmann U, Reith W, Haas D, Wanders RJ, Duran M, Marziniak M (2006) Mutations in the gene encoding peroxisomal sterol carrier protein X (SCPx) cause leukencephalopathy with dystonia and motor neuropathy. Am J Hum Genet 78:1046–1052

Ferdinandusse S, Falkenberg KD, Koster J, Mooyer PA, Jones R, van Roermund CWT, Pizzino A, Schrader M, Wanders RJA, Vanderver A, Waterham HR (2017) ACBD5 deficiency causes a defect in peroxisomal very long-chain fatty acid metabolism. J Med Genet 54:330–337

Goto-Inoue N, Hayasaka T, Zaima N, Setou M (2011) Imaging mass spectrometry for lipidomics. Biochim Biophys Acta 1811:961–969

Hama K, Nagai T, Nishizawa C, Ikeda K, Morita M, Satoh N, Nakanishi H, Imanaka T, Shimozawa N, Taguchi R, Inoue K, Yokoyama K (2013) Molecular species of phospholipids with very long chain fatty acids in skin fibroblasts of Zellweger syndrome. Lipids 48:1253–1267

Hama K, Fujiwara Y, Morita M, Yamazaki F, Nakashima Y, Takei S, Takashima S, Setou M, Shimozawa N, Imanaka T, Yokoyama K (2018) Profiling and imaging of phospholipids in brains of Abcd1-deficient mice. Lipids 53:85–102

Hanson RF, Szczepanik-VanLeeuwen P, Williams GC, Grabowski G, Sharp HL (1979) Defects of bile acid synthesis in Zellweger's syndrome. Science 203:1107–1108

Herzog K, Pras-Raves ML, Vervaart MA, Luyf AC, van Kampen AH, Wanders RJ, Waterham HR, Vaz FM (2016) Lipidomic analysis of fibroblasts from Zellweger spectrum disorder patients identifies disease-specific phospholipid ratios. J Lipid Res 57:1447–1454

Herzog K, Pras-Raves ML, Ferdinandusse S, Vervaart MAT, Luyf ACM, van Kampen AHC, Wanders RJA, Waterham HR, Vaz FM (2018) Plasma lipidomics as a diagnostic tool for peroxisomal disorders. J Inherit Metab Dis 41(3):489–498

Hubbard WC, Moser AB, Tortorelli S, Liu A, Jones D, Moser H (2006) Combined liquid chromatography-tandem mass spectrometry as an analytical method for high throughput screening for X-linked adrenoleukodystrophy and other peroxisomal disorders: preliminary findings. Mol Genet Metab 89:185–187

Hubbard WC, Moser AB, Liu AC, Jones RO, Steinberg SJ, Lorey F, Panny SR, Vogt RF Jr, Macaya D, Turgeon CT, Tortorelli S, Raymond GV (2009) Newborn screening for X-linked adrenoleukodystrophy (X-ALD): validation of a combined liquid chromatography-tandem mass spectrometric (LC-MS/MS) method. Mol Genet Metab 97:212–220

Jang J, Park S, Jin Hur H, Cho HJ, Hwang I, Pyo Kang Y, Im I, Lee H, Lee E, Yang W, Kang HC, Won Kwon S, Yu JW, Kim DW (2016) 25-hydroxycholesterol contributes to cerebral inflammation of X-linked adrenoleukodystrophy through activation of the NLRP3 inflammasome. Nat Commun 7:13129

Kemp S, Huffnagel IC, Linthorst GE, Wanders RJ, Engelen M (2016) Adrenoleukodystrophy - neuroendocrine pathogenesis and redefinition of natural history. Nat Rev Endocrinol 12:606–615

Lee DK, Long NP, Jung J, Kim TJ, Na E, Kang YP, Kwon SW, Jang J (2019) Integrative lipidomic and transcriptomic analysis of X-linked adrenoleukodystrophy reveals distinct lipidome signatures between adrenomyeloneuropathy and childhood cerebral adrenoleukodystrophy. Biochem Biophys Res Commun 508(2):563–569

Moser AB, Jones DS, Raymond GV, Moser HW (1999) Plasma and red blood cell fatty acids in peroxisomal disorders. Neurochem Res 24:187–197

Moser H, Smith KD, Watkins PA, Powers J, Moser AB (2001) X-linked adrenoleukodystrophy, the metabolic and molecular bases of inherited disease, vol II, 8th edn. McGraw-Hill, New York, pp 3257–3301

Mosser J, Douar AM, Sarde CO, Kioschis P, Feil R, Moser H, Poustka AM, Mandel JL, Aubourg P (1993) Putative X-linked adrenoleukodystrophy gene shares unexpected homology with ABC transporters. Nature 361:726–730

Nury T, Zarrouk A, Ragot K, Debbabi M, Riedinger JM, Vejux A, Aubourg P, Lizard G (2017) 7-Ketocholesterol is increased in the plasma of X-ALD patients and induces peroxisomal modifications in microglial cells: potential roles of 7-ketocholesterol in the pathophysiology of X-ALD. J Steroid Biochem Mol Biol 169:123–136

Ofman R, Hettema EH, Hogenhout EM, Caruso U, Muijsers AO, Wanders RJ (1998) Acyl-CoA:dihydroxyacetonephosphate acyltransferase: cloning of the human cDNA and resolution of the molecular basis in rhizomelic chondrodysplasia punctata type 2. Hum Mol Genet 7:847–853

Parmentier GG, Janssen GA, Eggermont EA, Eyssen HJ (1979) C27 bile acids in infants with coprostanic acidemia and occurrence of a 3 alpha,7 alpha,12 alpha-tridhydroxy-5 beta-C29 dicarboxylic bile acid as a major component in their serum. Eur J Biochem 102:173–183

Schutgens RB, Wanders RJ, Heymans HS, Schram AW, Tager JM, Schrakamp G, van den Bosch H (1987) Zellweger syndrome: biochemical procedures in diagnosis, prevention and treatment. J Inherit Metab Dis 10(Suppl 1):33–45

Shimozawa N (2007) Molecular and clinical aspects of peroxisomal diseases. J Inherit Metab Dis 30:193–197

Sugiura Y, Konishi Y, Zaima N, Kajihara S, Nakanishi H, Taguchi R, Setou M (2009) Visualization of the cell-selective distribution of PUFA-containing phosphatidylcholines in mouse brain by imaging mass spectrometry. J Lipid Res 50:1776–1788

Sun N, Wu Y, Nanba K, Sbiera S, Kircher S, Kunzke T, Aichler M, Berezowska S, Reibetanz J, Rainey WE, Fassnacht M, Walch A, Kroiss M (2018) High resolution tissue mass spectrometry imaging reveals a refined functional anatomy of the human adult adrenal gland. Endocrinology 159(3):1511–1524

Suzuki Y, Shimozawa N, Imamura A, Fukuda S, Zhang Z, Orii T, Kondo N (2001) Clinical, biochemical and genetic aspects and neuronal migration in peroxisome biogenesis disorders. J Inherit Metab Dis 24:151–165

Taguchi R, Houjou T, Nakanishi H, Yamazaki T, Ishida M, Imagawa M, Shimizu T (2005) Focused lipidomics by tandem mass spectrometry. J Chromatogr B Analyt Technol Biomed Life Sci 823:26–36

Takashima S, Toyoshi K, Itoh T, Kajiwara N, Honda A, Ohba A, Takemoto S, Yoshida S, Shimozawa N (2017) Detection of unusual very-long-chain fatty acid and ether lipid derivatives in the fibroblasts and plasma of patients with peroxisomal diseases using liquid chromatography-mass spectrometry. Mol Genet Metab 120(3):255–268

Takemoto Y, Suzuki Y, Horibe R, Shimozawa N, Wanders RJ, Kondo N (2003) Gas chromatography/mass spectrometry analysis of very long chain fatty acids, docosahexaenoic acid, phytanic acid and plasmalogen for the screening of peroxisomal disorders. Brain and Development 25:481–487

Theda C, Gibbons K, Defor TE, Donohue PK, Golden WC, Kline AD, Gulamali-Majid F, Panny SR, Hubbard WC, Jones RO, Liu AK, Moser AB, Raymond GV (2014) Newborn screening for X-linked adrenoleukodystrophy: further evidence high throughput screening is feasible. Mol Genet Metab 111:55–57

Vogel BH, Bradley SE, Adams DJ, D'Aco K, Erbe RW, Fong C, Iglesias A, Kronn D, Levy P, Morrissey M, Orsini J, Parton P, Pellegrino J, Saavedra-Matiz CA, Shur N, Wasserstein M, Raymond GV, Caggana M (2015) Newborn screening for X-linked adrenoleukodystrophy in New York State: diagnostic protocol, surveillance protocol and treatment guidelines. Mol Genet Metab 114:599–603

Wood PL, Khan MA, Smith T, Ehrmantraut G, Jin W, Cui W, Braverman NE, Goodenowe DB (2011) In vitro and in vivo plasmalogen replacement evaluations in rhizomelic chrondrodysplasia punctata and Pelizaeus-Merzbacher disease using PPI-1011, an ether lipid plasmalogen precursor. Lipids Health Dis 10:182

Yagita Y, Shinohara K, Abe Y, Nakagawa K, Al-Owain M, Alkuraya FS, Fujiki Y (2017) Deficiency of a retinal dystrophy protein, acyl-CoA binding domain-containing 5 (ACBD5), impairs peroxisomal beta-oxidation of very-long-chain fatty acids. J Biol Chem 292:691–705

Yore MM, Syed I, Moraes-Vieira PM, Zhang T, Herman MA, Homan EA, Patel RT, Lee J, Chen S, Peroni OD, Dhaneshwar AS, Hammarstedt A, Smith U, McGraw TE, Saghatelian A, Kahn BB (2014) Discovery of a class of endogenous mammalian lipids with anti-diabetic and anti-inflammatory effects. Cell 159:318–332

Chapter 12
Neurophysiology and Neuropsychology in Adrenoleukodystrophy (ALD)

Makiko Kaga

Abstract Adrenoleukodystrophy (ALD) is an X-inked degenerative disease caused by abnormal fatty acid metabolism due to *ABCD1 gene* mutation. ALD is known to appear as eight clinical types, which sometimes show transitional patterns; childhood cerebral, adolescent cerebral, adult cerebral, adrenomyeloneuropathy, Addison disease, brainstem-cerebellum, female symptomatic, and asymptomatic male types. Accordingly, the early diagnosis and hematopoietic stem cell transplantation treatment of ALD are indispensable, especially for patients with the childhood/adolescent cerebral types because of the devastating nature of their disease. The evaluation and monitoring of the disease onset, the severity and speed of the disease progression, and the nervous system involvement of the disease are necessary for the treatment of symptomatic and preclinical asymptomatic ALD patients. Neurophysiological and neuropsychological procedures are the best methods for this purpose, and the procedures should be noninvasive and performed successively. Electroencephalography, the evaluation of evoked potentials (the auditory brainstem response, visual evoked response and somatosensory evoked response), and event related potentials, and the combination of these tests are informative neurophysiological examinations. The use of the Wechsler Intelligence scale tests for each age range is essential to evaluate and follow ALD patients. The location of the main lesions of ALD with focal and diffuse permeation which cause a variety of concerned focal and diffuse symptoms must also be determined. Appropriate neuropsychological tests to evaluate visual, auditory, language and behavioral functions are also essential.

Keywords Adrenoleukodystrophy · Neurophysiology · Neuropsychology · Auditory brainstem response (ABR) · Visual evoked response (VEP) · Wechsler intelligence scale tests · Early diagnosis for early or preclinical treatment

M. Kaga (✉)
Tokyo Metropolitan Tobu Medical Center for Children with Developmental Disabilities, Tokyo, Japan

Department of Developmental Disorders, National Institute of Mental Health, National Center of Neurology and Psychiatry, Tokyo, Japan
e-mail: kaga@ncnp.go.jp

T. Imanaka, N. Shimozawa (eds.), *Peroxisomes: Biogenesis, Function, and Role in Human Disease*, https://doi.org/10.1007/978-981-15-1169-1_12

12.1 Neurophysiology and Neuropsychology as Noninvasive Methods for Early Identification of X Linked-Adrenoleukodystrophy (X-ALD) and Useful Follow-Up Procedure

A thorough overview of adrenoleukodystrophy (ALD) is provided by Shimozawa in the previous pages of this book. To summarize very, briefly, ALD is an X linked recessive, metabolic and progressive degenerative disease caused by *ABCD1 gene* mutation at Xq28. Clinical types are usually classified (1) Childhood cerebral type (CCALD), (2) Adolescent cerebral type, (3) Adrenomyeloneuropathy (AMN), (4) Adult cerebral type (ACALD), (5) Brainstem-cerebellum type, (6) Addison type, (7) preclinical asymptomatic male patient, (8) female clinical type.

Although it is hoped that the order-made gene therapy will become a treatment option in the near future, further clinical trial data are needed. At present, therefore, hematopoietic stem cell transplantation (HSCT) is the only realistic treatment for ALD. The origin of HSCT is based on human leukocyte antigen (HLA) match bone marrow transplantation (BMT) and subsequently by umbilical cord blood transplantation (UCBT). The results of HSCT have improved significantly due to the committed efforts of excellent hematologists. HSCT can now be performed at the clinical symptoms (and Loes score is less than 10 or 12*) or the appearance of MRI abnormalities in preclinical patients who are diagnosed as having a high titer of very long chain fatty acid (VLCFA) and mutation of *ABCD1* gene because of a family history. The recent introduction of universal newborn screening (UNS) in some states of United States of America and some European countries has identified such kind of patients. However, the type of ALD and the onset age of ALD at preclinical stage of ALD with definitive genetic abnormality cannot be predicted.

*The Loes score is often over 10 when patient are diagnosed as having clinically apparent ALD. Thus Japanese guideline (Kato S 2013) uses 12 as the upper limit of Loes score instead of 10 as the permitted score.

Lorenzo's oil was a rising star for the treatment of ALD at one time; it is regrettable that its effectiveness as an ALD treatment was denied. However, extensive work by Moser and his colleagues has indicated that in asymptomatic and MRI negative children may help delay the appearance of MRI lesions and reduce neurological impairment (Moser and Borel 1995; Moser et al. 2005). Patients with ALD and without evidence of clinical disease are thus candidates for HSCT as an early treatment. However, sometimes the disease progress is very rapid, or it is difficult to secure donor. In such conditions, an HSCT is problematic despite of relatively recent introduce of UCBT, which alleviates ALD, it is thus very important to identify the clinical start of ALD. The diagnosis can be made when definite lesions in MRI or apparent symptoms such as motor paralysis and central visual impairment.

For patients without definitive evidence of the emergence of ALD, an MRI examination is usually recommended every 6 months, but since the disease sometimes progresses too rapidly to perform an HSCT, other tools (preferably noninvasive) are needed to identify the appearance of ALD. Such tools can be provided by neurophysiological and neuropsychological examinations.

12.2 Neurophysiology of ALD

ALD has been explored as one of the degenerative diseases since the introduction of the concept of evoked potential. The electroencephalography (EEG) findings and pathological basis of the ALD disease process, measure that are available to make an early diagnosis, and an important way to follow-up patients are discussed next.

12.2.1 Electroencephalography (EEG)

There are a surprisingly paltry number of EEG studies in ALD. Most of the relevant neurophysiological study of ALD focused on evoked responses. In 1979, Ochs et al. conducted EEG examinations of siblings with ALD. The older brother was severely affected, bedridden and had communication difficulty but his younger brother had only hyperpigmentation of his skin and no neurological symptoms. Ochs et al. described the older brother's EEG findings as moderately and diffusely slow basic activity and continuous delta activity with minimal theta components, whereas the younger brother's EEG findings were normal. They also described the general differences in EEG findings among patients with white-matter and with those of gray-matter disease. In the former group, the reported findings were high amplitude irregular delta activity whereas the gray-matter diseases showed bisynchronous paroxysmal discharges consisting of rhythmic delta activity or bisynchronous spike-wave activity (quoted from Gloor et al. in 1968).

In 1996, Wada et al. described the EEG findings of an 8-year-old boy and his younger brother, both of whom had been diagnosed with ALD. Their EEG abnormalities (i.e., slow wave bursts in the occipital lesion) were the first symptoms of their ALD. In the younger brother's case, the EEG abnormality was confirmed before the appearance of the lesion on MRI. In this author's experience, slow waves were often confirmed in the leads of affected cerebral lesions revealed by MRI. This finding can be the predictive value of the future lesion location.

12.2.2 Evoked Potentials (EPs)

Ever since the early days of the clinical introduction of evoked potentials (Eps), the auditory brainstem response (ABR), the visual evoked potential (VEP) and short- or long- latency somatosensory evoked potentials (SSEPs or SEPs) have been reported in ALD patients. The effectiveness of multimodal EPs has been observed in clinical medicine. In the next section, each type of evoked potential and the multimodal recordings of Eps will be explained.

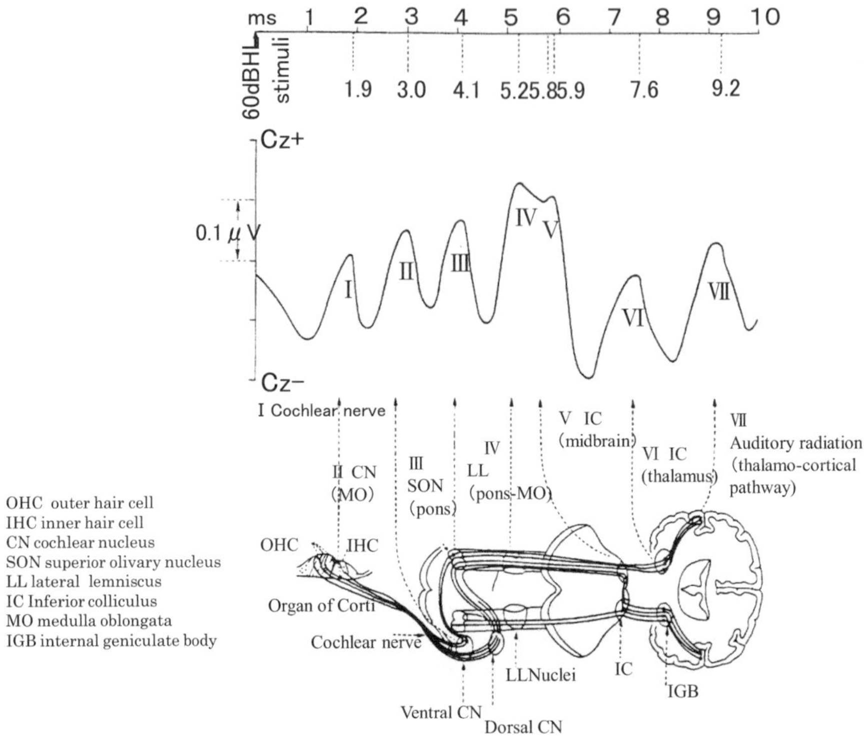

Fig. 12.1 Schematic diagram for an explanation of the origin of each wave of the ABR (Quoted from Stockard JJ, Stockard JE, Sharbrough FW. Detection and localization of occult lesions can be demonstrated by ABR. *Mayo Clin Proc.* 1977;52:761–9)

12.2.2.1 Auditory Brainstem Response (ABR)

The auditory brainstem response (ABR) was discovered independently by Jewett et al. (1970) and Sohmer et al. (1970). The importance and usefulness of ABR became clear very soon after the crucial discovery of this potential, based on its stability and reproducibility and the conclusive evidence of the origin of each wave. Stockard's schematic diagram is famous for an explanation of the origin of each wave of the ABR (Fig. 12.1, Stockard et al. 1977). The threshold of wave V is almost the same as that of subjective audiometry. The measurement of the wave forms, latency, amplitude and threshold of ABR are not affected by sleep or general anesthesia and causes no pain and no side effects to the subjects examined. The ABR can be measured repeatedly or serially even in infants, children, unconscious patients and people who cannot fully cooperate with the examinations (e.g. those with intellectual or emotional disabilities). The ABR is thus an ideal objective audiometry tool and a strong neurological instrument.

In 1979, Ochs reported ABR in ten patients with leukodystrophy; five had Pelizaeus–Merzbacher disease (PMD), one had metachromatic leukodystrophy (MLD) and two had ALD. Two patients with ALD were siblings. The elder brother (aged 11) showed only wave I and questionable wave II. His younger brother whose only symptom was slight skin pigmentation showed almost normal ABR with slight prolongation of wave V. ABR abnormalities were demonstrated in leukodystrophies (PMD, MLD and ALD). The ABR findings established that brainstem white matter was definitely involved in this disease.

In 1980, Kaga K et al. reported a serial study of ABR and slow vertex potentials along with the pathological findings in a patient with ALD. Deterioration of ABR was apparent with the progression of his disease. At the beginning of the disease, the patient's ABR was normal despite the apparent clinical symptom, but abnormalities soon became definitive in parallel with his disease stage. This important case report explains the natural course of clinical symptom, ABR changes, and pathological correlation of the lesion location of auditory pathway in ALD.

In 2006, Inagaki et al. described the ABR, VEP and SSEP in 3 patients with ALD, 5 PMD, 4 patients with Alexander disease, 4 patients with metachromatic leukodystrophy (MLD) and 3 patients with Globoid cell leukodystrophy (GLD is also known as Krabbe disease) respectively. Their conclusion regarding the ABR in patients with ALD were as follows; At the early stage of ALD, waves I, III and V are normal, but as the disease progresses, the interpeak latency (IPL) of waves I-V and I-III elongate from normal to abnormal. In addition to elongation of the wave I-V IPL a decreased amplitude ratio of V/I becomes evident. Finally, later components of the ABR (waves III and V) disappear.

An example of the serial change of the ABR in ALD is shown in the Fig. 12.2 (Inagaki et al. 2006). In an individual with ALD, early components of the ABR in ALD remain normal for a relatively long time. A prolonged I-V IPL then becomes apparent. A decreased V/I amplitude ratio eventually becomes clear and prolonged, and disappeared later components are confirmed in occipital type of ABR.

The following reports focused on adrenomyeloneuropathy (AMN). Grimes reported the bilateral abnormalities in the ABR of three patients with AMN: no response in one patient and prolonged IPL in two patients.

In 2006, Pillion et al. studied the ABR in 96 patients with ALD. The five types of disease were childhood/adolescent cerebral type, pure AMN, AMN plus cerebral, pure Addison's disease and a group of symptomatic females with ALD. ABR abnormalities were often exhibited by these patients, as were a delayed wave V and prolonged IPL. An abnormal ABR was observed most often in the patients with AMN and cerebral symptoms. Age was the most sensitive predictor of ABR abnormalities. Pillion et al. stated that an abnormal ABR was caused by the abnormalities in fiber tracts in lesions of the lateral lemniscus and inferior colliculus. This may be accurate, but physiological abnormalities can be detected more easily because a prolonged IPL may be apt to arise from the longer route abnormalities in AMN compared with pure cerebral types.

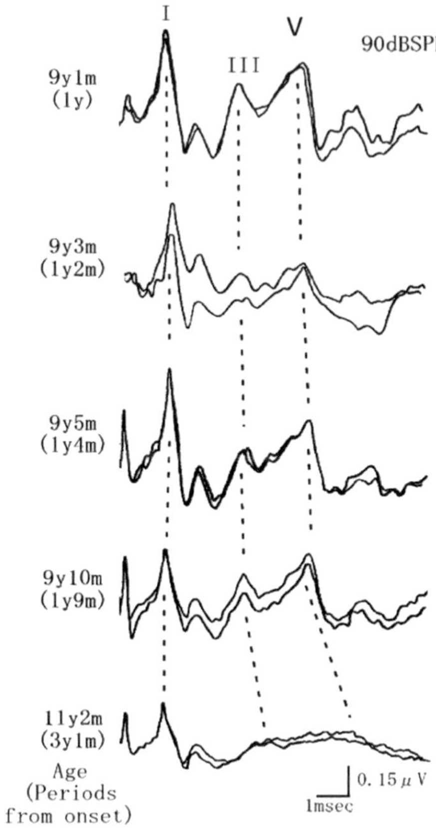

Fig. 12.2 Serial change of ABR in a patient with ALD. The patient's motor and mental development was normal until 8 years and 1 month old. His first symptom was defect of visual field. He admitted to a hospital aged 8 years and 11 months. He gradually lost his ability and became bedridden at last. His ABR was normal in shape, latency and amplitude after the onset of his disease (9 years and 1 month). However, interpeak latency of I-V gradually became prolonged and all waves reduced their amplitude. Three months from the onset, wave V disappeared finally. Figure noted in parenthesis is the time period from the onset of the disease (Quoted from Inagaki M, Kaga Y, Kaga M, Nihei K. Multimodal evoked potentials in patients with pediatric leukodystrophy. *Clin Neurophysiol.* Suppl 2006;59:251–63)

12.2.2.2 Visual Evoked Potential (VEP)

In 1993 Kaplan et al. examined 108 patients with ALD treated with Lorenzo's oil. The follow-up examination for 1 year later disclosed that reduced level of very long chain fatty acid (VLCFA) in the patient's serum but did not improve the N100 value of the patient's visual evoked potential (VEP). Inagaki provided the noteworthy case report of a patient with childhood cerebral-type ALD in 1995. After a bone marrow transplantation, the patient's VEP and SEP reappeared. Such cases highlight the importance and usefulness of evoked potentials. In 2006, Inagaki et al. reported an

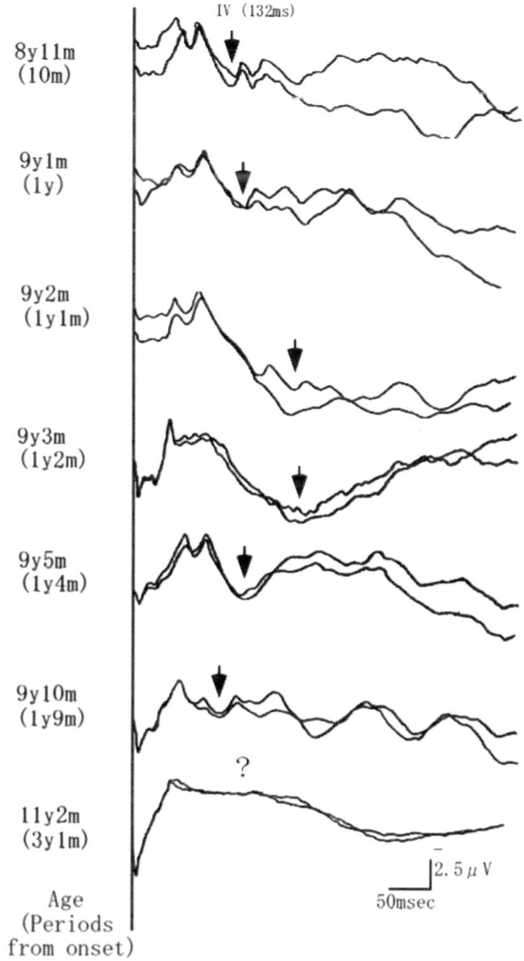

Fig. 12.3 Serial change of VEP in a patient with ALD. Serial change of VEP in the same patient with ALD of Fig. 12.2. The wave IV (P100 evoked by flash stimuli) latency of VEP was already prolonged at 10 months from the onset of the disease. Wave IV gradually diminished and could not be recorded at thee years of the onset. Figures noted in blankets are the period from his onset of the disease (Quoted from Inagaki M, Kaga Y, Kaga M, Nihei K. Multimodal evoked potentials in patients with pediatric leukodystrophy. *Clin Neurophysiol.* Suppl 2006;59:251–63)

update with serial change of VEP (Fig. 12.3, Inagaki et al. 2006) in the same patient described above in this section. The serial change of the patient's VEP demonstrated that the wave VI (P100 evoked by flash stimuli) latency of the VEP was already prolonged at 10 months from the disease onset. Wave IV could not be recorded at 3 years after the onset. Thus, serial recording reveals the pathophysiology and clinical course of ALD.

Kaplan et al. studied visual system abnormalities in 1995 in 59 patients with AMN using pattern reversal VEP, MRI, and clinical examinations, and they reported that visual pathways were affected in 63% of the patients. Their results seem to intensify the continuity of cerebral types of ALD and AMN.

In 2009, Furushima et al. reported the importance of P100 amplitude of the flash/pattern reversal VEP and P1 of visual event related potentials (V-ERP) in six asymptomatic boys with ALD along with their psychological profiles. The boys' full scale IQ values were all in the normal range, but in two patients, their performance IQ

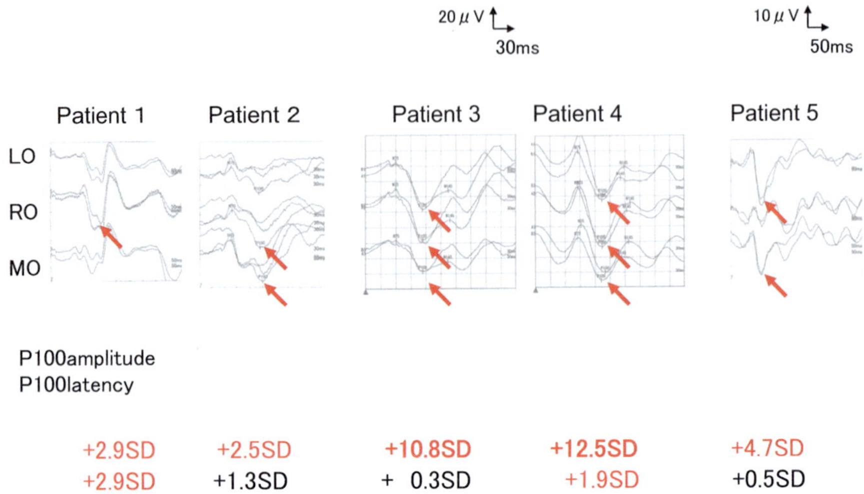

Fig. 12.4 High amplitude flash VEPs in 5 asymptomatic ALD patients

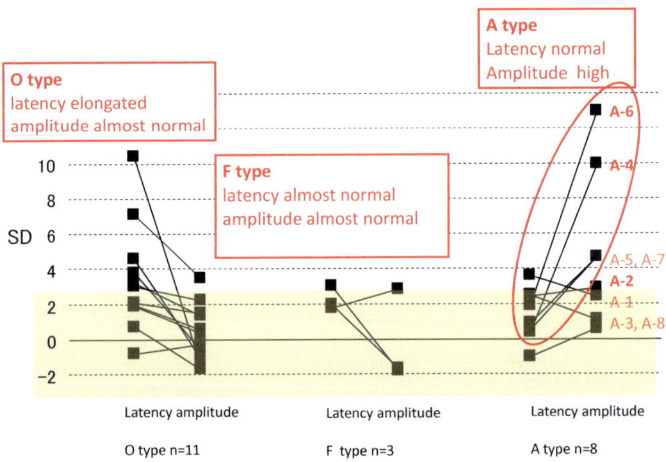

Fig. 12.5 Comparison of latency and amplitude of Flash VEP in different clinical types of 22 ALD patients. Vertical axis is Standard deviation (SD) of the amplitude of VEP. Horizontal axis is SD of the latency and the amplitude of 3 clinical types of patients with ALD. From the left to the right, 11 Occipital (O), 3 Frontal (F) type and 8 asymptomatic (A) type of patients. Data of average and standard deviation are based on healthy controls

(PIQ) was much lower than verbal IQ (VIQ). This finding will be discussed later in the Neuropsychology section. In the same study, regarding the VEP, the amplitude and latency values of P100 and P1 was intensified and prolonged in 8 asymptomatic patients with ALD compared to those of 22 healthy controls. Children generally show higher amplitudes in EEG compared to adults. However, the amplitude of the VEP in the asymptomatic patients with ALD was clearly high compared to the average and standard deviation (SD) of the age-matched control children.

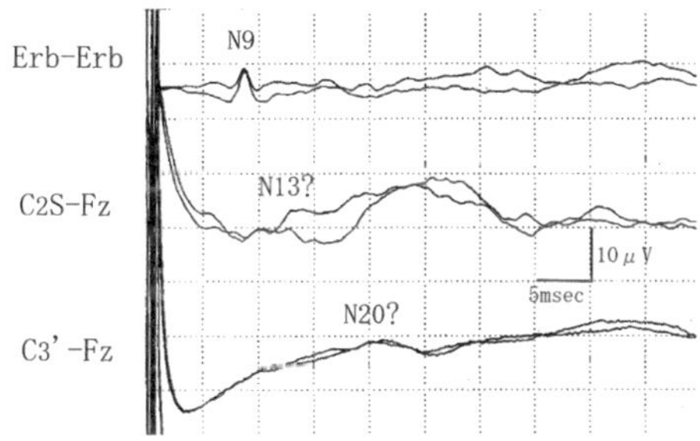

Fig. 12.6 Short latency somatosensory evoked potential (SSEP) in a patient with ALD. The patient was first noted by visual abnormality and seizure at the age of 7 years and 10 months. SSEP recorded at the age of 13 years 6 months is shown in this figure. Erb's potential (N9) is normal, brainstem (N13) to cortical potential (N20) are ambiguous. Central conduction is disturbed (Quoted from Inagaki M, Kaga Y, Kaga M, Nihei K. Multimodal evoked potentials in patients with pediatric leukodystrophy. *Clin Neurophysiol.* Suppl 2006;59:251–63)

Thus, a high-amplitude VEP (Fig. 12.4) may be an early sign of ALD prior to the appearance of a lesion MRI. The follow-up of asymptomatic children with the use of both electrophysiological and neuropsychological tests may serve as an aid for deciding the timing of therapeutic intervention. In fact, some of the children in the Furushima et al.'s study underwent an HSCT, and their VEPs returned to normal; no apparent symptoms of ALD were then observed.

In our 22 patients with ALD (11 occipital (O), 3 frontal (F), and 8 asymptomatic (A) types) showed different pattern of the latency and amplitude in VEPs (Fig. 12.5). In O type, prolonged latency with normal amplitude is most distinct. F type patients seemed to show almost normal latency and amplitude. In A types, patients who showed normal latency and high amplitude VEP are conspicuous. But there are patients with normal latency and normal amplitude. These findings suggest some patients will appear as O type and the other will commence from F type (Fig. 12.5).

12.2.2.3 Somatosensory Evoked Potential (SEP)

The origin of SSEP induced by stimulation of median nerve (or tibial nerve) has been defined by the experimental and clinical research. N9 is from brachial plexus. N13 and N14 originate from brainstem relay nucleus, and N18 to N20 is the contralateral somatosensory cortex. In 2006, Inagaki et al. described the SEP (Fig. 12.6, Inagaki et al. 2006), ABR, and VEP at 5 years and 8 months after the disease onset in a 13-year-old patient. His SSEP revealed a normal N9, which indicates normal peripheral nerve conduction and an ambiguous N13 and absent N20, i.e. abnormal function in the brainstem to sensory cortex.

The SEP has been well investigated in patients with AMN. For example, in 1997 Kaplan examined the SSEP induced by median and posterior tibial nerve stimulation in 83 patients with AMN. Ten of the 16 women (62.5%) had abnormal SSEPs involving central pathways, and 59 of the 67 men (88%) had abnormal median

SSEPs involving both peripheral and central pathways. Tibial SSEPs were abnormal in 14 of the 15 women and all 67 men with peripheral and central pathways. These differences among women and men may define the clinical severity of their ADL.

Restuccia et al. also reported in 1997 that 12 of 19 female ALD carriers exhibited abnormal SEP results reflecting central abnormalities with intact peripheral conduction. The patients with severe symptomatology in ALD/AMN showed severe abnormalities, and the patients with milder symptoms showed milder SEP abnormalities. In 2012, Cappa et al. reported an improvement of the SEP following a decrease in the plasma levels of serum 26:0, 26:0/22:0 ratio brought about by the dietary intake of Lorenzo's oil.

In 1992, Park et al. later provided a case report of a 33-year-old man with AMN and an abnormal SEP induced by posterior tibial stimulation. His main clinical feature was progressive spastic paraparesis and other autonomic dysfunction. In such a case, a detailed follow-up SEP study is necessary to clarify the clinical meaning of these abnormal findings.

12.2.2.4 Multimodal Evoked Potentials

In 1982, Markand et al. examined 3 patients with ALD, 12 patients with PMD, and three with MLD. The ABR, VEP and SEP findings in the ALD patients were as follows. All of the evoked responses were abnormal in all three patients with the exception of normal VEPs and SEPs observed in a patient with early ALD. In most of the patients, wave I with and without wave II were the only components of the ABRs that remained for long time. However, subsequent components (waves III-VII) were absent which indicates a rostro-caudal progression of the disease. The patient's VEPs were severely abnormal and the latency of major components was significantly delayed.

Inagaki et al.'s (2006) description of the serial recordings obtained from patients with various white matter disorders suggested that the degenerative process can also be observed by measuring the VEP. In their patients, median nerve stimulation evoked SSEP showed normal Erb's potential (N10) and absent or marked attenuation of cervical (N14) and early scalp components (N19 and P22). A significant delay in scalp components was also observed. Multimodal evoked responses provide more information regarding the functional integrity of several afferent systems in patients with white matter disorders.

Assessments of the individual evoked potentials are of course useful to evaluate clinical disease processes, and the results can contribute to auxiliary diagnostic procedure. Multimodal assessment can reveal the diseased nerve route pathway of a sensory modality.

In 2010, Matsumoto et al. reported multimodal evoked potentials in ten patients with AMN. An abnormal motor evoked potential (MEP) and SEP were observed in all ten patients. The ABR was abnormal in nine patients but an abnormal VEP was observed in only one patient. The brainstem latency of seven of the patients was observed; the cortical-brainstem conduction time was severely prolonged in three patients. These findings emphasize the importance of the discrepancy in the results of VEP, MEP, SEP and ABR as clinically useful electrophysiological features for the diagnosis of AMN.

12.2.2.5 Transcranial Magnetic Stimulation (TMS)

Hitomi et al. in 2003 described the improvement of motor conduction induced by transcranial magnetic stimulation (TMS) in a 20-year old AMN patient after HSCT treatment. His VEP and ABR were abnormal but ABR and VEP did not change after treatment. Thus the authors intensified the meaning of TMS in evaluating the effect of HSCT.

12.2.2.6 Event Related Potential (ERP)

Event related potential (ERP) is a transient brain electrical potential related with time course and is induced by the inner events such as cognition of some difference, decision making and others. P300 and mismatch negativity (MMN) are most often used ERP which will be described below. Other examples of ERP are contingent negative variation (CNV), N400 and others. CNV is a negative potential induced by anticipation or expectation. N400 is induced by relating with the meaning of words or sentences.

Mismatch negativity (MMN) is induced when an individual can detect a subtle difference of two or more types of stimuli under an 'ignore' condition. MMN was discovered by Naätänen et al. in 1978 and then thoroughly investigated by his group. The P300, first reported by Sutton in 1965 is induced when one can detect the difference between two types of stimuli in a conscious/attentive manner. By applying both methods (non-conscious or conscious) to induce the MMN response or P300 response, the authors used nonverbal and verbal stimuli to investigate patients with ALD who showed verbal auditory agnosia. And both ERPs induced by pair of tone bursts or pair of verbal sounds were compared. In both ERPs, amplitudes are higher in response to tone bursts than to verbal sounds. Both ERPs (i.e. MMN and P300) induced by a pair of tone bursts or a pair of verbal sounds were compared. The amplitude of the latency of both was higher and shorter response to tone bursts than to the verbal sounds. These results suggest the clinical severity of the ALD patient's auditory verbal agnosia. The MMN and P300 wave forms in a patient with ALD whose main symptom was auditory agnosia are shown in Fig. 12.7.

12.2.2.7 Summary of Neurophysiological Examination in ALD

Neurophysiological examinations are noninvasive methods, and those examinations can thus be performed repeatedly without harm to patients. This is an advantage that allows the monitoring features of degenerative disease. In addition, the VEP and EEG findings (and perhaps more features) may clearly change before the onset of

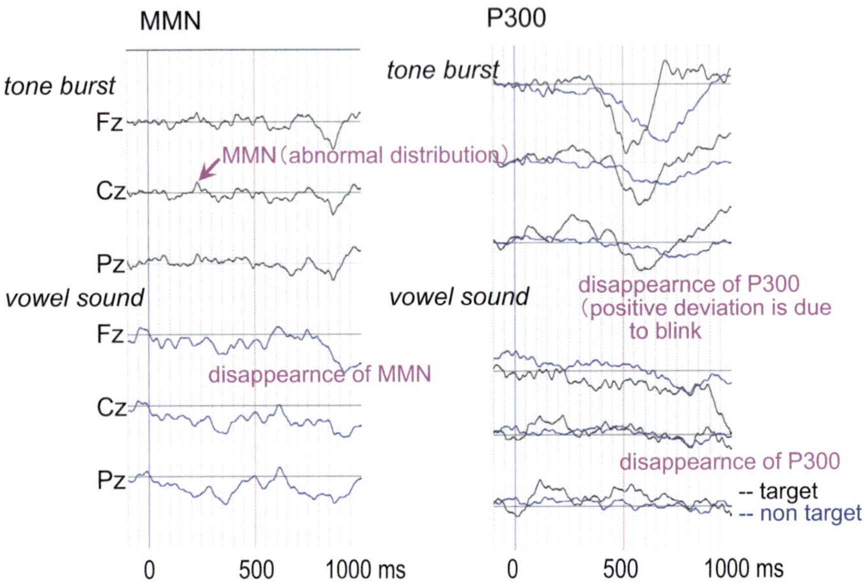

Fig. 12.7 Mismatch negativity (MMN) and P300 to tone and vowel sounds in a patient with ALD with auditory agnosia. Mismatch negativity (MMN) and P300 to tone bursts ([1000 Hz] versus [2000 Hz] and vowel sounds ([a] versus [ae]) in a patient with ALD aged 11 years. He showed typical auditory agnosia at the onset of his disease. MMN to tone bursts was clear at Cz lead, but no MMN to vowel sounds. (MMN is usually recorded best at frontal leads.) No P300 was recorded both to tone bursts and vowel sounds. He could not discriminate clinically the difference of both kinds of stimuli for himself

clinical disease. Neurophysiological examinations are therefore desirable. As an example, an increase in VEP amplitude identified by neurophysiological testing may be the critical sign of the disease onset in asymptomatic ALD patients. This we can have tools to help the decision of deciding HSCT treatment for the anxious patients and their families. Thus the author intends to consider more positive introduction of neurophysiological examination.

12.3 Neuropsychology of X-ALD

12.3.1 What Is Neuropsychology?

The scope of neuropsychology research includes the study of the relationships between higher cortical activities and brain structures. The case of patients with a brain injury and organic symptoms can teach researchers a great deal about the brain. Our understanding of aphasia, agnosia and apraxia has been greatly increased by the constant advances in neuropsychology methods. A recent example is the trend of using neuroimaging techniques together with clinicopathological methods. The application of neuropsychology tests to ALD are summarized next.

12.3.2 Application of Neuropsychology to ALD

ALD is a metabolic neurodegenerative disease but its course and appearance as symptoms are not the same in all patients. The speed of progression and the order of the lesion to be located in the white matter of the brain often differ among ALD patients. The classification and early symptoms of ALD are described by Shimozawa in the previous chapter, but a rough classification of the cerebral type of ALD considers two types, the frontal and occipital types. Symptoms of frontal type can appear as inattention, reduced executive function, and behavioral change/alteration in personality due to frontal lobe dysfunction. Symptoms of the occipital type often appear as the central visual impairment along with gait disturbance or a variety of symptoms. The occipital type of ALD is more frequent than the frontal type; the reported rates are 80–85% and 15–20% of ALD cases, respectively. Like these localized brain symptoms can be identified because the progress of ALD is not even and patients' overall intelligence does not become compromised uniformly compared with other types of leukodystrophies and many gray matter degenerative diseases. Thus cognitive tests have often disclosed a visual processing disorder in patients with occipital type ALD.

12.3.3 Neuropsychological Tests

Over the past century, many types of neuropsychological tests were devised and used to identify cognitive functions. The following functions are among those considered when patients' overall cognitive state is being assessed: general intelligence, language function, visual function (copy, memory, visuospatial cognition, visual motor function, etc.), auditory function (repetition, recognition of verbal and nonverbal sounds, auditory spatial cognition, memory, etc.), frontal function (e.g., attention, executive function, Stroop test results, comprehension of no-go tasks), behavior, recognition of body and face images, space, emotion, memory (long term, short term and working memory), adaptive behavior and academic skills.

12.3.3.1 Wechsler Intelligence Scale Tests

The Wechsler Intelligence Scale Tests are essential to evaluate all the patients suspected of having ALD. When serial testing is necessary, the interval between these tests should be ≥1 year. The Wechsler intelligence Scale Tests are divided into three types by the chronological age of the subjects being tested; adults (the Wechsler Adult Intelligence Scale or WAIS), children (the Wechsler Intelligence Scale for Children or WISC) and preschoolers and toddlers (the Wechsler Preschool and Primary Scale for Intelligence or WPPSI). These tests are standardized in each area of the world, used internationally, and revised periodically to better assess intelligence based on separate cognitive functions, i.e., language (verbal comprehension index [VCI]), perceptual reasoning index ([PRI], visual comprehension and information processing), working memory index [WMI] and processing speed index [PSI] in the WISC-IV[the fourth edition of the WISC]). For comparison, VCI and PRI have almost all the same meaning with Verbal IQ (VIQ) and Perceptive IQ (PIQ), respectively in WISC-III (Wechsler Intelligence Scale for Children, the third edition).

Subcategories of WAIS and WPPSI are somewhat different from WISC-IV but fundamental designs of the three tests are not significantly different. Large segments of cognition can be evaluated by using the Wechsler Intelligence tests. Several newer tests have been added to some test batteries to enable more precise evaluations, or to confirm the findings obtained with Wechsler tests, or to identify the necessary cognitive function.

12.3.3.2 Other Neuropsychological Tests

- Draw a man test (body image, eye hand coordination, vision and visual image) This test helps reveal an individual's body image and eye hand coordination. A picture drawn by a patient with central visual impairment is shown in Fig. 12.8.
- Peabody Picture vocabulary test (language ability, concrete and abstract words)
- Frostig visual perception development test (visual perception and eye–hand coordination)
- Raven colored matrix test (visual reasoning)
- Rey's complex figure test (visual cognition, eye–hand coordination and visual memory)
- Kaufman assessment battery for children (academic skills, sequential and simultaneous processing scale)
- Standard language tests for aphasia (by Japan society of Higher Brain Function)
- Token test (comprehension of words and sentences)
- Word fluency test (frontal function)
- Stroop test (frontal function)
- Wisconsin card sorting test (frontal function)
- Rey's auditory verbal learning test (AVLT) (auditory memory)

12.3.3.3 Neuropsychology in Children with ALD

In 2000, Riva et al. reported that neuropsychological testing may predict early progression of asymptomatic adrenoleukodystrophy. They investigated seven symptomatic and eight asymptomatic pediatric patients. All symptomatic boys showed lesions in occipital white matter, corpus callosum and visual pathway. Five of the seven symptomatic patients had lesions in the auditory pathway. MRI revealed frontal white matter lesions in three of the symptomatic patients. Among the eight asymptomatic boys, six developed apparent ALD within 16–30 months, and seven showed parietooccipital white matter lesion Only one patient was free of MRI lesions.

Fig. 12.8 Draw a man test in a patient with ALD aged 12 years. He had Occipital type of ALD. His body image was impaired. Proportion of the head, body shoulder and trunk are unbalanced. Left leg/his trousers, both shoulders/arms and right upper half body/his trousers are separated. A picture drawn by a patient with central visual impairment is shown in Fig. 12.8 also affects this phenomenon

The neuropsychological test results of the symptomatic patients revealed marked abnormalities in full IQ (FIQ), verbal IQ (VIQ) and performance IQ (PIQ) by WISC-III. Most of their other test results were abnormal.

The asymptomatic patients showed relatively normal neuropsychological test results, but verbal fluency and naming functions were severely impaired. These results seem to derive from frontal lesions. Riva et al. noted that MRI demonstrated frontal lobe-related lesions in only three out of eight asymptomatic patients.

Cox et al. (2006) examined 52 asymptomatic boys with ALD (6.7 ± 3.6 years old)and reported that the boys showed normal overall cognitive function when they showed no clinical or MRI signs of the disease. They attempted to determine the patients' full cognitive functions just before the onset of their disease. The tests included Wechsler intelligence tests (Fig. 12.9). The test results indicated completely normal full cognitive functions in 48 of the 52 patients. The remaining four patients showed somewhat abnormal results at their asymptomatic stage. Briefly, (a) One patient (age 5.2 years) showed low FIQ/VIQ values and depressed language and adaptive behavior. (b) One patient (age 12.8 years) showed lower-limit PIQ and low executive function. (c) One patient (age 2.6 years) showed lowered visual cognition. (d) One patient (age 2.4 years) showed low daily adaptive function revealed by the Vinland Adaptive Behavior Scale. Moreover, they found negative correlation between patient's age and their visual cognition. The authors (Cox et al.) also identified a negative correlation between the patients' age and their visual cognition, The findings suggests that as patients grew older, their visual cognition declined. In other words, the possibilities of neuropsychological abnormalities come to the surface as an individual with ALD ages.

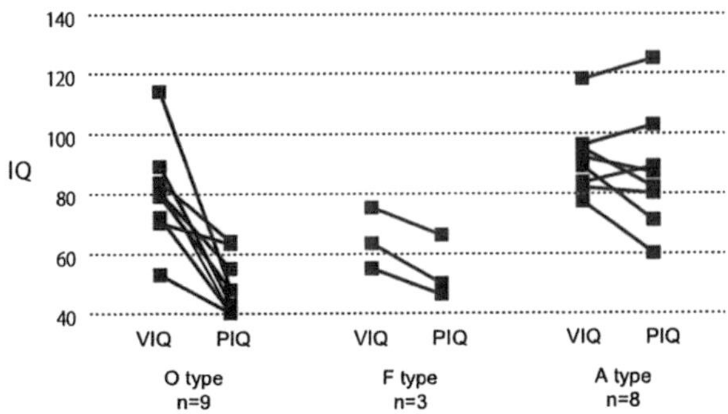

Fig. 12.9 IQ value of Wechsler Intelligence scale in patients with 20 ALD patients. O type: occipital type. F type: frontal type. A type: asymptomatic type. VIQ is verbal IQ and PIQ is performance IQ of Wechsler Intelligence scale for children-III (Quoted from Kaga M, Furushima W, Inagaki M, Nakamura M. Early neuropsychological signs of childhood adrenoleukodystrophy (ALD). *Brain Dev*. 2009; 31:558–61)

In 2009, Kaga et al. reported neuropsychological test results of eight asymptomatic patients with the childhood cerebral type of ALD (A type) and 14 symptomatic ALD patients before HSCT treatment. The group of symptomatic patients was composed of 11 patients with the occipital type (O type) and three patients with the frontal type (F type). The age ranges of the complete series of 22 patients were as follows: A type, 5–14 years, O type, 3–11 years; and F types, 10–11 years. No clinical and radiological signs of disease were confirmed in A-type patients when they were examined. Precise neuropsychological tests were done in all patients.

In addition, the distribution of the verbal IQ (VIQ) and performance IQ (PIQ) of the WISC-III results (or WPPSI in one patient) are shown in Fig. 12.8.The PIQ test of the WISC III is almost the same as Perceptual Reasoning Index (PRI) of the WISC-IV. Eight of the nine O-type patients tended to show marked discrepancy of VIQ and PIQ. In the three F-type patients, the discrepancy was rather low compared with the O-type patients. The A-type patients, overall IQ values were higher than those of the symptomatic patients, and they showed almost normal FIQ (7/8). The PIQ result was significantly lower than VIQ in three patients. There were three patients who showed a definitive difference between their VIQ and PIQ values (VIQ≫PIQ).

Both symptomatic and asymptomatic patients with a wide discrepancy in their VIQ and PIQ results (≥10) in the Kaga M et al. study showed abnormal results in tests designed to detect visual cognition. The Gestalt Closure in the Kaufman Assessment Battery for Children and the Constancy of Shape in the Frostig Developmental Test of Visual Perception seemed to be effective for detecting visual cognitive dysfunction in the asymptomatic ALD children whose difference of VIQ and PIQ was within the normal range.

These findings may indicate that the pattern of this discrepancy can predict each patient's main lesion of disease at onset (Kaga et al. 2009; Furushima et al. 2009).

Along with the high-amplitude VEP, neuropsychological abnormalities can be early neuropsychological signs of childhood adrenoleukodystrophy in asymptomatic boys. Prior to the disease onset, the cognitive function of asymptomatic patients appears to be normal, but the neuropsychological and neurophysiological abnormalities identified by the tests described above may predict that clinical disease is not very far off for the patients.

12.3.3.4 Auditory Agnosia in Children with ALD (Furushima et al. 2015)

Central visual impairment is often stressed in ALD patients because their lesions often begin from occipital white matter. Though hearing impairment is not rare as an early symptom of childhood ALD, it is apt to be missed; overlooking this symptom can contribute to a delayed diagnosis. The precise evaluation of patients' auditory function can aid the diagnosis. In ALD, hearing impairment is of central brain origin and not from ear disease, indicating that the symptoms are derived from a bilateral dysfunction of auditory cortex in temporal lobes. ALD is essentially a white matter disease, and the sites of the organic lesions usually involve the route of the retrocochlear nerve, lateral lemniscus, internal geniculate body, and auditory radiation. Typical symptoms are seemingly ignoring other people's statements, frequent ask-back, and hearing impairment-like attitudes (Furushima et al. 2015). Individuals with these symptoms are often misdiagnosed at first as malingering, rebelling against authority and/or "acting up." It is also important to note that children can have auditory agnosia due to their underlying disease, although this is rare.

12.3.3.5 Summary of Neuropsychological Tests in ALD

The careful follow-up of asymptomatic boys with ALD by administering neuropsychological tests along with some of the above-described neurophysiological tests can help determine the timing of therapeutic intervention. In ALD, neuropsychological abnormalities have seemed to precede the appearance of clinical and MRI alterations.

References

Cappa M, Bizzarri C, Petroni A, Carta G, Cordeddu L, Valeriani M, Vollono C, De Pasquale L, Blasevich M, Banni S (2012) A mixture of oleic, erucic and conjugated linoleic acids modulates cerebrospinal fluid inflammatory markers and improve somatosensorial evoked potential in X-linked adrenoleukodystrophy female carriers. J Inherit Metab Dis 35:899–907

Cox CS, Dubey P, Raymond GV, Mahmood A, Moser AB, Moser HW (2006) Cognitive evaluation of neurologically asymptomatic boys with X-linked adrenoleukodystrophy. Arch Neurol 63:69–73

Furushima W, Inagaki M, Gunji A, Inoue Y, Kaga M, Mizutani S (2009) Early signs of visual perception and evoked potentials in radiologically asymptomatic boys with X-linked adrenoleukodystrophy. J Child Neurol 24:927–935

Furushima W, Kaga M, Nakamura M, Gunji A, Inagaki M (2015) Auditory agnosia as a clinical symptom of childhood adrenoleukodystrophy. Brain and Development 37:690–697. https://doi.org/10.1016/j.braindev.2014.10.011. Epub 2014 Dec 26

Hitomi T, Mezaki T, Tsujii T, Kinoshita M, Tomimoto H, Ikeda A, Shimohama S, Okazaki T, Uchiyama T, Shibasaki H (2003) Improvement of central motor conduction after bone marrow transplantation in adrenoleukodystrophy. J Neurol Neurosurg Psychiatry 74:373–375

Inagaki M, Maegaki Y, Ohtani K, Asano J, Suzuki Y (1995) Reappearance of visual and somatosensory evoked potentials in a patient with childhood adrenoleukodystrophy after bone marrow transplantation and dietary erucic acid therapy. Acta Paediatr Jpn 37:125–128

Inagaki M, Kaga Y, Kaga M, Nihei K (2006) Multimodal evoked potentials in patients with pediatric leukodystrophy. Suppl Clin Neurophysiol 59:251–263

Jewett DL, Romano MN, Williston JS (1970) Human auditory evoked potentials: possible brain stem components detected on the scalp. Science 167(3924):1517–1518

Kaga K, Tokoro Y, Tanaka Y, Ushijima H (1980) The progress of adrenoleukodystrophy as revealed by auditory brainstem evoked responses and brainstem histology. Arch Otorhinolaryngol 228(1):17–27

Kaga M, Furushima W, Inagaki M, Nakamura M (2009) Early neuropsychological signs of childhood adrenoleukodystrophy (ALD). Brain Dev 31:558–561

Kaplan PW, Tusa RJ, Shankroff J, Heller J, Moser HW (1993) Visual evoked potentials in adrenoleukodystrophy: a trial with glycerol trioleate and Lorenzo oil. Ann Neurol 34:169–174

Kaplan PW, Kruse B, Tusa RJ, Shankroff J, Rignani J, Moser HW (1995) Visual system abnormalities in adrenomyeloneuropathy. Ann Neurol 37:550–552

Kaplan PW, Tusa RJ, Rignani J, Moser HW (1997) Somatosensory evoked potentials in adrenomyeloneuropathy. Neurology 48:1662–1667

Kato S (2013) Guideline for HSCT to patients with ALD. Research report to Ministry of Health Welfare and Labor, Japan

Markand ON, Garg BP, DeMyer WE, Warren C, Worth RM (1982) Brain stem auditory, visual and somatosensory evoked potentials in leukodystrophies. Electroencephalogr Clin Neurophysiol 54:39–48

Matsumoto H, Hanajima R, Terao Y, Hamada M, Yugeta A, Shirota Y, Yuasa K, Sato F, Matsukawa T, Takahashi Y, Goto J, Tsuji S, Ugawa Y (2010) Efferent and afferent evoked potentials in patients with adrenomyeloneuropathy. Clin Neurol Neurosurg 112:131–136

Moser HW, Borel J (1995) Dietary management of X-linked adrenoleukodystrophy. Annu Rev Nutr 15:379–397

Moser HW, Raymond GV, Lu SE, Muenz LR, Moser AB, Xu J, Jones RO, Loes DJ, Melhem ER, Dubey P, Bezman L, Brereton NH, Odone A (2005) Follow-up of 89 asymptomatic patients with adrenoleukodystrophy treated with Lorenzo's oil. Arch Neurol 62:1073–1080

Näätänen R, Gaillard AW, Mäntysalo S (1978) Early selective-attention effect on evoked potential reinterpreted. Acta Psychol 42(4):313–329

Ochs R, Markand ON, Demyer WE (1979) Brainstem auditory evoked responses in leukodystrophies. Neurology 29:1089–1093

Pillion JP, Kharkar S, Mahmood A, Moser H, Shimizu H (2006) Auditory brainstem response findings and peripheral auditory sensitivity in adrenoleukodystrophy. J Neurol Sci 247:130–137

Restuccia D, Di Lazzaro V, Valeriani M, Oliviero A, Le Pera D, Colosimo C, Burdi N, Cappa M, Bertini E, Di Biase A, Tonali P (1997) Neurophysiological abnormalities in adrenoleukodystrophy carriers. Evidence of different degrees of central nervous system involvement. Brain 120(Pt 7):1139–1148

Riva D, Bova SM, Bruzzone MG (2000) Neuropsychological testing may predict early progression of asymptomatic adrenoleukodytrophy. Neurology 54:1651–1655

Sohmer H, Feinmesser M (1970) Cochlear and cortical audiometry conveniently recorded in the same subject. Isr J Med Sci 6:219–223

Stockard JJ, Stockard JE, Sharbrough FW (1977) Detection and localization of occult lesions with brainstem auditory responses. Mayo Clin Proc 52:761–769

Sutton S, Braren M, Zubin J, John ER (1965) Evoked-potential correlates of stimulus uncertainty. Science 150:1187–1188

Wada N, Takeuchi Y, Fujiwara F, Odani I, Sawada T (1996) Adrenoleukodystrophy indicated by abnormal electro-encephalography. Acta Paediatr Jpn 38:352–356

MIX
Papier aus verantwortungsvollen Quellen
Paper from responsible sources
FSC® C105338

If you have any concerns about our products,
you can contact us on
ProductSafety@springernature.com

In case Publisher is established outside the EU,
the EU authorized representative is:
Springer Nature Customer Service Center GmbH
Europaplatz 3, 69115 Heidelberg, Germany

Printed by Libri Plureos GmbH
in Hamburg, Germany